Amir Muzur, Hans-Martin Sass (Eds.)

Fritz Jahr and the Foundations of Global Bioethics

Ethik in der Praxis/Practical Ethics

Studien/Studies

herausgegeben von

Hans-Martin Sass
Ruhr-Universität Bochum,
Potomac Institute for Policy Studies,
Arlington VA

Schriftleitung: Arnd T. May

Band 37

LIT

Fritz Jahr and the Foundations of Global Bioethics

The Future of Integrative Bioethics

edited by

Amir Muzur and Hans-Martin Sass

LIT

Cover Picture: Erwin J. Löhr

Supported by a grant (2012-FJF3B) of the Institut für Ethik in der Praxis e. V., Recklinghausen Germany.

Bibliographic information published by the Deutsche Nationalbibliothek
The Deutsche Nationalbibliothek lists this publication in the Deutsche Nationalbibliografie; detailed bibliographic data are available in the Internet at http://dnb.d-nb.de.

ISBN 978-3-643-90112-5

A catalogue record for this book is available from the British Library

Klosbachstr. 107
CH-8032 Zürich
Tel. +41 (0) 44-251 75 05
Fax +41 (0) 44-251 75 06
e-Mail: zuerich@lit-verlag.ch
http://www.lit-verlag.ch

LIT VERLAG Dr. W. Hopf
Berlin 2012
Fresnostr. 2
D-48159 Münster
Tel. +49 (0) 2 51-620 320
Fax +49 (0) 2 51-23 19 72
e-Mail: lit@lit-verlag.de
http://www.lit-verlag.de

Distribution:
In Germany: LIT Verlag Fresnostr. 2, D-48159 Münster
Tel. +49 (0) 2 51-620 32 22, Fax +49 (0) 2 51-922 60 99, e-mail: vertrieb@lit-verlag.de

In Austria: Medienlogistik Pichler-ÖBZ, e-mail: mlo@medien-logistik.at

In Switzerland: B + M Buch- und Medienvertrieb, e-mail: order@buch-medien.ch

In the UK: Global Book Marketing, e-mail: mo@centralbooks.com

TABLE OF CONTENT

Section 2
Roots of Integrative Bioethics

Section 3
Stems of Integrative Bioethics

Section 4
Branches of Integrative Bioethics

INTRODUCTION

The foundations of bioethics seem to be deeply rooted in human nature and human culture and thus bioethics and ethics have been and are global phenomena associated with human strive, human failure and human culture. Different cultures interact differently with their natural and social environments, with animals and plants. Holistic worldviews see humans as part of the living universe; spiritual worldviews sense visible and invisible life together as a cosmos, in which human and other life grows and strives. Anthropocentric worldviews reflect an antagonistic and combatitive relationship of humans and human cultures with nature as raw nature, as something wild and dagenrous which needs to be stewarded, dominated, cultivated, even eradicated. Some cultures focus on dominating and cultivating the world of *bios* for survival, others envision the survival of humankind and themselves in a more natural interaction with nature. Different models in the dialectics and interactions between egoism and altruism express the multitude of these approaches, as Fritz Jahr has suggested. Issues of compassion play a role in all cultures, – and they definitely have played a role for individuals and groups, no matter to which culture or tradition they belonged.

While those different attitudes and visions quite often have developed independently from each other on various continents and societies and in different times, the newly developing pluralistic and global societies of the 21th century interact more closely with each other and might actually develop new interactive and integrative models in ethics and bioethics. Fritz Jahr, in overcoming Kant's anthropocentric Categorical Imperative had put forward a new vision and command for an integrative model of ethics, the Bioethical Imperative 'Respect every living being on principle as an end in itself and treat it, if possible, as such!' (1)[1] For him the basic human capacity, principle and virtue in bioethics was *compassion*, while slightly more than four decades later for Van Rensselaer Potter it was *survival.*[2] Both developed their new model on the basis of close interpretation of modern scientific knowledge in terms of culture and ethics for a more harmonious and civilized global society

---

[1] All numbers in brackets refer to the respective number of author and article in the Table of Content.

[2] Van Rensselaer Potter (1970) ‚Bioethics', Perspectives in Biology and Medicine, autumn 1970 (adapted from Chapter 1 of Potter's *Bioethics. Bridge to the Future*, Englewood Cliffs; Prentice Hall 1971).

of the future. Jahr in particular made use of different traditions, including those beyond his own European and Christian culture and society.[3]

Today we are witnessing, on a global scale, a growing discontent with our cultural and moral handling of many technological and scientific advancements, also *angst,* that we are not prepared for the changes these advancements are bringing. For Van Rensselaer Potter (1911-2001), that time arrived sooner than for others and resulted in the early 1970s in a series of publications launching a new discipline – bioethics – predestined to help regain humane values to the too-rapid pace of 'progress' in the interest of *survival* and as a *bridge to the future*.[4] For the next thirty years[5], however, although the name of the new 'science' did survive and spread, Potter was not adequatly credited for it. In the USA, André Hellegers institutionalised medical ethics under the name bioethics and a series of authors oriented it more toward a narrowed-down concept of 'new' medical ethics. In the 1990s, nevertheless, Potter and his original ideas, now shaped as 'global bioethics', started to receive more attention all over the world and provoked the formation of a unique and promising network. By that time, bioethics had also been introduced to Europe (first to Spain, then to France, Croatia, etc.), somewhere, nonetheless (like in Germany), encountering quite a resistance[6]. And then, shortly before the new millennium, bioethics entered into a new phase. The long forgotten works of Fritz Jahr (1895-1953) were rediscovered. A concept of bioethics, similar but not identical to Potter's, had been shaped by a humble German theologian and teacher some fifty years earlier, already in 1927.[7] The recapture of this early tradition of bioethics, being based on the

---

[3] According to a report in *Chemical & Engineering News*, October 11, 1971, about the forming of the Kennedy Institute of Bioethics, its Director André Hellegers had stressed the integrative character of the new institute: "The institute, although located at a Catholic university, will be universal in approach and will include Protestant, Jewish, and humanist thinkers. Studies will not be inhibited by Catholic doctrine on abortions, contraception, or other problems. Close cooperation with Georgetown's hospital will keep research abreast of the 'real world.'"

[4] Gensobella M (2009) "Biodiversity and the global bioethics of Van Rensellar Potter", 1. EMUNI Research Souk 2009, Unity and Diversity of Euro-Mediterranean Identities, Jan 9, 2009.

[5] Reich W (1994) "The word Bioethics: its legacy and those who shaped it" Kennedy Inst of Ethics J 4:319-335; (2005); "The word Bioethics: the struggle over its earliest meanings", Kennedy Inst of Ethics J 5:19-34.

[6] Sass HM (1992) "Bioethics in German speaking countries", Bioethics Yearbook 2, Dordrecht: Kluwer 211-231; (1995); "Bioethics in German speaking countries", Bioethics Yearbook 4, Dordrecht: Kluwer 247-268; (1996); "Bioetika u Europi", Društvena istraživanja 23/24: 629-649.

[7] Engels EM (1999) Article "Bioethik" in: Metzler Lexikon Religion. Stuttgart, Weimar: Metzler, 159-164; Sass HM (2008) "Fritz Jahr's 1927 Concept of Bioethics" Kennedy Inst of Ethics J 17:279-295.

Bioethical Imperative, has slowly penetrated the world and signals e new integrating and integrative step forward into a global future. Not knowing about Fritz Jahr at that time, Edmund Pellegrino stated already in 1988: 'It is without risk to predict that changes, caused by bioethics in each and every country of the world, one day will be judged by historians as key events in medical history and in human history'.[8]

It may very well be said that the definitions of bioethics are as numerous as bioethicists themselves. Eve-Marie Engels defines: bioethics is 'a major field of applied ethics aiming at a normative understanding of the space and limits of human acting with respect to the living nature'.[9] Others, understanding bioethics as a discipline devoted to the study of 'moral problems of the life sciences', try to discern theoretical bioethics, clinical ethics, regulatory and policy bioethics, and cultural bioethics.[10] One of the most comprehensive and forward looking definitions of bioethics as a yet unfinished business on a global scale has been offered by Hrvoje Jurić, comparing it to 'an open point of meeting and dialogue of various sciences, activities, and worldviews.'[11] A first International Conference 'Fritz Jahr and European Roots of Bioethics: the Establishing of International Scholars' Network' was held in Rijeka and Opatija (Croatia) in March 2011 (www.eurobionethics.com). Some contributions in this volume were presented at that conference, which had been supported by the Kennedy Institute of Ethics at Georgetown University in Washington DC and by the Zentrum für Medizinische Ethik in Bochum. Other international conferences are scheduled for 2012 in Saõ Paolo (Brazil) and 2013 in Asia. The book, you have in your hands, is the first dedicated to integrative bioethics and its foundations on a global scale.

We have grouped the contributions in this volume in a natural way in presenting a sort of bioethics biography by orienting ourselves and the reader in processes well established by all form of *bios*. Fritz Jahr might have approved of such a sequence, demonstrating anf following the biography of a moral and cultural intuition and its implementation in many traditions, cultures, academia, and in a growing global dialoge. First, we present one of the *seeds* of modern integrative bioethics, i.e. 16 articles by Jahr (1-16) in English translation by EM Miller and HM Sass. Then there are articles

---

[8] Pellegrino E (1988) „Preface“, Sass HM et al., „Bioethik in den USA“, Berlin: Springer, 17.

[9] Engels EM (1999) Article “Bioethik” in: Metzler Lexikon Religion. Stuttgart, Weimar: Metzler, 159-164.

[10] Callahan D (2004) Article “Bioethics”, in: Encyclopaedia of Bioethics. New York: Macmillan, 278-287.

[11] Jurić H (2007) „Uporišta za integrativnu bioetiku u djelu Van Rensselaera Pottera“, in: Integrativna bioetika i izazovi suvremene civilizacije. Sarajevo: Bioetičko društvo u BiH, 77-99.

discussing the *roots* (17-27) of modern integrative bioethics, followed by articles on major *stems* (28-36) of developing contemporary global bioethics. The final section presents new and growing important *branches* (37-44) of integrative and integrating bioethics with an outlook into the future.

A Seed of Integrative Bioethics. The 16 articles by Jahr represent just one set of seeds for contemporary and future integrative bioethics. Jahr mentions already 1927 in his seminal article 'Bio-Ethik' the names of Buddha and Francis of Assisi among others as seeds in their respective cultures for the development of the Bioethical Imperativ and for a culture of compassion and respect for all forms of life (1). Other articles elaborate on details in animal and plant ethics (3), the importance of a day of rest for all forms of live and the secular bioethical origin of the Sunday (11; 16). Actual issues of the 21th century in implementing bioethical attitudes and practice are the role of modern mass media (4) and of character formation (7), also the difference between faith and deeds in Christian life (14; also 13). Egoism and altruism are identified as interacting and integrating capacities and attitudes (6; also 9), appreciating cultivated forms of egoism as virtuous and indispensable in all forms of life fighting for survival and for living a good and cultivated life. Discussions on sexual ethics (5) in the culturally challenged 1920s may serve as case studies for addressing comparable issues in our times. Quite unique and definitely in the center of Jahr's understanding of integrative bioethics is the inclusion of spiritual and unseen worlds (2; 10; 15) integrated into the visible and invisible cycles of life; previous and later generations than ours have found and might again find easier access to these visions than many of us today. Of course, doubts about religious traditions have always played a role in the autobiography of morality and culture (8; 12); Jahr uses the narrative of Richard Wagner's *Parsifal* for Jahr as an opportunity to address the virtue of compassion in recognition of the many ways of suffering life (10).

Roots of Integrative Bioethics. Eleni Kalokairinou (17), Nicosia and Tessaloniki, traces the roots of modern integrative bioethics back to ancient Greek physicians-philosophers such as Hippocrates, Galen and Celsus; Ivana Zagorac (18), Zagreb, focusses on Francis of Assisi, while Igor Eterović (19), Rijeka, discusses the relationship between Kant's Categorical Imperative and Jahr's Bioethics Imperative. Eve-Marie Engels (20), Tübingen, underlines the importance of Darwin's theory for Jahr's conception of bioethics. Fernando Lolas-Stepke (21), Santiago de Chile, discusses Jahr's vision in the light of Victor von Weizsäcker's anthropology and cultural theory. José Roberto Goldim (22), Porto Alegre, compares Albert Schweitzer to Jahr; both had published in *Ethik. Sexual- und Gesellschaftsethik* and must have known each other's position, only Schweitzer becoming prominent during his lifetime, different to Jahr. Marija Selak (23), Zagreb, presents Karl Löwith as one of the few 20th century European thinkers with a broad global view and a critique of narrow-minded anthropocentrism, while Hrvoje Jurić (24), Zagreb, discusses Hans Jonas' quest for responsibility in a philosophy of biology and life. Van Rensselaer Potter's (25), Madison WI, last e-mail communication to his friends in a bioethics network from 2001, which included Ivan Šegota of Rijeka, Croatia, together with the text of Potter's video-address to the participants of the 2001 Lošinj (Croatia) bioethics symposium, are the only contributions not commissioned for this volume, but demonstrating Potter's views of a growing global interest and movement in integrative bioethics. Irene M. Miller (26), Harrisville NH, reflects on Jahr's very personal life and reasons for his late recognition by presenting an early CV and two yet unpublished letters by Jahr. Amir Muzur and Iva Rinčić (27), Rijeka, present first detective findings on Jahr's life and on the birth and early years of recognizing his Bioethics Imperative on a growing global scale.

Stems of Integrative Bioethics. H. Tristram Engelhardt, Jr. (28), Houston, focusses on unavoidable and real cultural wars in the coming global age of multiculturalism in the first article in this section, while Hans-Martin Sass (36), Arlington VA and Bochum, discusses the many yet further to be detailed faces of the Bioethics Imperative at the end of this section. The next four contributions focus on the European stem of developing integrative bioethics. Ante Čović (29), Zagreb, had been a visionary promotor of integrative bioethics in the multifacetted societies of Southeast Europe. Matti Häyry and Tuija Takala (30), Manchester and Helsinki, discuss Jahr's Bioethics Imperative as a mid level principle and report on previous European attempts to establish a more genuine European vision in bioethics. Judge Christian Byk (31), Paris, reflects on and summarizes the role of biolaw for European bioethics following Fritz Jahr. Iva Rinčić and Amir Muzur (32), Rijeka, review the various institutional settings for bioethics in the European scene. Michael Cheng-tek Tai (33), Taichung, Taiwan, reviews Jahr's vision of compassion as a virtue and a principle in the light of Asian traditions and confronts it with a model, he has proposed earlier[12]. Darryl Macer (34), Bangkok, compares Jahr's position to Asian reasoning and his own concept of bioethics as love for life. James Giordano, Roland Bendikter, Nicola Boris Kohl (35), Arlington VA, Oxford UK, Munich Germany, present scientific facts and insights of comparative and integrative neuroscience essential for developing an integrative neuroethics.

Branches of Integrative Bioethics. Jayapaul Azariah (37), Chennai (India) confronts Jahr's and Asian bioethics critically to contemporary bioethics as strongly influenced by American culture. Arnd May (38), Halle an der Saale

[12] Tai MC (2008) "The Way of Asian Bioethics". Taipei: Princeton Intern Publ.

discusses his experiences in clinical ethics consultation as living entities. Nada Gosić (39), Rijeka, underlines the importance of Jahr's model of character formation in general education and in bioethics education. Geni Maria Hoss (40), Curitiba, Paraná, Brazil, analyses the role Jahr's Bioethics Imperative can play in Christian ethics and environmental concern and action. Hans-Martin Sass (41), Arlington VA and Bochum, discusses modern multiverse models in physics and biology in the light of Jahr's call for respect of all forms of life. Natacha Salomé Lima (42), Buenos Aires, starts a dialogue among Jahr, Schweitzer and Freud on issues in psychoanalysis. Hans-Martin Sass (43), Arlington VA and Bochum, uses Jahr's hermeneutics of the 5th Commandment as a role model to support the importance of being in touch with traditions and to question contemporary bioethics and medical, making no or limited use of the treasures in the life history of cultures. The 2011 Rijeka Declaration on the Future of Bioethics (44) is just another branch of the strongly developing future global and integrative bioethics; other declarations, conferences, reflections and activities will follow in the future of yet unfinished integrative bioethics.

We write this introduction 85 years after the publication of Jahr's editorial *Bio-Ethik* in the January edition 1927 of the journal *Kosmos*. The global future of integrative and integrating Bio-Ethics has just begun. We thank Iva Rinčić and Arnd May for valuable work in reviewing and editing the manuscripts for print.

January 2012
Rijeka, Croatia; Bochum, Germany; Arlington VA, USA

Amir Muzur, Hans-Martin Sass

Section 1

A Seed of Integrative Bioethics: Articles by Fritz Jahr

BIO-ETHICS, 1927

Reviewing the ethical relations of humans towards animals and plants

The strict distinction between animal and human being [Mensch], dominant in our European culture until the end of the 18th century, cannot be supported anymore. Up to the French Revolution, the heart of the European human being was struggling for a unity of religious, philosophical, and scientific knowledge; but such a unity had to be abandoned under the pressure of more information.

It will always be the credit of modern natural sciences to finally render an unbiased study of the world [Weltgeschehen]. We would not be seekers of truth today, if we would have given up the results of animal experimentation, blood research etc.. On the other hand, we cannot deny that precisely these scientific triumphs of the human spirit have infringed upon the dominant position of the human being in the world in general. Philosophy, formerly prescribing leading ideals for the natural sciences, now has to build her systems on the basis of specific knowledge from the natural sciences, – and it was only a poetic-philosophical [dichterphilosophische] interpretation of Darwin's insight, when Nietzsche considered humans to be a somewhat inferior stage towards a higher stage in evolution, as a 'rope extended between animal and superman [Übermensch].

What resulted from this revolution? First, the fundamental equalization of human being and animal as an object in psychology. Today, it [psychology] does not restrict itself to human beings, but applies the same methods to animals as well; and, as documented by comparative anatomical-zoological research, quite instructive comparisons between human soul and animal soul have been done. Yes, even beginnings of plant psychology are visible, – the most prominent representatives are G. Th. Fechner in the past, R. H. France, Ad. Wagner and the Indian Bose at present, so that modern research in psychology covers all living beings in research. Given these circumstances, it is only logical when R. Eisler speaks of Bio-Psychik (science of the soul of all that lives).

From Bio-Psychik it is only a step to Bio-Ethics, i.e. the assumption of moral obligations not only towards humans, but towards all forms of life. In reality, bio-ethics is not just a discovery of modern times. An especially attractive example from the past is the figure of St. Francis of Assisi (1182-1226) with his great love towards animals, his warm sympathy for all forms of life, centuries before Rousseau's romanticism for the entire nature.

When the unity of the European weltanschauung broke down at the end of the Baroque period, European intellectual life for the first time was able to receive foreign worlds of thought [Gedankenwelten] without prejudice. Already Herder's comprehensive spirit – probably the most sensitive in those days for things to come – expected of humans, based on the image of an all encompassing deity, that they project themselves into each and every creature and sense with it the way it needs. Such a reasoning already reminds us of the Indian philosophy, which by the way of England just had been discovered. But only during the time of Romanticism has India really influenced European intellectual life, and especially in Germany, its most important province. The teaching of reincarnation, as developed in India, has influenced the reasoning of Indian schools of philosophy, especially the school of Sankya. An offspring of this school is the yoga teaching, drawing the most rigorous consequences from those thought processes. The yoga repentant [Jogabüßer] under no circumstances is allowed to live at the cost of co-creatures; above all, he shall under no circumstances kill any animal, and only under certain settings enjoy vegetable foods. He has to wear a veil over his mouth in order not to inhale even a small living being; for the same reason he has to filter drinking water and shall not take a bath. The passion to not harm a living being in the process of self-preservation even leads some Indian repentant to eat horse manure. If in this context Buddha is mentioned, one has to stress that especially this religious leader refused such fanatic self-harm of the school of yoga. Buddha forbade, that food be based on animal products, but fully allows vegetable based foods. How much Buddha himself and his teachings totally believed in re-incarnation of the soul, is very well demonstrated for us Europeans by the collection of Buddhist stories collected by Jatakas, stories ascribed to Buddha and narrating about his early life. He claims that he has lived as a human being before, but also remembers his former lives as an elephant, a gazelle, a crab etc.. Even more beautiful than in Francis of Assisi, these narratives express the thought, that a human being in essence is related to all creatures.

Such sequences of reasoning caused similar thoughts in European intellectual life, even if not in such a strict version. Theologian Schleiermacher (1768-1834) declared it to be immoral to destroy life and formation [Leben und Gestaltung], as they are, if there is no reasonable cause to do so. Similarly, the philosopher Krause, a contemporary of Schleiermacher, requests to respect each and every living being and not to destroy it without reason. Because, they all, plants and animals, also humans, have similar rights, but not Equal Right, depending on the requirements for reaching their specific destiny. The philosopher Schopenhauer, who claimed as special importance of his ethics as based primarily on the sentiment of compassion, required towards animals as well, openly referred to the Indian intellectual world

[Gedankenwelt]. Via Richard Wagner, who was strongly influenced by Schopenhauer and a compassionate animal lover and friend of animal protection, those thoughts have become a common value for a broadest group of people.

Thus, in regard to animals such a rule has become evident, at least as far as needless torture is concerned. With plants it is different, so. For some, it seems at first unreasonable to have certain ethical obligations towards plants. But already [Apostle] Paul directed our compassion towards animals and plants. Comparable are the illuminated sentimental [verklärt stimungsvollen] interpretations in Richard Wagner's 3rd act of 'Parsifal'. In pious devotion, humans at least on Good Friday avoid hurting stalks and flowers in the fields by walking more carefully. But also in the thoughts of plant ethics by a sober philosopher such as Eduard von Hartmann, who passed away 20 years ago, we find similar thoughts. In an article on flower-luxury he describes a cut flower: 'She is an organism deadly hurt, but only her colors not yet destroyed, a head still there, but separated from the torso. – Whenever I see a rose in a glass of water or tied into a bouquet, I cannot fight the unpleasant thought that a human being has murdered a flower life for the sole purpose to enjoy his/her eyes while dying, heartless enough eyes, not to sense an unnatural death under the appearance of life.'

A majority of people naturally is not as sensitive as Ed. von Hartmann. However, everyone knows quite well, that plants are living beings, and that cutting flowers hurts them; but the thought that the flower might sense it, is far away. The concept of a plant-soul so far has not taken hold in us. Additional, we know that flowers also die and dry out, while they are on the plant, and therefore one does not take issue with cutting flowers, in particular when they were cultivated for that specific purpose.

Thus, we start from a totally different point of view than the Indian fanatics, who do not want to hurt any living entity. Also, our regulations by law and police protect specific plants and flowers in certain areas (such as plants in the Alps) are based on totally different assumptions. The police state [Polizeistaat] intends to protect those plants from becoming extinct in those areas, also to be enjoyed by other people in the future. Whenever there are plants abundant, the state does not intervene to protect them as an end in themselves.

Also, our concept of animal protection rests on a decidedly different foundation than the attitude of the Indians. When we read in the novel 'Holy Hate' [Der heilige Hass] by Richard Voss, that a Rodyia-boy, i.e. a member of a despised caste, does even not want to kill a snake, because 'also the snakes are our brothers and sisters', we do not accept such a reasoning; we actually hold it to be our duty to kill harmful animals, if we can. We have our farm animals been killed by the butcher and the harmless prey by the hunter, because we want to eat meat, which in our areas some do not want to do without, while in tropical countries vegetarian food is abundantly available. Our animal protection, thus, has a utilitarian aspect, which is daringly disregarded by the Indians, while we content ourselves with avoidance of unnecessary suffering. Unfortunately, legal regulations against prevention or punishment of those cruelties are not strong enough in all civilized countries [Kulturländern] yet. But, we are on the road of progress and animal protection get more and more support in wider circles, such as no decent human being [anständiger Mensch] will without criticism accept, that a thoughtless lout [Flegel] without any afterthought beheads flowers with a stick while on the hike or that children break flowers only to through them away after a few steps. Our self-education, in this regard, already has made considerable progress, but we have to go further, so that the guiding rule for our actions may be the bio-ethical demand: *'Respect every living being on principle as an end in itself and treat it, if possible, as such!'*

Bio-Ethics, 1927. Bio-Ethik. Eine Umschau über die ethischen Beziehungen des Menschen zu Tier und Pflanze. *Kosmos. Handweiser für Naturfreunde* 1927, 24(1): 2-4

DEATH AND THE ANIMALS, 1928

Contemplating the 5th Commandment

'Thou shalt not kill', thus are we Christians warned by the 5th commandment. Does not killing always relate to something alive? Since the commandment does not expressly forbid the killing of humans only, should we therefore not also apply it to other living beings, especially the animals? Perhaps the artist-painter Fidus was right on target with his painting of 'You shall not kill' [Du sollst nicht töten]: a child, reflecting innocence and purity, spreads its own protective arms in front of a deer to save it from the hunter's deadly bullet!?

Are animals so close to us that we should acknowledge and treat them as our 'neighbors'? – No doubt, considerable differences do exist between man and the animals, and modern sciences acknowledge this. However, that does not prevent biology and the life-sciences to utilize related trends in practical application, especially since many of these trends had also been acknowledged by Darwin. To mention here: animal experiments, blood testing, blood-protein research, transplants of animal tissue to man, as well as others. On a spiritual plain, interesting parallels between man and animal have emerged as well, so that both share not only physiologic but also psychological 'nearness' [nahestehen].

We should not find this disquieting; to the contrary, we should be proud that the discoveries of adventurous human spirits in recent times already are contained in the core of the Holy Scriptures. Genesis, the 1st book of Moses, speaks of a 'soul' of animals (Gen. 9:16). The Preacher Salomon also presupposes a soul in them, similar to humans and asks doubtfully: 'Who knows, whether the spirit of man goes upward, and the spirit of the beast goes down into the earth?' (Ec. 3: 21)

In any case, all creatures, be it man or beast, long for delivery from death and transiency, as taught by the apostle Paul in the epistle to the Roman congregation (Ro 8, 18-23), a sign that already of old a commonality between man and animal was recognized. Small wonder, that later on Franz von Assisi called all living creatures his sisters and brothers, and that Herder, too, saw in animals 'man's older brothers'. If that be so, one can grasp that when God made the covenant with humankind, he did likewise with the animals, as written in the 1st book of Moses (Gen 9:9-10. 17), and in Hosea (Hos 2:18-20). Even in the kingdom to come there will be a place for them, as written in Isaiah: 'Wolves shall dwell with lambs, and the leopard shall lie with the young goat, and the calf and the lion and the fattened calf together; and a little child shall lead them. The cow and the bear shall graze; their young shall lie together; and the lion shall eat straw like the ox. The nursing child shall play over the hole of the cobra, and the weaned child shall put his hand on the adder's den.' (Isa. 11:6-8)

In his table-talks no one lesser than Luther himself shared the belief of an acceptance of animals in the kingdom to come.

Given that the natural sciences as well as the Holy Scriptures hold animals in such high regard, it follows that we Christians have ethical obligations to them. We should indeed apply the 5th commandment to animals. Schleiermacher, most noted theologian of the newer times, expressly considered it immoral to destroy life and formation wherever established, including animals, without reasonable cause.

However, the fulfillment of the 5th commandment extended to animals seems utopian: Slaughter and killing of animals is hardly avoidable, even if done only for needed supply of a growing population. The struggle for life [Kampf ums Dasein] makes this a necessity. However much we may regret it, this same principle also influences our ethical behaviors towards our fellow man. In our entire life and activities, be it in politics, in economics, in the office, in the workshop, in the fields, the basic goals are in no way primarily directed by love, but rather by struggle with some kind of competitors of ours. Mostly we are not quite aware of this, as long as the struggle does not breach the limits of the law. In such struggle for life we humans deliberately and consciously use human power, human health, human life, not only in times of war, but also in 'peaceful' life such as in cultural development, especially in some of the industries. In spite of all this no one considers the 5th commandment a utopian charge. As our attitude towards animals – as determined by struggle for life – basically does not fall outside our attitude towards man, the commandment can and must be valid here as well, an ideal and a point of reference for our moral strife.

But what effect will this ideal have in reality, given the mentioned restrictions in sparing the lives of animals? – Schleiermacher requests to kill animals only then, when a reasonable cause demands it. Besides, animal protection laws and societies give us further guidelines on how to be 'merciful to our beasts'.

The admonition to spare animal-life is absolutely valid, without any regard to being to our advantage, as ethics in general does not and may not ask those questions. Richard Wagner, as does Schopenhauer, stresses this point especially: 'Everyone, who at the sight of a tortured animal suffers outrage, is so moved singly by compassion; and whoever unites with others for the protection of animals is solely motivated by compassion, an empathy by nature indifferent and inconsiderate of any calculated gain.'

In view of this principle it is interesting and useful to raise the following question: 'What effects will this expansion of our ethical duties beyond man to animals have on our relationship to our fellow man? Should we fear that the attention to the needs of our fellow man could be diminished because of our obligation to the animals?' – Exactly the opposite occurs: When we nurture in our breast a feeling heart toward the animals, we will not withhold compassion and help from suffering humankind. He, who has love so great that it reaches beyond the human-only realm and can see holiness in the poorest creature, he will also see and highly regard the divine in his poorest and lowest brothers, he will not limit himself to a certain societal class, interest group, or certain party. On the other hand, cruelty to animals is a sign of an unrefined character, which could also be dangerous to the human environment. Many great thinkers attest to this fact and the philosopher Kant also states that the compassionate and merciful dealings with animals are a duty man owes to himself.

Above all, the sparing of animal life as far as possible is a duty to God; if we want to honor the creator, we must regard with awe and respect his creatures, the animals as well, the more that we know he loves them too (Jon. 4:11) and bore them in mind as he commanded: 'Thou shalt not kill!'

Death and the Animals, 1928. Der Tod und die Tiere. Eine Betrachtung zum 5. Gebot, *Mut und Kraft*, Halle 1928, 5(1): 5-6

ANIMAL PROTECTION AND ETHICS, 1928

Compassion with animals shows up as an empirically given phenomenon of the human soul. This fact has been voiced, among others, by the poet and philosopher Herder in his 'Ideas in History of Humankind' [Ideen zur Geschichte der Menschheit]. The fact, that this phenomenon is more or less present in a normal human soul, is also assumed by the German Penal Code when expressing in paragraph 360(13) that cruelty to animals is a public offense. Exceptions, which of course are there, cannot change the truth of such a psychological observation, similarly as the existence of blind people can be used to argue that the capability of seeing is not an essential part of humans. Such compassion now is the central motive of the idea of animal protection [Tierschutz]. Thus, it does not calculate whether or not one has any benefit from it; e.g. Richard Wagner in an open letter to Ernst von Weber, under the influence of Schopenhauer's 'On the Foundation of Morality' [über das Fundament der Moral], argues: 'Everyone, who reacts indignant when witnessing the suffering of an animal, is motivated solely by compassion, and who works together with others in protecting animals does so similarly only out of compassion, naturally not of calculations of utility by an otherwise unconcerned and determined compassion.'

And if someone does not accept the unconditional nature of Animal Ethics (as one may call it, following Brenzinger, who was the first to publish a scientific Animal Ethics [Tierethik] from an ethical-legal position), one nevertheless may not suppress the question of the relationship between animal protection and ethics, respectively seek to answer such a question. In other words: what are the consequences in our relationships towards our fellow humans when we extend our moral obligations beyond humans towards animals? Don't we have to fear, that we will turn away our awareness from the misery of the latter towards the first? The philosopher Eduard von Hartmann, who, by the way, is not hostile against animals, voices those concerns in his article 'Our Relationship towards Animals' [Unsere Stellung zu den Tieren]. He gives the example of a 'dry old maid', feeding meats and sweets to her pug while letting her employees go hungry. He also finds love towards animal among embittered misanthropes, cold and cruel judges of heretics and bloodthirsty heroes of revolutions. While there are indeed those cases, Hartmann's arguments only target *false* animal love. This kind of false love also may be used towards humans. It is expressed in disgusting pampering, in unjustifiable preferences, in cronyism [Vetternwirtschaft], and unfortunately widely spread otherwise. But if such false love of people is no good argument against ethics, so is occasionally occurring false love of animals no demonstration against the justification of animal protection.

This is the issue: If we have a compassionate heart towards animals, then we will not withhold our compassion and help towards suffering humans. If someone's love is great enough to go beyond the borders of human-only and sees the sanctity even in the most miserable creature, he or she will find this sanctity as well in the most poor and lowest fellow human, will hold it high and will not reduce it to class of society, interest group, one party or what else may be considered. On the other hand, senseless cruelty towards animals is an indication of an unrefined character becoming dangerous towards the human environment as well. Among other thinkers, philosopher Kant expressively has hinted at this fact of highest importance for social ethics, when in 'Metaphysische Anfangsgründe der Tugendlehre' he calls the careful and compassionate treatment of animals a human obligation towards oneself. The word from Count Leo Tolstoi: 'From killing animals to killing humans, it is only one step' might be overly strong; but his position expresses Kant's concept and understanding. This is also the case for R. von Hippel, a jurist, who has collected and organized most of the relevant historical and statistical materials.

But effective and successful animal protection is only possible, when enough knowledge and at least some understanding of nature are present. The reason is, that we only can protect animals in reality, when we are somewhat knowledgeable about their physiological and psychological properties and life conditions. Therefore one of the main goals of the animal protection movement is the promotion of such a kind of knowledge and a better understanding of nature, – awakening, broadening, and deepening it. Such an interest in nature will by itself not limit itself to animals, but on the other side also towards plants and (what is most important in this content) towards humans as well. If such a goal will only be reached in part, then we can, for sure, expect a positive influence on humans and their way of life, i.e. in the attitude of a normal and healthy naturalness, which has nothing to do with a limitless life of over-excited, unhealthy, and thereby unnatural urges, which quite often wrongly are considered to be natural. The fact that the promotion of knowledge and understanding of nature and of a true love for nature will also have a positive effect on sexual ethics does not need to be demonstrated additionally.

If, indeed, it is true that a correctly understood and executed animal protection works positively on ethics, then it is also true that it has a value in public education and public knowledge; and this may not be underrated at all. On the other hand, everyone, who is active in animal protection, will support as much as possible general ethical activities, which, as already mentioned, cannot ignore or be silent about animal ethics, because this is also indirectly supportive of animal protection.

The fact of a close interrelationship between animal protection and ethics finally is based on the reality that we not only have moral obligations to fellow humans, but also to animals, even to plants – in short: to all forms of life – , so that we can speak about 'Bio-Ethics'.

As such, Bio-Ethics is not just an idea of modern times. – Already Montaigne, a skeptic, was the first Frenchman – so far being the first representative of a modern ethos of sentiment – who dared to reason, that all living beings have entitlement of being treated based on ethical principles: we owe justice towards humans, mildness and compassion towards all other creatures who will have a benefit from it. So he wrote in his 'Essays' in 1588. – Precisely in the same sense, Herder expects from humans to follow the example of a God, who transfigures with sentiment every living entity and is empathetic with it as much as it can feel and need. He expressively includes plants in this. – Highlights are reached by the theologian Schleiermacher and philosopher K. Chr. F. Krause. The first one, in his 'Philosophische Ethik', declares it to be immoral to destroy life and forms of life, wherever they are, i.e. including animals and plants, without a reasonable cause associated with such an act. The latter, a contemporary of Schleiermacher, requests in his 'Rechtsphilosophie' that every living being be treasured as such and not be destroyed without reason: because they all, plants and animals, also humans, are of equal standing [gleichberechtigt]. But they are not identical, and each only in a way which is a necessary requirement to reach its destination. So we read in Krause's 'Abriss der Philosophie des Rechts'. – A note in the diary of the poet Hebbel reminds us of the intuition of Herder, according to which not only humans, but everything living and moving [was lebt und webt] sees an inscrutable divine light, which one can access only via love.

It must be mentioned in this context that one has tried and still is trying – the longer the more – to support the bioethical thought by biological and biopsychological arguments, and not without success.

At first hand, it might appear utopian to realize such moral obligations towards all living entities. But we may not overlook that such moral obligations towards a living entity in practice are determined by its 'needs' [Bedürfnisse] (Herder), respectively its 'destiny' [Bestimmung] (Krause). So, the needs of animals seem to be lower in quantity and less complicated in content than those of humans. This is even more true for plants, so that practical moral obligations, which are already there towards animals (if not basically, but practically), create less difficulties. Additionally we have to take into account the principle of struggle for life and existence, a principle which in some way also modifies our obligations towards fellow humans, even if we might feel unhappy about it. Our entire life and activity in politics, in business, in administration, in the laboratory, in the workshop, in the fields is – as Naumann has underlined – in its reasoning and goals not focusing on love in the first place, quite often rather focused on struggle with some sort of fellow competitors. Quite often we don't recognize it, as long as such a struggle is without hate and in an open and legally accepted way. As much as we cannot avoid the struggle with fellow humans, similarly the struggle for survival [Kampf ums Dasein] with other living entities is unavoidable. Nevertheless, neither in the first nor in the second case, we will lose the idea of moral obligations as a principle. The paragraphs of animal protection laws in the penal codes of civilized countries and the activities of animal protection societies give testimony, in which ways animal protection becomes practical. In the field of plant ethics, our intuition will hinder us from killing plants right and left with a cane during a stroll, or from picking flowers only in order to throw them away after a short while, or when we detest the blind acts of destruction by raffian lads breaking the crowns of young trees along the road or in the woods.

From all this follows as guidance for our moral acts the *bio-ethical Imperative: Respect every living being, including animals, as an end in itself, and treat it, if possible, as such!* And if someone does not accept the validity of this principle, as far as it is concerned with animals and plants, then, in repeating what already was said, one nevertheless should follow it in recognition of the moral obligation toward human society in general.

Animal Protection and Ethics, 1928. Tierschutz und Ethik in ihren Beziehungen zueinander. *Ethik. Sexual- und Gesellschaftsethik. Organ des ‚Ethikbundes'* 1928, 4(6/7): 100-102

SOCIAL AND SEXUAL ETHICS IN THE DAILY PRESS, 1928

All ethical reflection gets a real purpose only when work is done, either theoretical or practical. Even as practical confirmation is most important – especially by role model – nevertheless theoretical enlightening information [Aufklärung] by word and text should not be underestimated. – As far as the spoken word, i.e. the form of private influence from person to person, is concerned, also lectures and presentations, the success by means of speaking and talking should not be overestimated, in particular since these lectures sometimes do not draw too many people. Radio probably has a much greater influence.

This leads to the printed word as far as theoretical enlightening information is concerned. At first, we have to mention the *professional press*. The professional press has the great advantage that its opens its pages exclusively to issues of controversial ethical problems, being basically more or less independent from those people, who are neutral or even hostile to these concerns. But a certain deficiency is connected with such an advantage. The professional journal reaches only those, who already are members of such a movement or otherwise close to it. Therefore, an immediate influence on the masses is missing. – Such a deficiency is even greater as far as books are concerned; on one side the purchase price is deterring, on the other side, not everyone has time and leisure to read a not so easy to digest and sometimes quite voluminous publication. Additionally, the book market is not short of publication on ethical issues, so it is difficult to make the right selection.

Thus, the most important issue is still to review the *daily press* in regard to its importance for ethics. From the point of view of the 'Ethikbund', newspapers are not without deficiencies. The daily press by far and large has not the same space available as a professional journal or even a book. Also, the daily press sometimes has to be careful to handle certain issue in recognition of its sponsors, parties, advertiser or subscribers. – This deficiency however is balanced with great advantages. First of all and most importantly, there is wide distribution of the daily press. Immense amounts of paper are used daily to produce newspapers. Millions of eyes review on a daily basis the multimillion lines of newspapers. Not even the most widely read professional journal nor a book can compete with that. Such an incredible distribution is matched with a similar incredible influence, whatever else one thinks about the usefulness of newspapers. Events going on in the city, in the county, in the fatherland, activities in the wide world, – where do we get information about those if not in the daily press? It also analyses the big political and economical issues and developments, – things we need to know in order to recognize and to fulfill our social obligations; – we also get information about food prices, how to dress, what the costs of certain garments are, what other products are recommendable, which is not in direct but in indirect relationship to ethics. And as the newspaper knows, men don't live by bread and other daily stuff alone; it reports about intellectual and spiritual issues as well, and so ethics cannot be missed. Finally: as newspapers will only on and off bring articles of ethical content, not too long and preferably in a feuilleton style, the creation of interest and understanding among lay readers – also from this perspective the importance of the daily press for ethics, including social ethics and sexual ethics – should not be underestimated.

Thus, the relationship between newspapers and ethics must be underlined very strongly and in any case. This is independently from being a supporter or enemy of the daily press. Also irrelevant is whether one considers the press to be an expression of public opinion or rather being the 'producer' or modifier of such a public opinion in the first place. – The newspaper friend will use it as an instrument of ethical work, in particular because in this way he will not only have an opportunity to express ethical reasoning but also to distribute it widely. – But even the sworn enemy of the newspaper will not refrain from at least to work with or support it, even if only for the purpose (if this is justifiable or not, is irrelevant in this regard) to influence it positively, The *possibility* of such an influence is always given, if one recognizes the newspaper as an expression of public opinion. And if one considers the pages in the newspaper to be a medium to establish or at least strongly influence public opinion, then from an ethical point of view, it becomes even an obligation, to take part in this type of character formation [Gesinnungsbildung], – actively and with best knowledge and conscience.

To start with the most suitable newspapers is a matter of pragmatism. Also important is holding close contacts with correspondents. Best advice can be sought from experts in both areas: ethics and newspaper expertise. To use this opportunity is of particular importance for the good cause, to which social ethics and sexual ethics definitely belong, and must be recommended strongly.

Social and Sexual Ethics in the Daily Press, 1928. Soziale und sexuelle Ethik in der Tageszeitung. *Ethik. Sexual- und Gesellschaftsethik. Organ des ‚Ethikbundes'* 1928, 4(10/11): 149-150

WAYS TO SEXUAL ETHICS, 1928

The sexual problem is one of the hottest problems in present-day ethics. One should not miss, that in olden days, yes, even in ancient times the same topic had surfaced among the most different peoples. Only, it was not given the broad latitude as in younger days, especially our present.

As a means, perhaps the only means, to stave off and prevent sexual-ethics harm to youth as well as adults, since the 18th century until today, was adequate rational information [Aufklärung] about sex life and its physiological base. For example, Rousseau, whose 150th anniversary of death we celebrate this year on the 2nd of July, in his 4th book of the education novel 'Emile, ou de l'education' advocates reasonable sexual information for young people above the age of sixteen. Rousseau regards it such an essential part of reasoned enlightenment, that he even planned visits for his imagined pupil Emil with syphilis patients and brothels, all for clearness' sake. As a significant friend and advocate of Rousseau's ideas, Basedow, a contemporary and acquaintance of Goethe, gained notoriety in Germany. He likewise advocated sex education [Aufklärung], as did many youth-educators with and after him. In these circles the suggestion originated to demonstrate suitable cadaver resections for better visualization of sex education. We see clearly that reasoned teaching on sexual matters is just as valuable today as in the 'times of 'Enlightment' [Aufklärungszeit], as Kant calls it. Numerous lectures, exhibits, films, writings and sundries attest to this (some among them worthless and even harmful).

How to judge this diffuse rationality-oriented attitude towards issues in sexual ethics? – Without doubt, thought and rationality can and must be applied to all matters, including ethical as well as sexual ethics, nay – it is necessary. To deny this necessity also means to deny the pursuit of the problem in question, as well as any progress, which might result from it. Above all, that would be the worst possible mistake. In Goethe's 'Faust' Mephistopheles rightfully says: 'If you despise rationality and knowledge, mankind's highest power, – then I have you already, for sure!'

So far, all should be clear. Only, we may not overlook that man is not exclusively a thinking being, rationality is not representative of all the other impulses of the mind. The omission of this long established psychological fact is widespread and it is often forgotten that man is endowed with strong drives and sentiments. And because of the strength of these inner drives, it is extremely difficult to subordinate sexual ethics to rationality alone: we have to recognize that the sexual drive may be especially powerful. In general, these drives are so strong, that man commonly gives in to them, even if reason tells him that he will be harmed. As an example of the real drive for self-preservation: a drowning man will grasp the last straw, as the proverb says, or, also his savior, if there is one, even though he knows it will complicate his rescue, might even make it impossible. The drive for self-preservation works blindly in this case and reason has no power to sway. A similarly impelling force of resistance exists in sex drive, especially if strongly endowed or deranged through illness.

To regard the last extreme case: A homosexually oriented male in his soul might regard his sexual orientation as wrong, unnatural, or ill; so for the remainder of his life he is unhappy. Even though his rationality might regard this as abnormal, he cannot alter that drive. Often he cannot even totally suppress acting on it. Similarly, this also applies to both sexes concerning simple or mutual masturbation, a widespread bad custom. Even the natural, more or less healthy sex-drive will often triumph over reason and teachings. (It goes without saying that a phlegmatic person is in less danger than a sanguine person.) In such a situation it is evident that on occasion well-meant rational teachings can have the opposite of the intended results, especially in individuals who can easily be aroused. – An example of old is found in the Bible: Even though matters of the sex life are easily and freely discussed, any enticing details are missing. Double meanings and sultry atmospheres are absent. Yet until the present day, lecturing to youth from such passages is considered dangerous. And similarly today, we find illustration of this with public educational films: While regular educational films are poorly attended; sex education films are always well attended. This, without a doubt, puts into question that the interest is purely that of reason, but much more – if not always consciously – it is prompted by sex drive.

It is nothing new, that rationality alone is not the only momentum driving the demands of promotion of sex ethics. Formerly mentioned Rousseau and Basedow are outspoken opponents of this exaggerated 'reasonable-ing' [Vernünftelns]. Already in yonder times attempts were made to directly influence, from childhood on, the so-called 'moral feeling' [moralisches Gefühl] in sex ethics. Of course, this was aimed at the furthering of sex ethics in adults. However, focusing on practical issues and effects, let's consider the following.

What practical consequences would result from the significant impact of human drives and their influence and impact on sex ethics? It is self-evident, to explicitly repeat the already stated, that one cannot do without rational sex education. It is only our reason that can convince us to consider seriously not only the rational aspect, but the physiological and, in the most modern sense, especially the psychological aspect as well. One of the means to this is to care for active religiosity. Much is expected of the awakening and strengthening of 'moral sentiments' [moralische Gefühle], as they were called during the time of enlightenment; in our specific field it would entail a cultivation of a sense of shame. To accomplish this, it is essential to remove damaging influences. We must keep a keen eye on the widely offered, mostly illustrated literature on sexual topics. This attention must not only be directed towards new publications, but include classical writings. I only want to mention here the (unedited) Tales of a '1001 Nights', 'Decameron' by Boccaccio and 'The Adventures of Casanova'. That these writings were not isolated in the past is proven by the repeated serious warnings of already mentioned pedagogue Basedow against those unsuitable, sensually arousing readings. Most important would be a purposeful lifestyle. Included in this is a balanced diet with properly sized portions at timely intervals, hygienic principles applied to clothing, living quarters (especially sleeping arrangements), that work and recreation are in due proportions to each other (not too much or too little of either), and that instead of social and worldly pleasures, there should be physical exercise and enjoyment in an ever more neglected nature (sports, hiking, gardening, etc.). Guard the golden rule of: Early to bed and early to rise! A further address of important issues (for example the question of alcohol) cannot be made here at this time. Those are best addressed by a competent and experienced physician.

Of special importance in childhood and youth is a direct influence in sex ethics, not accessed from rational education alone; for one must always balance the danger of moral harm against the possibility of a successful impact in education. Here starts the development of sex ethics in the soon to be adult, no proof needed.

Ways to Sexual Ethics, 1928. Wege zum sexuellen Ethos. *Ethik. Sexual- und Gesellschaftsethik. Organ des ‚Ethikbundes'* 1928, 4(10/11): 161-163

EGOISM AND ALTRUISM, 1929

Two basic moral problems: opposition and alliance in social life.

There are two highly important basic issues in individual ethics and social ethics: (I) the egoistic or egocentric, and (II) the altruistic position, also called the sense for justice, compassion, pity, love (not in the sexual sense) etc.

Philosophical and theological academic ethics does not always use these parallel terms in similar ways. It may be allowed for reasons of simplicity to use 'egoistic' respective 'egocentric' as a psychological fact, the result of which is the struggle for life, and 'altruism', 'love', etc. representing the opposite attitude of feeling, willing, and reasoning, connected with appropriate practical consequences.

I. First to the egocentric attitude, i.e. the interest in me myself. We should not neglect that such an attitude originally is an intuitive, instinctive drive. Only secondly, if at all, it becomes a conscientious reasoning form of thinking in regard to a potentially justified struggle-for-life attitude and the best methods in such a struggle. We recognize already the struggle for life among plants, by means of heavy multiplication, unpleasant odor or taste, sting hair, thorns etc., used against predatory animals. – Animals, on the other hand, know how to neutralize plant's self-defense properties. But then, they have to defend themselves against humans with a number of inherited tricks, often without a chance. – Humans, then, use plants and animals for their own personal goals, a fact which does not need to be demonstrated. And it is irrelevant that some farm plants [Nutzpflanzen] (such as crops and potatoes) and farm animals [Nutztiere] (cattle, goats, swine etc.) owe their wide distribution to humans. The last motive, anyway, was and is the self-interest and the struggle for life of the individual and collective I [Ich].

Such egoism is so big, that it does not stop short from an individual's own class. It is well known, that plants take away foods and light from each other and that animals use others as food. In principle, we see the same in humans. This is particularly true of economic life [Wirtschaftsleben], as has recently been pointed out by Naumann. Is it not the drive of the business man to outdo his competition or at least to be equally successful; his dealings with customers serve the same purpose. He would be a poor business man, besides, if he did otherwise! Similarly, national economy and world economy cannot and will not do without making good use of the struggle between different professions and different states. – Even intellectual intentions, at least as far as they are primarily professional, are not an exemption. How does a young person (respectively parents or guardians) act, when determining which intellectually satisfying job for life to choose? – He asks about his main interests and his best opportunities to have a good career, also in consideration and regard to the family. Quite rarely is it asked seriously: 'How can I serve humanity [Allgemeinheit] best with my properties and gifts?' Not even Kant could keep a utilitarian aspect totally away from his 'Categorical Imperative', as Schopenhauer shows in detail in his 'Über das Fundament der Moral'.

Again, it must be pointed out in this context, that these reported facts are not intended to be a demeaning critique. The egocentric attitude and the struggle for survival are extremely important for the creation and development of civilization, respectively culture. Under such an aspect, the consequences are a most blessed result for the community and the individual, even if not planned in the first place.

II. Who, however, only values egoism, such as Stirner in his book 'Der Einzige und sein Eigentum' and Nietzsche with his 'MasterHuman' model [Herrenmenschentum], 'beyond good and bad', does not recognize that there is an altruism, also presenting a natural gift of normal human soul-life [Seelenleben]. Therefore, the empathy for right and fairness, compassion, sympathy, love – or what else we want to call it – first of all has to be recognized as a psychologically given fact and valued as such. If we don't do that or even suppress it, then we suppress and dominate human nature. And if there is someone arguing, that everything is based on genetics and education, then the following question has to be answered: 'How come, altruism is inherited similarly to other human character traits [Seeleneigenschaft]?'. Respectively: 'Why do we target the education of the youth and the entire humankind in this direction and not in any other?', The most satisfying answer still is, that altruism is an empirically tested psychological fact of human soul, consequently we have to take this into account. 'What, now, is altruism?' – It is the fact, that my own I [Ich] retracts totally behind something else, under certain conditions up to the point of self-destruction, and that egotistic motives do not come into force at all. An impressive example is animal protection out of sheer compassion, as understood by Schopenhauer and Richard Wagner and in the understanding of contemporary animal protection societies and animal protection legislation. – Those, who follow their altruistic inclinations, also have a personal satisfaction in doing so; this is another demonstration, that it might as well be an egotistic inclination. On the contrary: he, who with Kant wants only to accept reason or with Eduard von Hartmann cool-logical justice as the single correct motive for ethos, is as close to egoism and additionally sacrifices this phenomenon of human soul life.

Of course, similarly egoism is not without altruistic aspects, so is altruism not thinkable without any egoism. One example for many: Even though Christianity very correctly may be called the religion of love – a love disinterested in loving oneself as any religion of redemption, which can be defined best with the word: 'Work out your own salvation' (Phil. 2:12). Important are also the words of Jesus: 'So, whatever you wish that men would do to you, do to them' (Mt 7:12; Luke 6:31), also: 'You shall love your neighbor as yourself' (Luke 10:27, following Lev 19:18). So, egoism is accepted as existing, and it is presupposed to be self-evident that everyone wants only good things from his neighbor and that everyone cares for himself. Do we want to regret this? – Surely, not easily. How would it be possible to provide for another being, i.e. to care for it in its struggle for survival, i.e. in his egotistic attitude, to protect it and to support it?! In particular, the interrelationship between altruism and egoism shows in the relationship of the individual as part of a community [Gesamtheit], may it be a professional organization, a political party, a village or an urban community, a national organization etc. Without doubt, there are people who submit their entire existence unto such a community [Gesamtheit], even sacrifice for it and in this way behave purely altruistic. As far as this community is concerned, it normally is resolute in struggling with other organizations or persons, i.e. it represents a collective egoism. Such a collective egoism, however, works well for the individual members of such a community in question and is in relation to it altruism as well. If such a positive effect of a specific organization is understood and used (and it has to be this way, otherwise such an organization under question would be useless), then the support of the community will play an important role for the individual as well. (F. Paulsen in his Ethics has expressed this and related issues in more detail).

As far as the evaluation of these two psychological, respectively ethical facts, as discussed above, is concerned, we have to mention the following: Only ethical skepticism seriously supports egcentric attitudes; but these are relatively few exemptions. Much larger is the number of those, who unconsciously make their own self the center of their interest. That this attitude is unavoidable and not necessarily to be criticized has been shown above. However, an exaggeration of egoism associated with endangering the ethos is obvious. These circumstances make it understandable that altruism carries a much higher recognition in public. Even those, who call love a 'sentimentality', routinely refrain from admitting their psychological shortcomings (and, indeed, these they are). Most people, however, possess, depending on their predisposition [Veranlagung], a serious will and confidence to be 'just', 'good', 'altruistic'. The emotional, more or less unconscious higher esteem for altruism over egoism can be recognized by referring to witnesses in the history of ideas [Zeit- und Geistperioden], the first being the ideal, hoped to be realized finally in spite of the latter. Already the old Jewish prophet Isaiah refers to a future kingdom of peace, where the struggle for survival among animals and between animals and humans will have come to an end (Jes 11:6-9). Jesus similarly preaches about the coming of the kingdom of God, this means also the kingdom of love. It is only understandable when Paul in the first letter to the Corinthians, chapter 13, ascribes love a higher standing than faith and hope (verse 13) and attributes eternal existence to it (verse 8). In recent times, Kant, in postulating a moral demonstration of God's existence [Gottesbeweis], shows that he as well has the ideal of an exclusive rule of the good in the future. It should be mentioned that the favored idea in socialism about a 'State of the Future', is an idea which without doubt bears a witness to ethical idealism and optimism.

What are now the most important consequences for social ethics?

1. Egoism and altruism are not necessarily incompatible ethical adversaries.

2. Egoistic mentality as a natural phenomenon thus is also a basic human right. If such a right is used reasonably (natural and healthy lifestyle, especially in regard nutrition, dressing, living and working conditions, also in a regulated and sufficient pay, not to forget also in the struggle for survival which can be regulated largely in justice and fairness), then it works supportive in wide circles and is – at least in its consequences – altruistic.

3. The psychological fact of egoistic attitude may also not be overlooked; and should not be without recognition. This requirement is easier to fulfill since its observation does not always result in personal harm. For example, what is spent in social welfare and in support and improvement of national competitiveness [Volkskraft], returns with interest income, since the state and the economy have a greatest interest to have trustworthy public servants, good workers, financially well off consumers, good development of the youth, and in general the wellbeing of the entire nation.

4. Now, egoism unfortunately quite often is overdone, a fact which is not the case with altruism. Therefore we have to care as much as possible for the belief in and the ideal of love, and to look optimistically at its future realization, e.g. with the faithful words of Goethe's 'Faust': 'Whoever strives hard enough, we will be allowed to redeem him', to redeem also from an exaggorated egoism. Caring for this ideal, of course, will never be without practical consequences. We may mention all activities in

social care, in particular the support of the economically weak, not taking into account whether or not it will pay back (i.e. the care for the old and fragile; again also animal protection). The Christian churches, by the way, give enough references to work in the service of love.

Egoism and Altruism: Opposition and Alliance, 1929. Zwei ethische Grundprobleme in ihrem Gegensatz und in ihrer Vereinigung im sozialen Leben. *Ethik. Sexual- und Gesellschaftsethik* 1929, 6: 341-346

CHARACTER DICTATE OR FREEDOM OF THOUGHT? 1930

Thoughts about a liberal model of character education

Character [Gesinnung], in one way or the other, always is based on morally grounded judgment. Different from character is science, which does not accept value judgments as a basis of work. Of course, science can provide material for a character formation. And without doubt, this is very welcome. – But whenever it is done, it may also happen that no consequences are drawn or – whenever they are drawn – those may differ from person to person. (Germans and Frenchmen consider themselves to be traditional enemies. Some accept this without critique. Others probe deeper, but come to different conclusions: some want to solve the issue by means of war, others by means of agreement and understanding).

On the other hand, people try to influence science and culture and not always without success; they neglect scientifically proven facts of reality and start from a preconceived opinion. This sometimes happens even without a clear understanding of such a process. The mentioned transition from conviction to science (if such a transition is possible at all) only happens by eliminating objective reasoning in general or at least for a specific issue. Additionally, such a step becomes *a decorum* by covering it with an undeserved scientific cover.

Instead of presenting additional reflections, I want to review the practice of character education as it was, is today by far and large, and how it should be. Let's look at first to the traditional disciplines of character education: religion, German literature, and history.

Traditional religion classes from the beginning have a firm understanding that religion, more than Christianity or even a denomination, has to form character, – first in class, but then as well in later life. However, widely forgotten is that faith, in particular in Christian understanding, cannot be made obvious; also, the shortcomings of one's own conviction and the merits of other denominations are not admitted. The bible and other religious books preferably are used to select the passages which one likes, contradictions are not mentioned or manipulated according to the rule of Mephistopheles in Goethe's Faust: '*Always be vigorous in hermeneutics. If you don't understand, you obscure*'. Of course, we find similar misdeeds by those who are of different opinion; it is a misdeed, to breed arbitrary characters this way, which renders freedom of thought factually impossible (cf. article 135 of the German Constitution, which in no unclear terms guarantees all citizens of the Reich full freedom of faith and conscience).

German literature classes and history classes – especially since the times of the Romantic – are intended to instill the love of the fatherland and of the people, thereafter of the state and of the dynasty. Here are a few well known examples from a vast number of materials: 'Right or wrong my country' says the Brit, while the German thinks, that his country has to heal the world, and the Frenchman calls his country 'La Grande Nation'. Whether these character traits are desirable or not, shall not be judged here. Let's assume for a moment that those character traits are wanted. But then, what about the methods of character formation or character education? – One starts with an already existing character trait, but hides the subjective form of such a character under a so-called scientific cover and strictly rejects an unprejudiced logical review of such a questionable moral fiber. See, how Goethe lets his Faust say: '*You are right; I don't find a dot of spirit, everything is broken in by training [Dressur]*'. So, not freedom of thought, but trained, dictatorship of opinion (it has to be made very clear, that dictatorship may come from the political and pedagogical right, the middle, and the left). There are groups who accept these methods because of their effectiveness, irrespectively of their subjectivity and misconceived psychology. Thus it is very doubtful, if one can ever be content with the withdrawal of freedom of thought and its replacement by character dictate [Gesinnungsdiktatur].

But how should it be? – The answer is self-evident from the preceding: Not brake-in by training, but liberalization, respectively 'democratization', of character formation. In classroom teaching it will work this way: – 1. Do not teach predetermined subjective disposition [Gesinnung]. – 2. Strictly avoid the cover-up of a predetermined opinion with so-called objectivity and with so-called interactive teaching [Arbeitsunterricht]. 3. Methodologically it is not acceptable to present only what is suitable and to suppress unsuitable facts, or to deny or to manipulate them at will. – 4. Always consider different character attitudes [Gesinnungseinstellungen]. – 5. The benefits and shortcomings of different opinions and attitudes must be discussed. (Neither through a rosy glass nor through a black one). – 6. When you present your personal opinion, it must be in an impartial form. Also, one should forget to discuss problems associated with one's own position. – 7. Instead of presenting biased character formation [Gesinnungsmacherei], students should be given the opportunity to form their own opinion, respectively objective information should be given, so they may form their own character at a later date. – 8. 'Reason and science, people's highest authority' never should be bypassed in the formation or review of an already existing character. Wrong is a rule, established by a newspaper in Munich: 'Attitude first, thereafter reason'. By the way, it would be satisfactory, if reason were to be allowed an objective review of character attitude. – 9. One should not claim, that youth is only ready for authoritarian methods, not for methods of freedom, – a position which might be contradicted by some. But, let it be! Sowing is always earlier than harvesting. The 'Guidelines for Curricula at High Schools in Prussia' draw these consequences in methodological recommendations for various class teachings, expressively indicating that religious teaching in class has to be content with presenting material only, to be followed by student's self-determination *later*. – 10. And if a new expected character is not developing, we should not forget, that had happened under the old method even more often. Additionally: a character acquired and formed by the self is better than one just accepted from others, also better than a childish or immature attitude towards character issues.

It is important to recognize that all disciplines in school teachings will support character formation. Mathematics, physics and chemistry teach exact precision and strict verification by avoiding any hint of moralizing. Biology will improve much underdeveloped understanding of nature, get rid of old preconceptions, instill a love of nature, and respectively improve it. Music, painting, handcrafts and needlework favorably develop taste including aesthetics in character formation. – Gymnastics and sports satisfy human drives to wrestle in life. Additionally, they educate in camaraderie including the other team. Both establish a reluctance to scuffle, hunger for war, and hatred among nations. In such a context, teaching foreign languages with cultural and historical aspects [Sachunterricht] is of eminent importance in character formation for the young, i.e. for the next generation of adults.

Character Dictate or Freedom of Thought?, 1930. Gesinnungsdiktatur oder Gedankenfreiheit? Gedanken über eine liberale Gestaltung des Gesinnungsunterrichts. *Die neue Erziehung. Monatsschrift für entschiedene Schulreform und freiheitliche Schulpolitik*, 1930, 12: 200-202

OUR DOUBTS ABOUT GOD, 1933

1. The truth cannot be known. Also, what we commonly call science does not give us true knowledge or actual recognition of truth. This I conclude from skepticism in old and new times, and from agnostic, positivist, and pragmatic schools of thought. When I ask myself: 'What is truth?' (Pontius Pilate in John 18:38), I have to answer myself: 'I know, that I don't know' (Socrates), and I see that we *can* know nothing (Goethe; Faust).

2. One thing, however, I do know with certainty is whether I am happy or not.

3. My happiness does not depend on money or possessions or other outward things, at least not in its deciding factors. Ultimately it always depends on something from within, something spiritual.

4. The greatest happiness of all things spiritual for me is always intertwined with love.

5. As love brings the greatest happiness, yes, even the greatest blessedness, my strivings cannot be selfish, but will unfailingly include essentially the wellbeing of others.

6. I judge my faith by what I know experientially, from a practical point of view. Two forms of belief are of interest to me: (a) Steadily changing opinions of the day, at that particular time called 'common sense' [gesunder Menschenverstand] and based on then available scientific views and perceptions (which are only just another form of belief). – (b) The Christian religion.

7. (Ad 6a). What has called itself science at different times and in different places has surely quite often contributed to good fortunes for everyone, and thus to mine as well. In those cases, faith in dubious scientific knowledge is quite justified. However, in the most difficult areas and most fateful moments of my aspirations for happiness, the presumed recognition of truth, the so-called scholarly truth, most often abandons me. At times it even harms me, forcing me by civilization and technology into a realm of superficiality, away from the inner and spiritual life, on which my happiness and the happiness of the world depend. In this case, the true and good faith in presumed knowledge and truth becomes a false faith, a harmful superstition.

8. (Ad 6b). The Christian religion is as well a faith only. It possesses no less believability as the faith based on reasonable or scientific perceptions. But above all: This faith makes me happy; for it gives me spiritual gifts, it is a

religion of love. And as a religion of love it enables me to be a living and useful member of society at large.

9. The more ‘unadulterated’ and ‘massive’, the more ‘positive’ and decisive the Christian faith is; that is to say: the less dependent it is on ever changing opinions of the day, even as they are paraded as exact scientific knowledge; the greater is the probability that it will bring happiness, – the more veracious it is.

10. ’Our doubts about God?’ – They have become null and void.

Our Doubts about God, 1933. Unsere Zweifel an Gott. Subjektive Gedanken beim Thema eines Anderen (Nr. 1, Jg. X der ‚Ethik‘). *Ethik. Sexual- und Gesellschaftsethik.* 1933, 10: 115-116

THREE STUDIES ON THE FIFTH COMMANDMENT, 1934

The 5th Commandment as an Expression of the Moral Law:

How do we do good? – The so called 'Golden Rule' gives answer to this question: Whatever you want the people do to you, the same do to them (Matthews 7:12; Luke 6:31). Kant's 'Categorical Imperative': Act only according to the maxim whereby you can, at the same time, will that it should become a universal law, - this basically means the same. – But are these and similar formulations not only just a formal criterion for 'good' action. The motive, disregarding such a criterion, could just be blatant egoism, a so called contract on reciprocity: Do nothing to me, so that, in return, I will do nothing to you (Schopenhauer hints at this in his 'Grundlage der Moral').

If we are aware that love is the fulfillment of the moral law (Romans 18:10), we indeed are already one step ahead: We know the motive. But we do not yet know the concrete content of the moral law; we don't know what to do or not to do specifically. Here, Schopenhauer, serves as a helping hand: He says that the best, the most concrete specification of the moral act is the sentence: Neminem laede, imo omnes, quantum potes juva! (Don't hurt anyone, but help everyone, as far as you possibly can!)

More than two millennia before Schopenhauer, the 5th commandment already provides such an insight, and, in fact, in a broader perspective than benefit or harm, namely under the perspective of sanctity of life and life's manifestations. Therefore the command: 'You shall not kill!'. We know from Jesus, that the 5th commandment does not only prohibit killing, but prohibits all wrong deeds against others, even the bad word, even the bad thought. This means: He not only forbids the malicious or careless destruction of life, but also everything that, in one way or another, may hamper or trouble life. Luther in his Catechism has made it clear, that the 5th Commandment has to be understood not only in a negative, but as well in a positive way. – Consequently, the 5th commandment is a very good expression of what it means to be morally and practically good.

The Duty of Self-Preservation

When talking about moral duties, normally we mean duties towards other people in the first place. Routinely we do not consider that each person has moral duties towards oneself as well, and that those duties are of immense importance. Christian religion expressively mentions those moral duties of everyone towards oneself. That basically applies to the 5th commandment as well: 'You shall not kill'. In this sense – 'You shall not harm or hurt anyone's body or life, rather help and support him/her in all distresses of body and life, wherever you can'[1] – in the first place, meaning the life of our 'neighbor'. In a later consequence, however, it means: in a Christian perspective every human life as such is morally 'sacred' – including one's own life. Preservation of life – and one's own life not excluded – is a duty. And destruction and harm – again, including one's own life – is a moral sin. 'Don't you know, that you are God's temple and that God's spirit dwells in you? You shall keep God's temple sacred and not destroy it.' (following 1. Corinthian 3:16-17)

How should these moral duties, expressed in the 5th commandment towards one's own life, be applied in real life's practice? By not taking one's own life, not shortening it, not harming or endangering it, not weakening one's health by unchastity, excesses in eating and drinking, heavy anger, frivolous foolhardiness and daredevilry, etc. . Particularly important is the protection of sexual virtue and the avoidance of abuse of alcoholic drinks. – As far as the first one is concerned, the judgment of the New Testament is particularly clear: 'If you have loose sex, you sinfully harm your own life' (following 1. Corinthian 6:18). But not only is it a duty to oneself to abstain from fornication, but also avoid anything, which might lead to unchastity: indecent looks, unclean or double talk, dancing, dresses etc. – As far as alcoholism is concerned, the Christian attitude is based on recognizing that 'wine kills many people' (Sirach 31:30), i.e. alcohol endangers life and brings great dangers to health.

Are the duties towards one's own life in conflict with duties towards the neighbor? – That is not necessarily the case. On the contrary: Whoever fulfills the duties towards oneself, avoids many forms of harm to other people. That can be shown in regard to the already mentioned issues of sex and alcohol: Who falls into dependency and unchastity, endangers and weakens oneself physically and spiritually. Venereal diseases threaten as well. Weakness and disease cause the victim to be more and more a burden to the community, harming everyone. If one has offspring, they also are harmed, as they may inherit a weak or sick nature, causing additional burdens and harm to the community. Whoever protects his own life in this regard, he fulfills one's duty also towards the community. Similar with alcohol: Those, who are dependent on consuming alcohol, may eventually expose themselves to severest physical and spiritual dangers. And thus does not only harm oneself, but one's family as well, one's offspring, one's country, and one's

[1] Cf. Luther's explanation of the 5th Commandment, German and Latin

race[2]. And again: If one protects oneself in this regard against harm, one does, at the same time, good to one's neighbor, actually to one's entire country.

The Bioethical Imperative

The 5th Commandment admonishes 'Thou shalt not kill'. Now, the term killing always means killing something which is alive. Living entities, however, are not only humans, but animals and plants as well. Because the 5th commandment does not expressively prohibits the killings of humans exclusively, should it not be applied towards animals and plants analogously? But are animals and plants so close to us, that we must recognize and treat them actually as our neighbors? – When we review publications in modern science, we find immediately similar studies of humans and animals as subjects in research, not only in physiology, but also in psychology. Such equal treatment today is not reserved, as already mentioned, for humans; similar methods are applied in the field of animals, and – as there is a comparative anatomical-zootomic research – similarly very interesting comparisons are made between the human soul and animal soul[3]. Yes, even the beginnings of plant psychology are recognizable – the most well-known among them are G. Th. Fechner[4] in the past, R. H. France[5], and Ad. Wagner[6] at present – thus modern psychology includes all living beings in its research. Given this, it is only consequent, that E. Eisler[7], in summarizing, speaks of a Bio-Psychik.

From Bio-Psychik, it is only a small step to Bio-Ethik, i.e. to the assumption of moral duties not only towards humans, but towards all living beings. In fact, bioethics is not a discovery of today. Montaigne[8] – the only early representative of modern ethics of sentiment – already grants all living beings an entitlement of being treated based on moral principles: We owe justice to humans; mildness and mercy towards all living beings, capable of having a benefit from that. Similarly, Herder[9] requires that humans – following the model of God in their sentiments – put themselves into the place of every living being and feel with it, as much as it requires. Those lines of reasoning

---

[2] Alcohol is 'a mean enemy of our race', cf. the brochure with this title by Wilhelm John, reviewed in no. 2 of 'Ethik'.

[3] Among recent publication in animal psychology especially recommendable are: Sommer, Tierpsychologie, Leipzig 1925. – Alverdes, Tierpsychologie, Leipzig 1925.

[4] G. Th. Fechner, Nana oder das Seelenleben der Pflanze [1848; 5th ed. 1921]

[5] R. H. France, Pflanzenpsychologie als Arbeitshypothese der Pflanzenphysiologie, Stuttgart 1909.

[6] Ad. Wagner, Die Vernunft der Pflanze, Dresden 1928.

[7] E. Eisler, Das Wirken der Seele, Stuttgart 1908.

[8] Montaigne, Essays.

[9] Herder, Ideen zur Geschichte der Philosophie der Menschheit.

are continued by the theologian Schleiermacher[10], who calls it immoral, to destroy life and formation – wherever they are, i.e. including animals and plants – without a reasonable argument for doing so. Therefore philosopher Krause[11], a contemporary of Schleiermacher, requests that every living being has to be valued as such and not be destroyed without reason. Because they all, plants and animals like humans, have an equal right; but not totally equal, each only as a precondition to reach its destiny. Schopenhauer[12] in particular refers to the Indian realm of reasoning, stressing compassion as the most important motive of his ethics, and requesting it also for animals. It was Richard Wagner, strongly influenced by Schopenhauer and a passionate animal friend, who made those thoughts commonly known.

As far as animals are concerned, the moral request has become obvious for a long time[13], at least in that form, not to harm animals without purpose. With plants it is different. However, in regard to new biological and biopsychic knowledge (see above), also regarding the various thoughts mentioned above from Montaigne, Herder, Schleiermacher and Krause, moral duties towards plants become visible. For purely sentimental-poetic argumentation such recognition is nothing new. Think of Goethe, who has Faust call plants his brothers, or of Richard Wagner's Parsival: In pious devotion people, at least on Good Friday, protect weeds and flowers in the meadow by walking carefully, in order not to hurt them. More seriously we have to consider the plant-ethical reflections of a quite matter-of-fact Eduard von Hartmann[14]. In an article on flower luxury he writes about a picked blossom: 'She is a deadly wounded organism, the colors of which are not harmed yet, a still living and smiling head, separated from its stem. – When, however, I put the rose into a glass of water, I cannot help myself but fighting the thought, that man has murdered a flower life, in order to enjoy the dying process by an eye, heartless enough to not sense the unnatural death under the appearance of life[15]'. The requirements of plant ethics, leading to such recognition, are quite clear.

As far as the potential realization of such moral duties to all living beings is concerned, it might seem utopian. But we may not ignore that moral obligations towards a living being related to its 'need' (Herder), respectively to its 'destiny' (Krause). It appears, that needs of animals seem much less in

---

[10] Schleiermacher, Philosophische Sittenlehre, Kirchmann 1870.
[11] K. Chr. Fr. Krause, Das System der Rechtsphilosophie, Roeder, Leipzig 1874.
[12] Schopenhauer, Über das Fundament der Moral.
[13] The most comprehensive book in this area still is Bregenzer, Tierethik, Bamberg 1894.
[14] Psychological preconditions are discussed in W. von Schnehen, Ed. Von Hartmann und die Pflanzenpsycholgie, Stuttgart 1908.
[15] Ed. Von Hartmann, Der Blumenluxus, 1885.

number, and their content less complex than those of people. This applies even more to plants, so that moral obligations to them should produce less complications than those to animals, as they are lower (if not conceptually, so nevertheless practically). Here also comes the principle of struggle for survival into play, a principle which also modifies our moral obligations towards fellow humans at no low scale. Within these limits there always will be enough possibilities for bioethical actions. Paragraphs for animal protection in penal codes of various cultivated nations[16] give guidance in this regard. Confer in particular the new German Reich Animal Law. As far as plant ethics is concerned, we are guided by our sentiment, which will hinder us to pick flowers and then throw them away carelessly shortly thereafter, or to deadhead plants with a walking stick, or when we find it disgusting to recognize the blind destructive impulse of rowdy lads in breaking the heads of small trees along the road. Also, excessive flower luxury – in learning from Ed. von Hartmann – is not morally refined and can be avoided.

In sum, the universal realm of authority of the 5th Commandment is clear and demands application to all forms of life. A transcription of the 5th Commandment results in the Bioethical Imperative: 'Respect every living being in principle as an end in itself and treat it, if possible, as such!'

Three Studies of the Fifth Commandment, 1934. Drei Studien zum 5. Gebot. *Ethik. Sexual- und Gesellschaftsethik*. 1934, 11: 183-187

[16] For the first time, material has been extensively collected and reviewed in R. von Hippel, Die Tierquälerei in der Strafgesetzgebung des In- und Auslandes, Berlin 1891.

FAITH IN THE HEREAFTER AND ETHICS IN CHRISTIANITY, 1934

A Post - Easter Contemplation

When we consciously listen to the word 'Christ'-ianity, we realize that the belief in Christ is an essential of Christianity. On closer examination we find that this faith concerns particularly the risen Christ. We don't have to prove that here. Only one single, but very instructive example is necessary: In Athens, when Paul preached among a gathering of philosophers, his core message was 'Jesus and the resurrection' (Acts 17:18, 31). And in a letter to the Corinthians he expresses the following: A Christianity without the risen Christ is no Christianity. (cf.1. Cor 15:14 and 17) The belief in the Easterly resurrection is at the same time a belief in the resurrection of those, who were saved in Christ. That means: Faith in Christ is a faith in the life-to-come.[1]

On the other hand, Christianity is highly valued as a religion of love, and rightfully so. Again, there is no need for proof here: In the New Testament, time and again, the commandment of brotherly love and love-of-your-neighbor, yes, even love-of-your-enemy confronts us, and from the very beginning until this day love is extended and practiced in the most dire situations. Especially praiseworthy in the history of charity is the work in the education of youth, the help given to the aging, the care for the poor, for the sick and frail, for the fallen and abandoned. This work also raised others into such useful service – their teacher long since forgotten and possibly even disdained.

During the last centuries, the Christian belief in the here-after has experienced more doubt and rejection than ever before. In contrast to this, the Christian ethic has held its ground and is widely respected, even when a belief in the Easter-message is wanting. One can understand this drift away from a belief in the here-after [Jenseits] and towards Christian ethics as a sign of the times. However, it would be inappropriate to separate the one from the other, because they do depend on each other.

On the one hand, the belief in the here-after is well-founded in ethics: when we know the commandment of God, which regards life as sacred[2], it is an obvious conclusion to believe that God would, in accordance with his commandment, regard and treat his creatures as holy and not let them expire in death. This is one of the ethic roots of the belief in the here-after. Another one follows: in the end, all good will finally find its just rewards, likewise all

[1] Cf. ‚Vom Leben nach dem Tode', 11(2), p. 50
[2] Cf. ‚Drei Studien zum 5. Gebot', 11(4), p. 183

evil will reap its rewards. Plainly, this is not the rule in this life and it becomes an axiom of faith in goodness and its final victory in the life-to-come.

Just as the belief in the here-after leans heavily on ethical considerations, so do ethics bear the stamp of a belief in the here-after: Ethics derives great strength and confidence from the expectation of the victory of good, which the belief-in-the-beyond confirms.

From this we learn: Whoever cannot rightly accommodate the Christian belief in the here-after – and that includes more Christians than the churches are willing to confess – but is cognizant of the value of Christian ethics, he may via ethics come to a deeper understanding and higher appreciation of the belief in the beyond.

Faith in the Hereafter and Ethics in Christianity, 1934. Jenseitsglaube und Ethik im Christentum. *Ethik. Sexual- und Gesellschaftsethik* 1934, 11: 217-218

THE ETHICAL-SOCIAL IMPORTANCE OF SUNDAY, 1934

The weekly rest and holiday in Caucasian-occidental western culture comes from the Bible. Yet, the true meaning of that day remains unclear to the followers. When asked, what do they answer? 'Sunday is to remember the resurrection of Jesus'. This is true, indeed. But Christians should remember such an important fact of faith all week long. Another answer might be: 'A work-free day on the weekend is a reminder of the completion of God's creation'. This is true as well. But the believer should not forget his Creator during the week. Or: 'On Sundays you are supposed to hear God's word and somehow serve him'. This is also correct, but one should not necessarily need a work-free day for that; it can also be done on weekdays. In some instances work is regarded as a service to God and the Sunday rest might even prove to be a hindrance in our service to God. What is important?: To not work and to not to ask others to work.

From an idealistic point of view, to be freed from work seems strange. Work is indispensable in three ways: it is a necessary instrument of love, or in more modern language, of a conscientious social duty. It is furthermore a practical necessity. And lastly: to work with sense and purpose is a joy, – to produce value (Werte schaffen) creates a sense of value (Wertgefühl).

Sadly, this threefold benefit of work is often lost, because love – the sense of social duty – is missing. Instead we find selfishness in greed and thirst for pleasure. Even though the government in Germany today tried a change, it remains true that humanity as a whole stays the same. Such selfishness only knows its own interests. Concerns for others are unimportant and conveniently ignored, unless these concerns serve one's own advantage as well. Such a feeling creates mistrust and combativeness, so the blessings of work are lost. In this way people turn work into some sort of fight, and that sadly means brother fights against brother.

Such a struggle does not lose its rancor just when it parades as 'friendly competition'. In this way people create in their work thorns and thistles, far worse than any field could grow. The beads of sweat created by this man-made fight burn and corrode far more than those, work alone might produce. Those man-made burdens are much heavier than those of other daily and natural events – to the believers these are god-sent fates (Gen 3:18, 19), which serve man for the best in the end (Rom 8:18, 28). Under these circumstances the joys of work may embitter people or even be completely taken away.

Here the weekly work-free day is of great help. When work is at rest, so are greed and envy; even though they still exist. When work is at rest, the fight for existence, for survival and the daily bread rests as well, – this fight which pits man against man instead of overcoming common burdens in a united way. On those days a Sunday-mood gives peace to all spirits. On such a day of peace one can hear more clearly the words of love, which otherwise are drowned out by the restless rush of the world. One's own heart can easily find the way to the other, also the people and spirits come closer to each other. Even if love at first has only the power to stop certain actions – then there is a multitude of actions that should cease out of compassion and love – it is a very great cultural benefit, – for the ethical culture, for the 'social' culture[1].

It is a special benefit if such a day of rest is not restricted to a small group of people. It is not only for the rich but also the poor, not only far the master but also for the servant and maid, not only for the entrepreneur but also for the workforce. Even the poor work-animal needs a day of rest. It is a blessing for all of humankind not just for those following the Bible; regardless of the Christian Bible it is a blessing to every living thing. Founding a day of rest for everyone, regardless of their rank, social or economic standing, or faith, from a historic point of view is an extraordinary achievement of greatest ethical and social value.

The fact that the believer recognizes that such a social institution, i.e. the day of rest is based on God's revelation, on God's love, and that respecting the day of rest as one of the first (in third or fourth place) commandments, shines a bright light on his faith. This light is even brighter for not being an afterthought or an adaptation from another faith, but an essential original thought from within. And even though it was formally ordered by government (Emperor Konstantin in the year 321 after Christ decreed the first Sunday law), the thought itself originated in the Bible and believers had long since observed it in their small circles. If the concept had not first arisen there and been established, humankind might up to this very day be without any holiday, any armistice in the harsh fight for existence – by people themselves robbed of any holiness –, day in day out, week for week, year in year out, 'until he returns to earth, from which he was taken' (Gen 3,19). Thus the social-ethical importance of the weekly holiday is significant and confirms at the same time the biblical source of the day of rest until this very day.

The Ethical-Social Importance of the Sunday, 1934. Die sittlich-soziale Bedeutung des Sonntags. *Ethik. Sexual- und Gesellschaftsethik* 1934, 11: 361-363

---

[1] Also: 'It is not only for ethical reasons, to respect the Sunday holiday, much more ... also for health ; cf. Abderhalden 'Die Bedeutung der Sonntagsruhe im Lande' [The Meaning of Sunday Rest], in 'Ethik', 10(4), p. 241

DOUBTS ABOUT JESUS? 1934[1]

In the person of Parsifal the ‘pure fool’ [reiner Tor] model is incarnated. Similar thoughts arise from two of the Beatitudes in the New Testament: ‘Blessed are the pure in heart! (Matth. 5:8); ‘Blessed are the poor in spirit!’ (Matth. 5:3). Further, we remember the story of Christ reminding his disciples to follow the example of children, innocent and free of intricacy (Mark 10:15). In harmony with this, Wagner opposes vehemently the intellectuality, which he sees as taking on an erroneous path. He notes especially the new directions of the natural sciences, which declare the world consists of ‘matter and energy’ exclusively, which contemptuously regard philosophy and religion as rudiments of the past, also deny any form of religious intuition. It would, however, be quite erroneous to regard Wagner’s premise as: ‘Despise any reason and science!’ His ire is not directed at the sciences, but at rationalistic one-sidedness and at materialistic distortions. Pure knowledge is of incomparably high value to him. And how can we achieve this? Through love which in Schopenhauer’s pessimistic worldview – shared by Wagner – can only manifest itself in compassion. ‘Knowledgeable by compassion’ [durch Mitleid wissend].

What about Parsival’s development in this regard? Parsifal’s first step on this road is to be able to feel pain for having killed a swan in haughtiness. (Wagner, a follower of Schopenhauer, considered animals and people identical in nature [wesensverwandt].) When led inside the Temple of the Grail, Parsifal is deeply moved to see the unhappy Grail-king Amfortas lying on his bed of pain. As he learns in Klingsors enchanted garden, he gains wisdom in understanding Amfortas’ pain, i.e. the greatest pain in spiritual anguish suffering from guilt; he gains this highest wisdom by empathy and compassion. Now: ‘through compassion did he become knowing’ and wise [‘durch Mitleid wissend’].

The meaning of the above is clear – also for religion: It is not intellectualism in a one-sided derailed form, but ‘the power that compassion imparts to love’ [des Mitleids Liebesmacht], which leads us to the knowledge of truth: ’Knowing through compassion’. It also applies to the truth of religion: ‘To love God is the sweetest wisdom’ (Jesus Sirach 1,14).

Can we likewise philosophically conceptualize that, which Wagner found through intuition in the arts? Love, clothed in compassion and directed to the wellbeing of others, yes, of all, is a curious fact of human ‘emotion’ [Gefühlsleben]. To feel and to exercise an all-embracing love represents

---

[1] A contemplation following Richard Wagner’s ‘Parsival’; cf. ‘Our Doubts about God’, 10(2), Nov./Dec. 1935, p. 115.

highest happiness and blessedness, as far as we know. The reality of such happiness stands up to all doubts; for, whoever feels happy, he is also happy. Here appearance and reality are identical. Does it even make sense to try to differentiate between them? Here, love is very well suited to be the foundation for further knowledge, the knowledge of truth in religious intuition.

The foregone seems to portray the religious worldview of 'Parsifal' as a Buddhist religion based on compassion, by way of Schopenhauer's intuits. Schopenhauer regards this kind of religiosity as to be contrary to Christianity. That, however, is incorrect. Yes, the highly valued compassion takes us to the very center of Christian thought. Christianity holds love (compassion being part of it) in the highest esteem, above what may itself call cognition, knowledge or wisdom (1 Cor. 13). Nietzsche, having a different understanding of Christianity, severely attacks Wagner and 'Parsifal' and herewith verifies that we are truly dealing with Christian principles. It remains true for the Christian religion: Love (shown as empathy, as compassion) is foundation and premise for any realization of the truth.

Exclusively and singularly it is love that determines our relationship to Jesus. His suffering is a 'love-offering' [Liebesopfer] out of mercy and compassion. After his death on the cross he mystically approaches his followers through 'compassion's power of love' [des Mitleids Liebesmacht] at the Holy Supper, which now has become a 'meal of love'. This is in keeping with the love between man saved and the savior, so that heavenly voices can sound in the Temple-of-the-Grail: 'Take my body, take my blood for the sake of our *love!'* Here, too, love means knowledge. And, whoever through love has found certainty that Jesus is the bearer of salvation [Bringer des Heils], he has conquered all doubt, 'blessed in love, blessed in faith'.

Doubts about Jesus?, 1934. Zweifel an Jesus. Eine Betrachtung nach Richard Wagner's ‚Parsifal'. *Ethik. Sexual- und Gesellschaftsethik* 1934, 11: 363-364

ETHICAL REFLECTIONS ON INNER-CHURCH QUARRELS, 1935

Christianity split into hundreds of directions, be it denominations, independent churches, sects, congregations and the likes. That surely is regrettable. Also, this splintering is not in to the spirit of Christ. However, such multifaceted expression makes sense from a psychological point of view. Genuine religiosity is a highly personal matter of the individual human soul. The more adherent and intensive the religious devotion, the more likely the deciding gradations of differences amongst the various people and groups will turn out, so that no two people's inner believes are alike. Perhaps the many forms of belief attest to the strength of the religion and, from that point of view, could be a good sign. It is mandatory that the richness of forms of belief must not cause religious unrest among the Christians. Such danger is very real and, from the past up to the present, the same thing always happens: The Christian churches, which are uniquely pioneering in their works of charity, a model for all of modern culture, show a terribly short supply of this love in the behaviors toward each other in matters of differing religious opinions. Instead of the peace, that Christ expects, we find intolerance, disharmony, quarrel and dispute – one could even call this the norm. It need not be and should not be so.

Take note. As the multifaceted Christian views can be a positive sign, an objective disagreement or a fight – if you wish to call it that – could be useful and necessary. Only, this "fight" among Christians must adhere to Christian standards and a Christian spirit. That takes a lot. I will discuss three important points of Christian bearings on this fight, which should be observed:

1. It is preferable to keep an eye on what unites us prior to that which divides us[1]. Central to any Christian church form is this important question: How do we differ from other Christians? In the evangelical state-church [Staatskirche] this question is: What makes us different from other protestant free-churches and sects? How does our view differ from the Catholics? These questions should be very similar in the free-churches and Catholics. Quite a few catechisms devote a special section to this. These subjects are part of school and university examinations. To know these differences is important – one should know what makes one's church special – but it is much more important to know what we hold in common. And the common ground consists of much more than a casual gleaning would show.

[1] Cf. the efforts of the 'German Christians' movement in Thuringia [Thüringer 'Kirchenbewegung Deutsche Christen] and a supporting review by Erich Friebel in 'Ethic', 11(5), p. 213 ff.:'Church of Love' [Die Kirche der Liebe].

2. When dealing with differences one should not exaggerate the other's conceptual weaknesses and remain silent about the positives.
   Everything done by man is incomplete and it is not at all difficult to find the shortcomings of it, especially if one tries. This applies to all religious conceptions and their expressions – including the Christian. In other words, it is always somehow possible to find weaknesses and a lack of completeness in the theories and practices of other Christians – with a tendency to exaggerate these. When at the same time you conveniently say nothing about their good points, it proves tactically effective: you can deal them a blow. Such tactics are no doubt very human. But this is totally incompatible with Christian ethics.

3. True mistakes or actual weaknesses in the other's belief should be rejected, however, in a most objective manner and without impassioned agitation.
   To speak metaphorically, one may not overlook a white or even change it into a black. Under no circumstances would truth and justice permit it. Clearly for the same reason a black may not be overlooked or covered with white. I must repeat: mistakes must be verified beyond any doubt. One has to be very cautious with negative critique; especially in the area of religion – also the Christian religion – the danger is great that one may judge from a presumed truth and justice, i.e. one misjudges. Indeed, if a true denial is justified and necessary, then it should always proceed in a most objective manner and without any impassioned agitation, that is to say: without any hint of coarseness, without animosity and odiousness, without arrogance and know-it-all, without strife and being over-critical, and without any desire to score mistakes.

Finally, one must become aware that one's own religion has shortcomings and weaknesses, for which a lenient judgment is taken for granted. Such a consciousness should suffice to take on a kinder judgment towards the shortcomings of others.

Ethical Reflections on inner-church Quarrels, 1935. Ethische Betrachtungen zu innerkirchlichen Glaubenskämpfen. *Ethik. Ethik. Sexual- und Gesellschaftsethik. Organ des Ethikbundes* 1935, 12: 58-61

FAITH AND WORKS: OPPOSITION AND ALLIANCE, 1935

The philosopher Kant states that a good will alone is good enough. This means: one can rightfully affix the designation of good only to the will. In everyday speech that would mean: All depends on the will, the heart, the character [Gesinnung]. Works of people are good, if attitudes, from which they come, are good.

However, one cannot look into another person's heart. Even in reviewing one's own work, one often errs about one's true motive: one usually believes in one's good intentions, but this quite often is far from being true. In practical life, most of the time we will not reflect from character onto deeds, rather from deeds onto character. (Matth 7:16,20: 'By their fruits you shall know them!') The same is in Kant's 'Categorical Imperative' [Kategorischer Imperativ]: 'Act in such a manner, that the maxims of your action can as well serve as a principle for a common law!' This is a call to good actions, whereby the good result of the deeds is not determined by character [Gesinnung], but by very real things, i.e. common goods. On the one hand: the intentions are what count, on the other: it is the actions. This appears contradictory, but is not. Of course, an honest good will is the condition for actions, if they indeed can be called good. But on the other hand, even if the deed logically comes after the intention, the deed is the yardstick for measuring intention. There are no truly good intentions if good actions do not follow, just as no truly good action can be without good intent. Intentions and deeds, or deeds and intentions are on equal footing and are inseparable in practice, even if not so in logic.

In applying these facts to Christianity, one would use faith for compassion – in place of actions one would use deeds – thus: faith and deeds, deeds and faith, both are practically inseparable. This fact has never been unknown to churches. At times, however the one or the other, faith or deeds, was regarded as more important, and this led to all kinds of misunderstandings among those, who differed in opinions about the greater importance of either. The more one side exaggerated; the surer the opposite side would exaggerate as well, this happening consciously or subconsciously. From a human point of view such fractions are quite understandable. But it is regrettable because both disunited parties could quite well have found common ground.

Take for example Catholic thinking in the Middle Ages. Never were they taught that character [Gesinnung] does not count and therefore, e.g., it is not necessary to believe in God. However, doing was more emphasized than faith, which is understandable: others can be deceived by a presumed or supposed kind character disposition – and even the self. That should be avoided, in those days and today as well. Even if good and right intentions are truly present, but too weak, easily disabled by fear or laziness or any other such hindrance, that likewise is not right. A focus on the importance of deeds does not disrespect the importance of character disposition. A misunderstanding did arise here: one could do this or that, e.g., give a sum of money for a letter of indulgence, a church-building or similar, regardless of one's character. Without any doubt, this was an erroneous concept. So-called good deeds were not really good deeds. For what derives from a lack of character [Gesinnungslosigkeit] or even an evil character [üble Gesinnung] cannot be called good. Such deeds are customarily called dead deeds. Dead deeds were a sin against the spirit of the Christian religion, as it was especially prevalent in the Middle Ages. An outward church-as-habit [Gewohnheitskirchentum] was the consequence.

Luther and his Reformation movement fought against this type of sin. Rightfully he said: action without character [Tun ohne Gesinnung], i.e. deeds without faith, are nothing. Character and faith alone are what counts. Luther in no way wanted to say that the deeds of people do not matter, or that humankind has no ethical duties. It was not necessary to address this, as it is self-evident: Just as genuine character intent, so will faith influence every action. Therefore it is important, that faith is there. This emphasis on faith was very important, as it exposed the wrongs of Middle Age piousness, i.e. dead deeds, and called for their avoidance. Thus, that erroneous course was blocked, indeed.

However, over time another error emerged into the opposite direction. The opinion, that faith has neither influence nor need to influence deeds, became accepted. That proved the most comfortable yet: One need not care about anything, need to feel no restraints in one's actions, and therefore had excuses for others and oneself: 'Since deeds do not matter, faith only counts, and that I have'. Whoever said this, forgot that such a faith – if he, who pretends this, really should have any – is really not a truly faithful. An intention which will not lead to actions, or at least to an honest try, is not a true in itself. Whereas the religious lives of the Middle Ages had dead deeds, the times after Luther suffered from a dead orthodoxy, i.e. a dead orthodox faith attitude [Rechtsgläubigkeit]: 'What good is it, though someone say, he has faith, and yet – no works? – Even so faith, if it has no works, is dead, being alone.'(James 2:14a,17). Also here an outward church-as-habit [Gewohnheitskirchentum] followed. It must be emphasized that this erring had nothing to do with Luther's teachings. But it happened and the situation became worse than what had been fought and overcome before. Any action, which lacks character intent, is a doubtful thing. Nevertheless, it is something palpable and may eventually be useful in a practical sense. But a character attitude – or rightly: an imagined character attitude – without actions or at least a willingness to act is empty talk, if not lying, and in any case useless. (A simile: When orders of a state, which do not meet with the approval of its citizens, are followed only because it is required then such an attitude is regrettable, but may be tolerable in extremis. However, it would be most unacceptable, if someone, expressing his free and joyful consent, would not even make an attempt to follow those orders.)

The above described dead faith and pursuant outward church-as-habit caused reactions on many sides in the 18.century – but not a relapse to the old mistake, which Luther had uncovered and fought. The Pietists have to be named first. (A.H. Francke and his foundation in Halle, Zinsendorf and the community of the Herrnhut Brethren are also known by non-theologians). Methodists appeared in protestant language regions [Sprachregion]. Pietists demanded instead of a 'dead' faith a 'living' faith associated with a pious Christian walk. It was called consecration [Heiligung]. Consecration means in other words, that one believed a real life faith cannot exist without an impact on life-style and actions. The consecration movement lent new life to the Christian walk. At approximately the same time, but from a different mentality the German Enlightenment [Aufklärung] followed the same direction. As an example a few lines of Gellert's poem 'The active Faith' [Der tätige Glaube]: 'Whoever does not keep God's word and says: I know God!, he deceives. Only then do I please God, when I practice obedience. A daily active Christianity is faith's fruit and glory.'

This is a reflection of enlightenment in religion, which highly values the practical-ethical side of Christianity, not unlike the sanctification of the Pietist's, demanding 'virtue'. (Sanctification and virtue are comparable, without being equal: Both stress deeds in contrast to dead faith.)

Also, thoughts and ideas emerged as a reaction to dead faith outside Christian concepts of faith. Kant was previously mentioned in the beginning of this article: when he says the will alone is good, it means: all depends on the heart and intent. As intent is characterized by will, action unmistakably is the aim. In the Categorical Imperative [Kategorischer Imperativ] Kant admonishes clearly to do good (see above!). Besides, when he discusses the 'maxim' of actions, it reveals that, as is necessary, he did not forget the importance of good intent, not repeating the mistakes of the Middle Ages before Luther's time.

Next to Kant and in this context Goethe needs mention. At the end of 'Faust', part 2 this word is indicative: 'Whoever strives and works hard, we can save him.' Firstly: The intent [Gesinnung] is what counts. In striving and acting it moves into the realm of the will and clearly aims at deeds. It is interesting to note the similarities of Goethe's words to those of Wesley, the founder of the Methodists, long before: 'Do not forget: the reward will be according to your work, not according to your success.' Goethe places decided emphasis on deeds in Faust's great monologue in the 1.part of the tragedy: 'In the Beginning was the Deed'. These words have remained an uncontested milestone for human conduct, also for the Christian consciousness[1].

Schleiermacher, the most well-known theologian since Luther, validated these insights and discussed the concept of sanctification (proving faith being effective in life) very positively and made it part of his theological teachings[2]. Over time, the situation has changed for the better. As one might say, Catholics do learn from Protestants. However that may be, it appears that some of what the churches were accused of in the Middle Ages has now been remedied, if not removed. Likewise the Protestants have abandoned the erroneous path of dead faith in favor of a living faith. In other words: Man is working in the direction of a Christianity of life and deed. May this way more and more become the only way.

Faith and Works: Opposition and Alliance, 1935. Glauben und Werke in ihrem Gegensatz und in ihrer Vereinigung. *Ethik. Ethik. Sexual- und Gesellschaftsethik. Organ des Ethikbundes, 1935, 12: 260-265*

---

[1] (22) Cf. in this context: Abderhalden in 'Ethik' 12(4), p. 151 'If I could as I would like to, then I would underline, what is basic, i.e. the Christianity of Deeds.'

[2] (23) Schleiermacher ‚Der Christliche Glaube etc …', paragraph 110-112.

THREE STAGES IN LIFE, 1938

A Contemplation following 2. Corinthians 5:1-10 and the Apostolic Creed

Much in philosophy (Weltanschauung) leads back to the Bible. This fact is quite often forgotten. And so it was with Gustav Theodor *Fechner's* (1801-1887) – scientist and philosopher – little book 'Büchlein vom Leben nach dem Tode' [Booklet on Life after Death] (1866). We find exactly the same thoughts in the 'Great Didactics' [Grosse Unterrichtslehre] (completed 1657) by Johann Amos *Comenius* (1592-1670), the well known pedagogue and upright Christian and confessor – he was Bishop of the Bohemian Brethren, a branch of the Herrenhuter Brüdergemeinde. Both Fechner and Comenius subdivide human life into three stages. Of special interest is the stage that goes beyond what we know as 'this life' to a 'life after death', i.e. the stage not only representing the last, but also the commonly acknowledged highest step.

There was one other – long, long before Fechner and Comenius – who perceived the division of human life into three parts, emphasizing the 'life after death': it is the apostle *Paul*. It is to Samuel Keller's credit, that he addressed these three stages anew in his book, 'Das Los der Toten' [The Lot of the Dead], earlier titled 'Auferstehung des Fleisches' [Resurrection of the Body], published after the war of 1914-1918.

Our consideration will focus on the sayings of Paul in 2. Corinthians 5:1-10. These correspond to the three stages of the life of Jesus, as does the Apostles' Creed. These stages are as follows.

First Stage

In the beginning life exists in the 'earthly house' of the body [*'in der Hütte*] (2. Cor5:1) (also 2. Peter1:13, 14), in the terrestrial housing (2. Cor 5:2) – in the very beginning even in the earthly housing of the mother's body 'in flesh' [im Fleische] (Gal 2:20; Phil 1:24; Col 2:1). The first stage in life is a preparation for the second (and lastly also the third). 'For we were not placed in this world for nothing, here we must ripen for another' (Matthias Claudius).

Jesus also had to live in an earthly hut and the motherly housing. He took on flesh and blood (John 1:14a) 'He was made in the likeness of men: and being found in fashion as a man, he humbled himself' (Phil 2:7). The Creed speaks of the latter part of his purely human life: *'Suffered under Pontius Pilate, was crucified, dead and buried'.*

Second Stage

Man is *unclothed* (2. Cor 5:4), or he is 'absent from the body' (2. Cor 5:8). It must be noted: To be unclothed does not in all cases mean to be stark naked. To follow the apostle's turn of speech – one can undress short of the undershirt, be in nightclothes and commonly is considered not dressed. Even if completely naked and bare one cannot shed one's skin' (also a type of dressing). Upon laying down the mortal form, there may be an 'inter-body' [Zwischenleib] or 'spirit-body' [Seelenleib], (expressions used by Keller).

With disappearance of earthly clothing, bodily weakness and imperfections disappear as well; Paul considers the release from this body to be a liberation. (Romans 7:24)

This state or place of 'being unclothed' is the middle stage and may well be called 'inter-place' [Zwischenort], 'inter-reign' [Zwischenbereich], or 'inter-world' [Zwischenwelt]. The New Testament (in the original Greek) calls it 'Hades', a word that could also be translated as underworld, world of the dead, world of the departed, shadow-world, soul world or spirit world. The conditions of that in-between world will correspond to how diligently we tried to please God (2. Cor 5:9), how we lived in this life-body, be it good or evil (2. Cor 5:10). We might find a state of rest (Rev 14:13), likened to sleep (1. Cor 11:30, 15:20; 1. Thes 4:13-15). If this going-to-sleep should signify death, and if the word death is actually used, it always refers to a state that shall end in awakening or resurrection. Besides, this condition does not preclude the hearing of Jesus' voice. (Joh 5:25; 1. Peter 4:6), – Or we find that the spirit-world could become a prison for us (1. Peter 3:19), or a place of torment (Lukas 16:23,24) – named 'Tartarus' (2. Peter 2:4 in the original Greek and Latin). Finally a place of blessedness may await us: 'Abraham's bosom (Luke 16:22), a 'paradise', an 'Elysium', which 'no eye has seen, nor ear heard, neither have entered into the heart of man, the things which God has prepared for them that love him' (1. Cor 2:9), a close communion with Jesus Christ, in contrast to the 'Tartarus', mentioned in Peter's 2. Letter. We remember the words of Jesus on the cross. 'Verily, I say unto thee, today thou shalt be with me in paradise'. (Luke 23:43). The apostle Paul looked forward to that same communion with Jesus: 'I desire to depart to be home with the Lord' (2. Cor 5:8; also Phil 1:23). We share this expectation and ask in song: 'Let go of me, let go of me, that my Jesus I may see! My soul is full of longing to embrace him forever and to stand before his throne' (Gustav Friedrich Ludwig Knak).

Johann Hermann's song 'Oh God, you faithful God' should also be mentioned here, the 7th verse: 'Take my soul hither for your gladness'. How did Jesus fare when he was disrobed of his body? The Creed states: 'He *descended into hell.'*

This is a difficult statement, even for believers: Is not hell the place of eternal damnation and what would Jesus be doing there? Now, Jesus was not really in 'hell' [Hölle]. This word appears only in the German translation of the Creed. In the deciding other-language experience it speaks of under-world, the world of the departed, or shadows, of the souls and spirits, as mentioned above. According to the scriptures and the Creed he was there from Good Friday afternoon to Easter Sunday morning and preached the gospel to the 'dead' (1. Peter4:6), and to 'the spirits in prison' (1. Peter 3:19).

Third Stage

Man has been '*layer-dressed' (überkleidet)* (2 Cor 5:4). The everyday dress is changed into a holiday dress, the hut becomes a palace, the natural body becomes a spiritual and 'transfigured' [verklärter] one (Phil 3:21). Whatever is sown in corruption, shall rise in incorruption. What is sown in dishonor, will rise in glory. What is sown in weakness, will rise in power (1 Cor 15:42, 43). Looking back we also remember Johann Hermann's song 'O God you holy God, in which the second part of the 7th verse and the 8th contain the prayer: 'In the holy Christian's grave, do grant the body a place of rest. On that day, when you will raise the dead, please extend your hand to my grave and let your voice be heard, awaken my body and when resurrected lead it to the chosen group'.

A good summary of the three stages of human life is contained in the liturgy of the Old-Prussian Union [Church] of 1895, and a newer one contains the formulary to be used at funerals: 'As it pleased God Almighty to call our brother (sister) from this life (1. Stage), we command his soul to the mercies of God (2. Stage) and lay his (her) body in God's earth – in the hope of a resurrection to eternal life through our Lord Jesus Christ (3. Stage)'.

This last mentioned point applies to us all, similar to Jesus, only he already was *'resurrected on the 3rd day from the dead.'* Eventually the whole world will be transfigured into a new heaven and a new earth (2. Peter3:13; Rev 21:1).

The passage from the first stage to the second is called *'death'*. Naturally this is contrary to the feelings of man. It is tolerated more or less with resignation in silent grief or noisy lament, – just as the baby at birth cries at leaving the mother's protective womb.

That is understandable: for leaving your quarters, even if it is only a ‘hut’, is an inconvenience as anyone knows, who has had to move. The laying down of our clothing is also a bit of trouble. Likewise, dying and being born are by nature associated with discomfort, not only the dying itself, but already the thought of it.

Can you not understand the desire to be ‘layer-dressed’ (überkleidet) without removing the other clothes, and ‘that the perishable could be swallowed up in the new life’ without laying down our desires, the hut, the cloth of life? (2. Cor5:4; 1. Cor15:51-53). According to the Holy Scriptures Henoch (Gen 5:24; Heb 11:5) and Elisha (2 Kings2:11) were taken up into the kingdom of God without tasting death. Every Christian hardly dares to share a hope for such special dispensation, but each would like to belong to those who could experience the return of the Lord in this life, they would instantly be ‘layer-dressed’ without being ‘undressed’ before. The apostle Paul anticipated this with great longing (2 Cor5:1-4; 1 Cor15:51-53). Such longings are in accordance with the Holy Scriptures. He who testifies to these things says, ‘Surely I come quickly!’ (Rev 22:20) This is answered with the request: ‘Even so, ‘come, Lord Jesus!’ (Rev 22:20)

But whoever is not called to this immediate passage, will not be far behind those who are. (1. Thes 4:15) For death, even though appearing to be an end to all life, in reality is only a deliverance from our weakness and imperfection (Rom 7:24), is the entry-gate to a complete state of being; dying is nothing more than ‘going home’ (2. Cor5:8).

And what about the kernel of wheat? (cf. 1. Cor 15:36-38). Does it not appear to be dead and get buried while being sown? But this dying, being buried and decaying is a mere appearance. In reality the corn swells and sprouts, hidden life-forces stir within it until it rises in a richer form of life than before. ‘We mournfully entrust even more precious seed into the lap of earth, that from the graves it may bloom into a happier lot’ (Schiller).

From the Christian point of view it is better to use the word ‘believe’ rather than ‘hope’: lending it a firm and certain reliance (Heb 11:1) so that mourning may lose its sharpest thorn.

The two parables of birth and sowing will remind us, that dying always appears like a catastrophic event, – at least for him, who has not yet experienced it before. And this includes all humankind on this earth. But whoever would be able to regard dying in retrospect, might perhaps find that it is not such a catastrophe, not nearly as forceful as it seemed! In 2. Cor 5:1 ff., Paul likens dying to a changing of clothes or moving out of a hut, giving a good image: In retrospect, we may find that, when our soul departs our body it may be no more uncomfortable than taking off a coat or clothing, which perhaps were rather uncomfortable and too small and now we feel much better.

It will likely be as if we stepped out of our hut or other building and now stride into a great, free nature. Or as if we exit from a narrow foyer with tarnished windows into a wide and bright hall with unfettered views. And that, in any case, is great improvement.

The passage from the second to the third life-stage is the *'resurrection'*, 'the resurrection of the flesh' (or body) reflected in the third article of the Apostolic Creed. It signifies the redemption of our body (Romans 8:23). To exactly know, how the ascension will occur is not given to us, as we can also not imagine how Jesus arose early on that Easter-morning. But one thing is firm in our faith, that 'Christ is risen from the dead, and became the first fruits of them that slept. In him shall all be made alive' (1. Cor 15:20,22; 6:14; 2. Cor 4:14). And 'we must all appear before the judgment seat of Christ, so that each one may receive what is due for what he has done in his body, whether it be good or evil.' (2. Cor 5:9,10). We refer to this in the Creed: Jesus is '*risen to heaven, sitting at the right hand of God, the almighty Father, from whence he will come to judge the quick and the dead'.*

Three Stages in Life, 1938. Drei Abschnitte des Lebens nach 2. Korinther. *Nach dem Gesetz und Zeugnis. Monatsschrift des Bibelbundes* 1938, 38: 182-188

THE SUNDAY – A SECULAR HOLIDAY, 1947

A Review of Paragraph 10 of the Draft of the Constitution

The work-free Sunday is not at all an achievement of Christian origin. Of course, the Jewish religion had a work-free day. But that was not the Sunday, rather the so-called Shabbat. Christians have observed the Shabbat as long as they lived in Jewish communities. But that was the case only in small Palestine and also for a few decades. Most Christians were Roman citizens and, as such, had no reason to observe a Jewish holiday. And there was no other, no weekly Roman day of rest from work. At least not in the beginnings. That was changed in the year 321 our time: the Sunday was introduced as a work-free day in the Roman Empire. Thus, this day definitely is a governmental, a secular institution.

Secular, therefore, up to today, are the names: Sunday is the day of the sun, – the Romans already named the seven days of the week after the seven planets: Sunday - day of the Sun, Monday - day of the Moon, Tuesday – day of Mars, Wednesday – day of Mercury, Thursday – day of Jupiter, Friday – day of Venus, Saturday – day of Saturn. Since we humans recognize the sun as the largest celestial phenomenon and her eminent importance for all life on earth, it made sense to give the Sunday this very special place among the days of the week.

Something else needs to be considered: Since the sun has been revered in many religions, the Sunday from its beginning had a kind of religious flavor, i.e. a pagan flavor. No wonder: Emperor *Konstantin*, who passed the Sunday law, was a pagan during all his life up to his last hour.

But, of course, the Christians had a great benefit from this day. They could have much more elaborated church services than before. So, for them the Sunday was a real 'donation Constantinii', a gift from *Konstantin*.

In summary, the Sunday was an institution of the state regardless of religion and confession. Today it is still the same. See the Constitution of the Weimar Republic, dated August 11, 1919, second part, third section, article 139 'The Sunday – as a day of rest from labor is protected by law'. The fact that the day of rest, indeed, was implemented, also is a fruit of the fight of organized labor for their right. Of course, the Sunday was supposed then and also today to be a benefit for Christians. This is indicated by article 139 being a part of the section on religions and churches.

Of actual interest is the 1946 draft of a Constitution for the German Democratic Republic prepared by the Socialist Unity Party of Germany [SED]. In article 16 the Sunday rest is protected by law.[1]

This direction is not so much a response to directions regarding religions and churches, but rather to those protecting the working public. The Sunday is associated with other holidays and with the first of May. In short: It represents itself definitely as a secular holiday, totally consistent with its history.

Thus, the Sunday is not so much a gift from Christianity to the world, rather a gift from the world to all Christians. But the Sunday is not for their exclusive and preferred use, rather for their lawful co-use [Mitbenutzung] according to article 33 of the draft of the Constitution.[2] Everyone without any difference in confession, belief (or unbelief), and weltanschauung may celebrate, whoever one wants to do it: in sport and play, in hiking and walking, in happy companionship, in peace and hobby, in going to church or any other religious activity.

The Sunday – a Secular Holiday, 1947 Der Sonntag – ein weltlicher Feiertag. Eine Betrachtung zu Artikel 10 des Verfassungsentwurfs. *Theoretische Zeitschrift des wissenschaftlichen Sozialismus* 1947, 2(6):607-608

---

[1] Draft of Constitution, article 16: … 'The Sunday, other holidays, and May 1 are days of rest and as such are protected by law …'

[2] Draft of Constitution article 33: 'Freedom of belief and conscience and undisturbed practice of religion is protected by the Republic …'

Section 2

Roots of Integrative Bioethics

TRACING THE ROOTS OF EUROPEAN BIOETHICS BACK TO THE ANCIENT GREEK PHILOSOPHERS-PHYSICIANS*

Eleni Kalokairinou[1]

If we look at the contemporary literature of medical ethics, we get the impression that Bioethics, an interdisciplinary science of about 35-40 years, has its origins in the United States. Gilbert Hottois, for instance, in his book, *Qu' est-ce que la Bioéthique?* argues that it was the American oncologist Van Rensselaer Potter who first used the term "Bioethics" in his article, "Bioethics, the science of survival", which was then included in his book, *Bioethics: Bridge to the Future* in 1971.[2] A number of publications following Potter's introduction of the term further support the idea that it was the American scientists' and philosophers' concern about the ethical dilemmas, raised by the development of medical sciences and technologies, which gives rise to this new interdisciplinary science called Bioethics.[3] But if we leave the term aside and, instead, concentrate on the kind of ethical problems which the development of the contemporary biomedical sciences raise, we will realize that, long before Potter, philosophers – physicians like Hippocrates, Galen and Celsus, philosophers like Plato, Aristotle, Immanuel Kant, Friedrich Schleiermacher, Hans Ionas, Albert Schweitzer and, of course, the German theologian and philosopher Fritz Jahr (1895-1953) investigated and attempted to answer the same questions which contemporary American bioethicists contend to have dealt with first. Our contention therefore is that Bioethics is a European discipline and that we must trace it to its roots if we wish to verify this fact.

In studying the origins of the European Bioethics it would be a serious omission if we did not turn to people like Hippocrates, Galen and the Roman Celsus who admittedly laid the foundations of the modern discipline known under the name of Bioethics. For, apart from their strict medical treatises, Hippocrates, Galen and their contemporary physicians composed certain

---

* The paper was presented at 1st International Conference „Fritz Jahr and the Foundations of European Bioethics", Rijeka/Opatija, March 11-12, 2011, and published in (2011) *Jahr* 2(4) 445-456.

[1] Eleni Kalokairinou, ekalo@edlit.auth.gr, Aristotle University of Thessaloniki, Department of Philosophy, 541 24 Thessaloniki, Greece.

[2] Hottois, G. (2004) *Qu'est-ce- que la Bioéthique?* Paris: J. Vrin, p. 10. See also Van Rensselaer Potter (1971) *Bioethics: Bridge to the Future.* Englewood Cliffs, New Jersey: Prentice-Hall.

[3] See, for instance, Beauchamp, T. (1999) "Ethical Theory and Bioethics" in: T. Beauchamp and L. Walters (eds.), *Contemporary Issues in Bioethics.* Belmont: Wadsworth.

deontological treatises to which can be traced almost all the principles of contemporary Bioethics. However, before one examines the content of the Ancient Greek deontology and the way in which it has influenced contemporary Bioethics, one has to consider the medical art or "science" as it was conceived and practised in antiquity.

Medicine, connected as it is to man and human nature, appears in a fairly advanced stage of human civilization.[4] In antiquity, when we talk about medicine we do not refer so much to a body of theoretical knowledge, as we do today, but, instead, to certain therapeutic practices. Similarly, the physician is not a scientist who possesses a fair amount of theoretical knowledge which he applies in life, but he is the practical healer who applies certain accepted practices for the healing of a disease or the cure of a wound. To be more precise, we should mention that these medical practices had a divine character. Before we say anything about the practical healers, we should be reminded that it was the soothsayers and augurs who, from the signs of the weather or the intestines of sacrificial animals, could conclude which practice –in the wide sense- could be followed for the cure of the disease or the expiation of the plague which had befallen a community or a royal House. Consequently, it was more the soothsayers' and the augurs' job than that of the practical healers' to find ways to purify the profane action and to expiate the plague. However, the idea of the divine origin of diseases began to give way. The Ancient Greeks soon realized that they were caught into an undesirable dualism and that they could not accept that all "normal phenomena were natural and all abnormal phenomena were divine".[5] They gradually reached the conclusion that all phenomena are natural and divine and that there are always certain elements of a phenomenon which cannot be explained. In this way, philosophy in the end replaces religion, as it tries to provide explanations for diseases which religion itself could not account for.

The kind of relation which exists between ancient medicine and philosophy is one of the most important problems that has engaged and still engages classicists and philosophers. Even though they all admit that ancient medicine and philosophy are related in a rather complicated manner, a number of classicists argue that it was ancient medicine that influenced ancient Greek philosophical thought. However, the dominant view nowadays is that it was

---

[4] In this paragraph I draw pretty closely to what I am saying in, Kalokairinou, E. «Επίμετρο. Η επίδραση της αρχαίας ελληνικής σκέψης στη σύγχρονη Βιοηθική» in M. G. Kuczewski and R. Polansky (eds.), *Βιοηθική: Αρχαία θέματα σε σύγχρονους προβληματισμούς* (2007). Transl. M. Katsimitses, editing of Greek translation and epimetron Eleni Kalokairinou, Athens: Travlos, p. 528-529.

[5] Hippocrates (1984) transl. W. H. S. Jones, The Loeb Classical Library, Cambridge: Massachusetts, London: Harvard University Press and William Heinemann, vol. I, General Introduction, p. x-xi.

the ancient Greek philosophers who laid the foundations of ancient medicine.[6] This view is mainly corroborated by the ancient Greek sources. Thus Aristotle writes in his treatise *On Sense and Sensible Objects*:

> It is further the duty of the natural philosopher to study the first principles of disease and health; for neither health nor disease can be properties of things deprived of life. Hence one may say that most natural philosophers, and those physicians who take a scientific interest in their art, have this in common: the former end by studying medicine, and the latter base their medical theories on the principles of natural science.[7]

Similarly, in the 1st century A.D., the Roman philosopher-physician Celsus in the prooemium of his work, *De Medicina* says:

> At first the science of healing was held to be part of philosophy, so that treatment of disease and contemplation of the nature of things began through the same authorities; clearly because healing was needed especially by those whose bodily strength had been weakened by restless thinking and night-watching. Hence we find that many who professed philosophy became expert in medicine, the most celebrated being Pythagoras, Empedocles and Democritus.[8]

The Milesian philosophers Thales, Anaximander and Anaximenes were mostly concerned with physics and astronomy and not so much with anthropology and medicine. However, things change as soon as the Pythagoreans were established in Croton of Italy, where there was a medical tradition. Alcmaeon of Croton is a Pythagorean or, at least, belonged to the Pythagorean circle and was the first philosopher who attempted to lay the theoretical principles of medicine and, then, to adapt them to experience. He breaks away from the prevailing view of his time according to which disease was conceived in ontological terms and, instead, he considers it as part of nature. In the extant fragment of his work, *Περὶ φύσεως (On Nature),* he argues that the body consists in a number of opposite elements or forces, i.e.

---

[6] On this claim see Frede, M. (1987) "Philosophy and Medicine in Antiquity" in: *Essays in Ancient Philosophy.* Oxford: Clarendon Press, pp. 225-242.

[7] Aristotle (1986) On sense and sensible objects 436a19-b1 in: *On the Soul, Parva Naturalia, On Breath,* transl. W. S. Hett, The Loeb Classical Library, Cambridge: Massachusetts, London: Harvard University Press and William Heinemann Ltd.

[8] Aulus Cornelius Celsus (1971) *De Medicina,* Prooemium § 6-7, transl. W. G. Spencer, The Loeb Classical Library, Cambridge: Massachusetts, London: William Heinemann Ltd and Harvard University Press.

cold-hot, moist-dry, sweet-bitter etc.[9] The harmonious mixing (κρᾶσις) and the balance (ἰσονομία) between these opposite forces of the body constitutes health, whereas the supremacy (μοναρχία) of any of these over the others causes disease.

The theory of the opposite constituents which Alcmaeon of Croton introduces was prevailing throughout ancient medicine. But, as Cornford points out, the various medical schools differed on the way each conceived of these ultimate elements.[10] Alcmaeon, as we have seen, considered these elements to be opposite powers. But when his theory is accordingly adopted by the medical school of Cos, thc powers are replaced by the "fluid substances, the *humours*".[11] This development took place gradually and we can trace it if we study carefully Hippocrates' treatise *On Ancient Medicine* – a treatise in which, as we shall see, the writer complains intensely for the intrusion of philosophy into medicine.[12] He maintains that these opposites are not substances but powers of secondary importance. He further argues that the body is composed of certain opposite *humours* which have properties or powers that influence health more than temperature does. Thus, in the Hippocratic school health is the harmonious blending of these humours (κρᾶσις), whereas the dominance of the one over the others (μοναρχία) is the sign of disease. In the treatise *Nature of Man,* which Aristotle attributes to Polybus, it is maintained that the humours are four: phlegm, blood, yellow bile and black bile.[13]

On the other hand, thinkers like Empedocles of Croton, who belonged to the Italian and the Sicilian school, followed a different line of thought. Empedocles, for instance, materialized these four ultimate constituents of the body, i.e. fire, air, water and earth, the ῥιζώματα, as he called them. These elements were taken to be the components not only of the human beings but also of all beings. The analogies with which these different elements are mixed determine not only the different kinds of beings but also the different individual human natures.[14] Given these four components, Philistion of Locri developed a theory of health and disease. Put briefly, it has as follows:

> Philistion holds that we consist of four 'forms' (ἰδεῶν), that is elements: fire, air, water, earth. Each of these has its own power: fire

---

[9] Diels, H. and W. Kranz (1989) *Die Fragmente der Vorsokratiker,* vol. I, Weidmann , 24, B4, [22].

[10] Cornford, F. M. (2000) *Plato's Cosmology,* The *Timaeus* of Plato translated with a running commentary. London: Routledge, p. 332.

[11] Cornford, p. 333. See, also, Hippocrates, vol. I, General Introducation, p. xlvi-xlviii.

[12] See, below, notes 28 and 29.

[13] Hippocrates, vol. I, General Introduction, p. xlviii-xlix. See, also, Cornford, p. 333.

[14] Diels, H. and W. Kranz, *Die Fragmente der Vorsokratiker*, vol. I, 31, B 110.

> the hot, air the cold, water the moist, earth the dry. Diseases arise in various ways, which fall roughly under three heads. (1) Some are due to the elements, when the hot or the cold comes to be in excess, or the hot becomes too weak and feeble. (2) Some are due to external causes of three kinds: (a) wounds; (b) excess of heat, cold, etc.; (c) change of hot to cold or cold to hot, or of nourishment to something inappropriate and corrupt. (3) Others are due to the condition of the body: thus, he says, 'when the whole body is breathing well and the breath is passing through without hindrance, there is health; for respiration takes place not only through mouth and nostrils, but all over the body…'[15]

Historians inform us that Philistion was a practising physician at Syracuse. It is almost certain that he influenced Diocles of Carystos in Euboea, who was later regarded as "a second Hippocrates".[16] Diocles, in his turn, practised in Athens and wrote medical treatises on almost every topic between 400-350 B.C.[17] Cornford observes that there is a lot of agreement on many issues between Diocles and Plato, something which leads us to conclude: (a) that they knew of each other's work, and (b) that they both had been influenced by Philistion's teaching.[18] In order to give further support to the above two claims, Cornford invokes Plato's *Second Letter* which, in his opinion, suggests that Philistion attended Dionysius II most probably during Plato's trip to Italy and that they must have met there.[19]

Plato is obviously influenced by Empedocles. In *Timaeus* he describes how the world was created, discusses the creation of man, presents the functions of the human body and the soul and, in the final part, offers an account of diseases. Following roughly Philistion's classification of diseases, he distinguishes three kinds of diseases. There are, first of all, the diseases that are due to the prevalence or the deficiency or even the misplacement of the ultimate constituents.[20] As Plato puts it:

> The origin of disease is plain, of course, to everybody. For seeing that there are four elements of which the body is compacted, – earth, fire, water and air, – when, contrary to nature, there occurs either an excess or a deficiency of these elements, or a transference thereof from their native region to an alien region; or again, seeing that fire and the rest have each more than one variety, every time that the body admits an

[15] See Cornford where he quotes the above fragment, p. 333.
[16] Cornford, p. 334.
[17] Cornford, p. 334.
[18] Cornford, p. 334.
[19] Cornford, p. 334, note 1.
[20] Cornford, p. 334.

> inappropriate variety, then these and all similar occurrences bring about internal disorders and disease.[21]

There are, secondly, the diseases of "the secondary tissues", as Cornford calls them.[22] Plato has in mind here the tissues which are composed of almost all the ultimate constituents. Such tissues are marrow, bone, sinew and flesh. This second type of disease appears when the normal process of nourishment is reversed. In this case, instead of building up in the tissues the appropriate substances which are in the blood in order to repair the waste and fight corruption, the flesh's composition is affected and as a consequence noxious substances are discharged back into the blood. This discharge causes certain kinds of poisonous humours which may be secreted and which may further affect and damage the bones and the marrow.[23] Plato describes the second type of diseases as follows.

> Again in the structures which are naturally secondary in order of construction, there is a second class of diseases to be noted... Now when each of these substances is produced in this order, health as a rule results; but if in the reverse order, disease. For whenever the flesh is decomposed and sends its decomposed matter back again into the veins, then, uniting with the air, the blood in the veins, which is large in volume and of every variety, is diversified by colours and bitter flavours, as well as by sharp and saline properties, and contains bile and serum and phlegm of every sort. For when all the substances become reversed and corrupted, they begin by destroying the blood itself, and then they themselves cease to supply any nourishment to the body.[24]

Thirdly, there are the diseases which are related to: (a) breath, (b) phlegm and (c) bile.[25] These are diseases which are mainly due to respiration problems, to the blockage of air inside the body. They are further due to the formation of noxious humours, such as phlegm and bile.

As may well be expected, Plato concludes his treatment of diseases in the *Timaeus* by discussing a further category, that of the diseases of the soul. These may be due either to the bad condition of the body or to the asymmetry

---

[21] Plato (1981) *Timaeus* 82 A in: *Timaeus, Critias, Cleitophon, Menexenus, Epistles* transl. by R. G. Bury, The Loeb Classical Library, Cambridge: Massachusetts, London: William Heinemann and Harvard University Press.

[22] Cornford, p. 335.

[23] Cornford, p. 335-6.

[24] Plato, *Timaeus*, 82 C - 83 A.

[25] Plato, Timaeus, 84 D; Cornford, p. 340.

which could exist between the soul and the body.[26] It is beyond our present purposes to examine the way Plato conceived of these diseases. However, it remains noteworthy that so long ago Plato was well aware of what we today would call mental illness.

Plato's pupil, Aristotle, though he did not follow his father's profession, esteemed medicine highly. Medicine is quite often employed by him as a model paradigm for developing his ethical and political ideas. The reader of the *Nicomachean Ethics* will soon realize the wide use of medical examples Aristotle makes in his discussion of ethical issues. Among his writings are included treatises which show his genuine interest in issues concerning man's physiology and pathology. Treatises like, *On the Soul, On Sense and Sensible Objects, On Memory and Recollection, On Sleep and Waking, On Dreams, On Prophecy in Sleep, On Length and Shortness of Life, On Youth and Old Age, On Life and Death, On Respiration* and others express his concern for medical and anthropological matters which he, as a philosopher, was in much more competent position to discuss than a mere physician. Aristotle's contribution to medicine has convinced almost everyone that philosophy and medicine were two inextricably related disciplines since neither philosophers can avoid studying medicine nor physicians can get their reasoning started unless they invoke the first principles of natural philosophy.[27] As he writes:

> As for health and disease it is the business not only of the physician but also of the natural philosopher to discuss their causes up to a point. But the way in which these two classes of inquirers differ and consider different problems must not escape us, since the facts prove that up to a point their activities have the same scope; for those physicians who have subtle and inquiring minds have something to say about natural science, and claim to derive their principles therefrom, and the most accomplished of those who deal with natural science tend to conclude with medical principles.[28]

Physicians and philosophers were very much convinced in the 4th century B.C. of the close relationship between philosophy and medicine. This relationship becomes even more obvious in the treatise attributed to Hippocrates. Hippocrates of Cos is a major physician of the 5th century B.C.

---

[26] Plato, *Timaeus*, 86 B - 87 B and 87 B - 89 D; Cornford, p. 343-352.

[27] On the relations between ancient medicine and philosophy see Kalokairinou, E. "Ancient Medicine and Philosophy: A philosopher's perspective" forthcoming in the proceedings of the conference, Medicine in the Ancient Mediterranean world, Nicosia 27-29 September 2008, ed. D. Michaelides, Oxbow Books, Oxford.

[28] Aristotle (1986) *On Respiration,* 480 b 22-31 in: *On the Soul, Parva Naturalia, On Breath,* transl. W. S. Hett, The Loeb Classical Library, Cambridge: Massachusetts, London: Harvard University Press and William Heinemann Ltd.

to whom are attributed more than sixty extant medical treatises. Classicists disagree as to whether or not all these treatises have been written by the same person; instead they prefer to talk of the treatises of the *Corpus Hippocraticum.* Leaving aside the issue of authorship, what is interesting is that while in certain treatises Hippocrates explains certain medical phenomena by arguing from given hypotheses or axioms to conclusions, as philosophers do, in certain other treatises this method is criticized. Thus, in the treatise *On Ancient Medicine* Hippocrates first criticizes those who deduce medical conclusions from first principles and then he puts forward his own view. He writes:

> All who, on attempting to speak or to write on medicine, have assumed for themselves a postulate as a basis for their discussion – heat, cold, moisture, dryness, or anything else that they may fancy – who narrow down the causal principle of diseases and of death among men, and make it the same in all cases, postulating one thing or two, all these obviously blunder in many points even to their statements, but they are most open to censure because they blunder in what is an art, and one which all men use on the most important occasions, and give the greatest honours to the good craftsmen and practitioners in it.[29]

And he adds:

> But my view is, first, that all that philosophers or physicians have said or written on natural science no more pertains to medicine than to painting.[30]

The first impression one gets from the above quotation is that in the treatise *On Ancient Medicine* Hippocrates attacks philosophy. This was how it was interpreted in antiquity. This interpretation was being held until recently. Celsus, for instance, in the prooemium of his work *De Medicina* writes that it was Hippocrates, a man of philosophical skill and medical talent, "who separated this branch of learning from the study of philosophy".[31] In light of further research, however, classicists, philosophers and physicians have come to conclude that this is not necessarily what Hippocrates has been doing. G. E. R. Lloyd in his article "Who is attacked in *On Ancient Medicine?*" is raising the question, whether the author of the treatise is attacking all the thinkers who reduced medical questions to philosophical questions of first principles, whether he is attacking a whole medical school, or just a particular

[29] Hippocrates, vol. I, *On Ancient Medicine*, I, 1-11.
[30] Hippocrates, vol. I, *On Ancient Medicine*, XX, 9-10.
[31] Celsus, *De Medicina,* Prooemium, § 7-8.

individual.[32] The conclusion which contemporary scholars and classicists tend to reach is that Hippocrates in the particular treatise is attacking a certain medical school, namely the Dogmatists, who behind the manifest symptoms of a disease, assumed the existence of the hidden causes of it, which to a great extent determined the kind of treatment to be applied to the particular patient. This does not mean that Hippocrates is combating philosophy as such, since the other medical schools of his days were also influenced by other philosophical schools. Thus the Empiricists, for instance, were influenced by the skeptic school, the Methodists were influenced by the atomic philosophers, whereas the fourth major school, the Pneumatists, were mainly eclectic and were equally influenced by the Stoic school and the theory of the four humours.[33]

It is no doubt that the ancient Greek physicians turned to philosophy in order to ask its support in the theory of knowledge, logic and natural philosophy. However, in the 5th century B.C. the character of philosophy changes. From cosmos- and nature-orientated which was so far, philosophy becomes man-orientated, it is focused on the study of man, it becomes primarily "anthropological". This is why in the 5th and 4th centuries B.C. philosophy's main object of research is man, and the branches of philosophy which mainly flourish then are moral and political philosophy. Philosophy influences medicine again but this time in a different manner.

We can find examples of the way philosophy influences medicine during this period in Hippocrates' deontological treatises, *The Oath (Ὅρκος), The Physician (Περὶ ἰητροῦ), Law (Νόμος), Decorum (Περὶ εὐσχημοσύνης), Precepts (Παραγγελίαι)* and *On Ancient Medicine (Περὶ ἀρχαίης ἰητρικῆς),* in Galen's brief treatise, *That the excellent physician is a philosopher (Ὅτι ὁ ἄριστος ἰητρὸς καὶ φιλόσοφος)* and in the Roman Celsus' treatises and in Sextus Empiricus' work.

If we study these treatises carefully, we will see that their author is not concerned so much with putting forward a theory of health and disease or a physiological theory of the functions of the human body. Instead, what interests him is to bring out the importance the physician's character has for the diagnosis and the cure of the disease. Put differently, the authors of these treatises do not see the physician merely as a mere "engineer", i.e. as a technocrat who knows how to apply specialized knowledge and practices in order to cure the disease. Instead, they see him as the good, wise man who

---

[32] Lloyd, G. E. R. (1963) "Who is attacked in *On Ancient Medicine?*" in: *Phronesis* 8, 108-126.

[33] Carrick, P. (2001) *Medical Ethics in the Ancient World.* Washington: Georgetown University Press, p. 41.

cares for and respects the patient as a human being. It is worth recalling what Hippocrates says on this matter in the most ancient text of medical deontology, the *Oath*:

> I will use treatment to help the sick according to my ability and judgment, but never with a view to injury and wrong-doing.[34]

And a few lines afterwards he adds:

> Into whatsoever houses I enter, I will enter to help the sick, and I will abstain from all intentional wrong-doing and harm, especially from abusing the bodies of man or woman, bond or free.[35]

The apprentice physician should not only be taught the medical art but he should also exercise his character so as to be well-disposed towards the patient. So, as the author of the *Oath* declares, the young physician swears to leave every injustice and harm aside (the contemporary principle of non-maleficence) and to enter the house of the patient with the aim to help the sick (the contemporary principle of beneficence).[36] And not only this. The young physician also swears to be trustworthy and never reveal what he sees or hears while practising his art, proving in this way to be the earliest initiator of what in contemporary medical deontology and bioethics we call the principle of confidentiality. Hippocrates writes in this respect: And whatsoever I shall see or hear in the course of my profession, as well as outside my profession in the intercourse with men, if it be what should not be published abroad, I will never divulge, holding such things to be holy secrets.[37]

The physician will approach his patient with the required respect, he will consider his case carefully and he will appreciate the difficult circumstances he and his family are in, showing in this way that he deserves his patient's trust who puts into his hands the most sacred thing he has, his life. As

---

[34] Hippocrates, vol. I, *The Oath,* 16-18.

[35] Hippocrates, vol. I, *The Oath,* 24-28.

[36] Hippocrates, vol. I, *The Oath,* 24-28. It is interesting to point out that the contemporary bioethicists who support the four-principles approach to Bioethics, otherwise known as principlism, among their basic principles include the two bioethical principles stated above by Hippoctates. Thus, the American T. L. Beauchamp and J. F. Childress in their book, *Principles of Biomedical Ethics* put forward the principle of respect for autonomy, the principle of beneficence, the principle of non-maleficence and the principle of justice. Whereas the British Raanan Gillon in his own work entitled, *Philosophical Medical Ethics,* also includes these two Hippocratic principles among the other bioethical principles he propounds.

[37] Hippocrates, vol. I, *The Oath,* 29-32.

Hippocrates writes in another, equally famous, deontological treatise, *The Physician*:

> The intimacy also between physician and patient is close. Patients in fact put themselves into the hands of their physician, and at every moment he meets women, maidens and possessions very precious indeed. So towards all these self-control must be used.[38]

In all these encounters with his patients and their families the physician should behave with continence and self-control. As Hippocrates puts it:

> Such then should the physician be, both in body and in soul.[39]

If what is of greatest importance is the patient's well being, then the physician should not try to exact his payment right from the start. Such a thing may lead the patient to believe that if the right agreement does not take place between the two, the physician will go away. On the contrary, the physician must be compassionate and must take into account the patient's financial situation. And if need be to offer his services for free, he should not hesitate to do it, bringing to mind the benefits he has already received, and his good name. He should not hesitate to offer his help to a stranger or to a needy. As he writes:

> For where there is love of man, there is also love of the art.[40]

Consequently, medical knowledge and skillfulness on their own do not contribute to the patient's cure, if the physician is not a good and charitable character. It is a happy coincidence if the physician is both good at his art as well as a good character. But where such a thing is not possible, then it is better if he is a good man and not particularly a good physician than the other way around. For, whereas the good character compensates for the deficient art, the bad character corrupts and damages the most perfect art.

It is becoming obvious now why, according to Galen, the man who was preparing to become a physician had to receive not only medical teaching and training, but he had also to study the liberal arts or what we would call today the humanities.[41] According to the Ancient Greeks, the medical teaching and training provided the students with the necessary knowledge and experience

---

[38] Hippocrates, vol. II, *The Physician,* 24-28.
[39] Hippocrates, vol. II, *The Physician,* 28-29.
[40] Hippocrates, vol. I, *Precepts,* VI, 6-7: «Ἢν γὰρ παρῇ φιλανθρωπίη, πάρεστι καὶ φιλοτεχνίη».
[41] Galen (1991) *On The therapeutic Method,* Books I and II, transl., introd. and comment. R. J. Hankinson. Oxford: Clarendon Press, Book I, 1.4-5, 3.15, 4.1-3, Book II, 6.14.

for treating the disease, in the same way as the teaching of an art, i.e. shipbuilding or the art of war, equipped the young with the necessary knowledge for building ships or winning a war. The liberal arts or the humanities, on the other hand, did not teach him a particular art. On the contrary, they addressed the student's character and contributed to the cultivation of his feelings and the development of his abilities and his virtues. By arousing his self-consciousness and his good will, the liberal arts urged him to perform prudent, just and brave acts and, in this way, to become himself prudent, just and brave, in a word *wise.* But, as he became wise, he at the same time became a better physician. It is in this sense that Hippocrates argues that the physician who is a philosopher amounts to being a god. As he puts it:

> For a physician who is a lover of wisdom is the equal of a god. Between wisdom and medicine there is no gulf fixed; in fact medicine possesses all the qualities that make for wisdom. It has disinterestedness, shamefastness, modesty, reserve, sound opinion, judgment, quiet, pugnacity, purity, sententious speech, knowledge of the things good and necessary for life, selling of that which cleanses, freedom from superstition, pre-excellence divine. What they have, they have in opposition to intemperance, vulgarity, greed, concupiscence, robbery, shamelessness.[42]

Today things, to be sure, are much more complicated. The bioethical principles which the classical deontologists propounded had to be further supplemented with more elaborate principles and rules so as to handle efficiently the complex problems which contemporary medical science and technology creates. Furthermore, our crowded contemporary societies could not just rely upon the physician's good character, as was the case in antiquity. They had to establish all the right social structures and mechanisms for protecting the patients and their families. Be that as it may, the truth remains that the basic principles and rules which are often invoked in serious discussions of bioethical issues are not modern and recent as one may at first think. Even though the term "Bioethics" was introduced in the 20th century, nevertheless the actual discipline of Bioethics, under any name whatever, was first conceived and widely practised some twenty-five centuries ago.

---

[42] Hippocrates, vol. II, *Decorum,* V, 1-13: «ἰητρὸς γὰρ φιλόσοφος ἰσόθεος· οὐ πολλὴ γὰρ διαφορὴ ἐπὶ τὰ ἕτερα· καὶ γὰρ ἔνι τὰ πρὸς σοφίην ἐν ἰητρικῇ πάντα, ἀφιλαργυρίη, ἐντροπή, ἐρυθρίησις, καταστολή, δόξα, κρίσις, ἡσυχίη, ἀπάντησις, καθαριότης, γνωμολογίη, εἴδησις τῶν πρὸς βίον χρηστῶν καὶ ἀναγκαίων, καθάρσιος ἀπεμπόλησις, ἀδεισιδαιμονίη, ὑπεροχὴ θείη. ἔχουσι γὰρ ἃ ἔχουσι πρὸς ἀκολασίην, πρὸς βαναυσίην, πρὸς ἀπληστίην, πρὸς ἐπιθυμίην, πρὸς ἀφαίρεσιν, πρὸς ἀναιδείην».

ST. FRANCIS OF ASSISI: BIOETHICS IN EUROPEAN MIDDLE AGES*

Ivana Zagorac[1]

In 1967 Lynn White, Jr., published a controversial text[2] which denotes Christianity as one – if not the only one – of the causes of the modern over-exploitation of nature. He identified the Middle Ages as the starting point of this misuse, which he found support for in (his misreading of) the *Holy Bible*. Moreover, he accused Christianity of being both the most arrogant of all the main religions, and the most anthropocentric one. According to White, man of the Middle Ages took seriously the Biblical implication that man rules over nature and that man is at the very top of the moral hierarchy. However, from this monochrome intonation of White's arguments,[3] one colourful character stands out. Namely, St. Francis of Assisi, a Christian saint from the 12th century. White declared St. Francis of Assisi to be the only source of light in darkness, an almost exotic phenomenon of the Christian Middle Ages, who deserves our admiration and could serve as a role model for moral rehabilitation even today. This "greatest spiritual revolutionary in the Western history", as praised by White, offered an alternative way to lived co-existence of the Christian religion, Man and Nature. The admirable life of St. Francis of Assisi inspired White to propose him for the saint patron of ecology, a title that was in the end conferred by Pope John Paul II.[4]

---

* The paper was presented at 1st International Conference "Fritz Jahr and the Foundations of European Bioethics", Rijeka/Opatija, March 11-12, 2011.

[1] Ivana Zagorac, izagorac@ffzg.hr, University of Zagreb - Faculty of Humanities and Social Sciences, Department of Philosophy, I. Lučića 3, Zagreb, Croatia.

[2] Lynn White, Jr., "The Historical Roots of Our Ecologic Crisis", *Science*, New Series, Vol. 155, No. 3767 (Mar. 10, 1967), pp. 1203–1207.

[3] Many others share White's view. Roderick Nash, for example, claims that St. Francis's early followers were completely insensitive to nature and fully devoted to its cultivation or, more specifically, to its subordination to human needs. (Roderick Nash, Wilderness and the American Mind, Yale University Press, New Haven–London 2001.) Such theses have been strongly objected to (cf. in the context of discussing the life of St. Francis of Assisi: Edward Armstrong, Saint Francis: Nature Mystic: The Derivation and Significance of the Nature Stories in the Franciscan Legend, Hermeneutics Studies in the History of Religions, Vol. 2, University of California Press, Berkeley 1973; H. Paul Santmire, The Travail of Nature: The Ambiguous Ecological Promise of Christian Theology, Fortress Press, 1985; Susan Power Bratton, "Oaks, Wolves and Love: Celtic Monks and Northern Forests", Journal of Forest History, Vol. 33, No. 1 (Jan., 1989), pp. 4–20, and Christianity, Wilderness, and Wildlife: The Original Desert Solitaire, University of Scranton Press, Scranton 1993, including an extensive bibliography).

[4] St. Francis of Assisi (born as Giovanni Francesco di Bernardone) was born in 1181/1182 and died on 3 or 4 October 1226. He was proclaimed a saint by Pope Gregory IX on 16 July 1228, less than two years after his death. On the occasion the Pope ordered that a basilica be built in Assisi dedicated to St. Francis. The Basilica of St. Francis in Assisi has

The life and teachings of St. Francis

What was so impressive about the life of St. Francis of Assisi that rendered him as one of the most lovable Christian saints and an equally lovable patron of ecology? St. Francis of Assisi gained fame across Europe during his lifetime, and was canonised less than two years after his death. The Franciscan Order was rising quickly, and until the death of its founder in 1226 reached almost the whole of Europe from Italy, spreading as far as the Mongol Empire.[5] Due to such expansion and interest in St. Francis's life, his biography was soon covered with layers of hagiographic elements.

Born in 1181 (or 1182) in the small town of Assisi in Italy, Francis was raised by a wealthy merchant family. However, the education he received was poor, and he struggled with basic reading and writing till the very end of his life. A lack of literacy skills reflected on his writings, which are relatively small in number,[6] making it difficult to reach his original thought. On the other hand, his life was well documented by his followers, providing great insight into Francis's vision of the world, nature and faith. Like many Christian saints from the Middle Ages, St. Francis also experienced a radical personal change, leaving his promising life of a merchant, and dreamt of becoming a knight for life according to the principles of the Christian tradition. But the change he underwent was much deeper and more radical than the usual pattern followed by other Christian saints. Disappointed with the multiplicity of incoherent interpretations of the *Holly Bible*, Francis formed his own interpretation of the interrelation between God, Nature and Man which was so bioethical that he could today rightly be proclaimed the saint patron of bioethics if one were sought.

---

remained the central place of his worship to date. In 1979 Pope John Paul II proclaimed St. Francis the patron saint of "those who promote ecology". Officially, the proposal was put forward by Cardinal Silvio Oddi (Apostolic Letter *Inter Sanctos*: AAS 71 (1979)). The Feast of St. Francis of Assisi is observed on 4 October, the day that World Animal Day is also observed. On that day or the Sunday closest to 4 October many Catholic churches organise the blessing of pets and animals. The largest cathedral in the world, namely the Cathedral of St. John the Divine in New York, is particularly famed for its blessing of the animals, when animals such as camels, elephants, deer or eagles can be seen being taken to church. Children are encouraged to bring their stuffed animals, and even photographs of pets are accepted.

[5] Edward Armstrong, *Saint Francis: Nature Mystic: The Derivation and Significance of the Nature Stories in the Franciscan Legend*, p. 7.

[6] The writings that are today attributed to St. Francis consist mainly of prayers, letters and the rules of the Order of St. Francis, and a testament from 1226. Regis J. Armstrong, J. A. Wayne Hellmann, William J. Short (eds.), *Francis of Assisi: Early Documents*, New City Press, New York 1999.

In spite of the layers of hagiographic exaggerations, St. Francis's empathy and compassion for both non-human creatures and non-living parts of nature are a fact supported by the writings of his contemporaries, most importantly by Thomas of Celano. In praising the Lord through all His creations, St. Francis practises his theological, sacramental vision of Cosmos. Having decided to live a life of poverty, he chose a life without any possessions. Renouncing possessions is also applied to animals, which – from St. Francis's point of view – we posses no more than we possess a forest, the sky, the sea or other people. For St. Francis, love and compassion are the media through which we can and do become open for communication with Divinity. All parts of nature must be treated with the same respect since love of God sees no difference between God's creations. A hierarchy of beings or a dualism of any kind – not even the one that would discriminate between the living and non-living parts of nature – have no place whatsoever in the philosophical concept of St. Francis. St. Francis transferred birds, rocks and worms from the category of 'things' to the level of ontic and moral equality with man within the framework of the same spiritual community. Egalitarianism of this kind necessarily implies one's acknowledgment of the intrinsic value that each and every part of nature has, as opposed to the well-established way of value attribution according to the standards of human needs. The bare fact of existence, seen as proof of God's love, was itself reason enough for St. Francis to include all Creations in his ethical considerations.

However, giving this theory reality, as practised by St. Francis, must have been interpreted as highly eccentric, especially to St. Francis's medieval contemporaries. Medieval theology was deeply dualistic, and saw a clear divide between humans and other living creatures, as well as between this and the other world. Preaching to birds, calling the Moon his sister and the Sun his brother, rescuing rabbits from hunting traps, or removing worms from footpaths, which is what St. Francis did, are all episodes that were reaching far beyond the typical understanding of one's Christian duties to God's creatures. St. Francis not only saw an ontological interrelation between all parts of nature, but also sought ways of expressing this unity through acts of moral benevolence. He does not see the human species or forests or bees as collective entities more or less worthy of moral consideration – he is dedicated to recognising the specific worth of every single entity, whether a human, a tree or an animal. He treats rock, water, fire, the Sun or the Moon with the same respect, perceiving them as much more than metaphorical decorations for suitable theological exemplifications.

St. Francis of Assisi widened the term 'community' from community based on shared religious beliefs, beyond community of all living creatures, to community of *all* creations.[7] Correspondingly, he chose yet another rather radical postulate to live by – the commitment to obedience by all the members of his Order to all the living and non-living parts of nature. This kind of emphasis on sensibility towards nature was unprecedented in both Christian and common secular life before St. Francis. For St. Francis living by this principle meant giving his own clothes as payment for animals that were to be slaughtered, or allowing mice to run over his body whilst lying severely ill in bed, or – as a particularly exotic element – letting his cloak be destroyed by brother Fire once fire broke out accidentally.

St. Francis's original idea has been preserved in an extraordinary poem, which is part of his small written legacy. It is entitled "The Canticle of the Sun" (or "Laudes Creaturarum") revealing a unique sensibility. In a radically animistic vision of the physical world, St. Francis addresses water, rock and fire as if they were sensible creatures, and preaches to flowers and birds in the same way that he would to people. He clearly hears their response: nature in turn replies with approval, confirming the rightfulness of his actions. Our world, as well as that of St. Francis, has been profoundly marked by a sharp division between us, and social and spiritual others. The revolutionariness of St. Francis's individualistic approach is especially evident in the historical context within which it developed. Christianity of St. Francis's time did not exclude nature from spiritual considerations, but it perceived nature only as a metaphorical tool for grasping the Lord's messages. The Middle Ages offered only two sources to learn from: the book of nature and the book of scripture. According to White,[8] the extreme anthropocentrism of late Antiquity and of the early Middle Ages saw the Universe as a riddle that had to be solved or as a cryptogram that needed to be decoded. An individual was held to be singular, but not self-worthy. The autonomy of individuals was annulated by the interpretative pluralism of reading through metaphors and allegories. For St. Francis, the constituents of nature also carry deep symbolic meanings, but the carriers of these meanings are at the same time acknowledged as subjects having their moral worth independent from an arbitrary human reading. St. Francis "forced man to abdicate his monarchy over the creation, and instituted a democracy of all of God's creatures".[9] It is the individual – and not the collective – that represents the core of moral acting, and the basic entity in St. Francis's alternative way of interpreting the teaching of respect for all Creations.

---

[7] J. Donald Hughes, "Francis of Assisi and the Diversity of Creation", *Environmental Ethics* 18 (3/1996), pp. 311–320.

[8] Lynn White, Jr., "Natural Science and Naturalistic Art in the Middle Ages", *The American Historical Review*, Vol. 52, No. 3 (Apr., 1947), pp. 421–435.

[9] Ibid., p. 433.

St. Francis of Assisi within the context of contemporary bioethical thought

A few interlinked elements are to be singled out in St. Francis's theory and practice. His worldview is undoubtedly imbued with (I) *profound religious experience*. On the other hand, it is precisely the way in which St. Francis lives his faith that sets him apart from the "mainstream" of Christian thought. The underlying characteristic of a lived faith according to St. Francis is accentuating the deep conviction that all creation is spiritually egalitarian. The very act of Creation, the belief that the origin of all Creations is in one God, makes all Creations participate in God. For St. Francis, spiritual egalitarianism, which presupposes appreciation of the inherent worth of all Creations, represents the (II) *annulment of ontic hierarchism*. Once all the constituents of nature are levelled with respect to worth, (III) *the boundaries of our moral concern expand*: due to the very fact that they exist, all Creations possess an inherent worth, which in turn implies having to be included in the moral concern of morally acting beings. Once all the constituents of nature are on a par, collective nouns, species and collective entities, such as living and non-living nature, no longer exist. St. Francis treats everything and everyone (IV) *individually*, in each and every constituent of nature – whether human or non-human, living or non-living – he finds his Brother or Sister.

St. Francis's relationship to nature is probably one of the key reasons for why he is today considered to be one of the most favourite saints. By perceiving the universal in the individual and the individual in the universal, and by acting in accordance with this vision, St. Francis has become the symbol of universal compassion. He has recently been frequently referred to within the context of new environmental ethics in general, and eco-theology, ecocentrism and biocentrism in particular. It is entirely needless to verify that St. Francis is miles away from anthropocentrism, which, as has been seen earlier, even the staunchest adversaries of anthropocentrism from his native religion acknowledge as his outstanding achievement. At the time of St. Francis of Assisi, he represented the antipode to the belief in the superiority of man which was grounded on the dualistic presupposition that man is distinct from nature, an attitude which has prevailed ever since. Contemporary ecologically orientated theories find their inspiring forerunner in St. Francis. However, even though these theories at first sight do present theoretical concepts that correspond with St. Francis's thought and practice, they soon turn out to be all too narrow.

Within the context of eco-theology, St. Francis represents a fresh and novel view on the relationship between God, nature and man. The key to his interpretation of the *Holy Scripture* is spiritual equality, erasing the boundaries between beings, and broadening the horizons of our moral concern to embrace all the constituents of nature. Religious rapture takes St. Francis far beyond the boundaries of the very religion that he draws his energy from. Although deeply immersed in religion, the theocentric orientation is foreign to St. Francis, almost as foreign as the anthropocentric one. By definition, theocentrism places God in the centre or, more specifically, at the top of a certain structure. But, St. Francis replaces structure, order and harmony of all Creations with a stance on spiritual autonomy. All Creations have the same Creator, they cannot be closer or further away from God, they cannot possess greater or lesser worth. There are no *others*. Deeply believing in the rightness of this principle of living elevates St. Francis to a level far beyond the frames of individual religions.[10] The implication of this attitude is truly simple, although perhaps paradoxical: St. Francis was simply too great a believer to stay within the boundaries of a defined religion. Accordingly, even a theocentric reading of St. Francis – including the reading by the contemporary theological-bioethical discourse which has attempted to rehabilitate and give reality to St. Francis's alternative, ecologically aware thought – ultimately proves to be insufficient.

St. Francis's nullification of the ontic hierarchy and his positioning of the individual as the basic entity of our moral concern are traits that disqualify him from being the forerunner of contemporary ecocentric theories. Ecocentrism, as understood by J. Baird Callicott,[11] presents itself as a holistic environmental ethics which includes in its moral concern holistic entities, such as species, ecosystems and biotic communities. The presupposition of their existence is inner harmony which enables each entity to function without impediment. Harmony also means order, acknowledging that there exists a sensitive structure and the need to protect the existing – biological, evolutionary or however established – hierarchy amongst beings.

---

[10] Referring to Matthew Fox, Michael W. Fox describes St. Francis's worldview as *panentheism*, considering it to be synonymous with St. Francis's sacramentalism. According to this definition, panentheism asserts that everything existing is in God, and that everything existing is God, as opposed to pantheism which claims that God is present in everything that exists. Fox adds: "This view is judged to be heretical by monotheistic fundamentalists who believe in a purely transcendent God" (p. 71). He holds that panentheism and monotheism represent complementary aspects of religious life. Irrespective of the way in which we label it, the possible panentheism of St. Francis's thought underlines all the difficulty in placing his thought within established concepts, whether they are or are not denominated as theological. Michael W. Fox, *The Boundless Circle: Caring for Creatures and Creations*, Quest Books, 1996.

[11] J. B. Callicott starts from the groundwork proposed in Aldo Leopold's land ethic. Callicott will himself name what I here call ecocentrism attributable to Callicott – holistic non-anthropocentric environmental philosophy.

Nevertheless, regardless of their position on the ladder, ecocentrism recognises that each and every entity does possess an inherent worth. However, as has been highlighted above, St. Francis's intention takes a significantly different – i.e. individualistic – direction.

On the other hand, biocentrism, as advocated by Paul W. Taylor,[12] and St. Francis's annulment of the ontic hierarchy of beings inspired by the *Holy Bible* have much in common. Taylor cites that the basics of all *human* ethics are a system of belief, respect, and a system of rules and standards. Any environmental ethics, such as biocentrism, should be organised adhering and according to the above structural presuppositions. For Taylor, the principle of acting in accord with acknowledging and inferred respect for the inherent worth of each living being lies at the very heart of biocentric ethics. As far as this point is concerned, St. Francis satisfies Taylor's presuppositions, unlike those of ecocentrism, and is thus qualified as a biocentric author. However, as has already been underlined, St. Francis takes a further step – he broadens our moral concern to include the non-living constituents of nature. Biocentrism, if understood to be ascribing worth to all *living* beings and practised accordingly, appears to be all too narrow for St. Francis. It is only in the field of implications that compatibility could be achieved between biocentrism and St. Francis's thought. In other words, if we take St. Francis's respect for all the constituents of nature which represent the *conditions* of life as lauding the diversity and interweavement of the living and the non-living world, and if we look at St. Francis's thought in the sense of inter-conditionedness, the definition of biocentrism could be stretched to include his views. On the other hand, recognising the inherent worth of every living being, biocentrism itself unreservedly implies the need to respect that which enables this life. The embodiment of moral value is another major point of contact: worth-wise biocentrism treats every living being individually, which is identical to St. Francis's practice. However, what a biocentric reading of St. Francis fails to see is that what is anchored in the groundwork of his thought is the annulment of all instrumentality. For St. Francis Brothers Sun and Wind, Sisters Moon and Water are on a par with the members of the Order of Friars Minor. In defending St. Francis's biocentrism, John Mizzoni[13] interprets St. Francis's above attitude within the context of giving thanks to God for helping living beings through these worthy objects. Even though St. Francis truly does recognise the correlation of the Sun, the Moon, water and air with the survival of living beings, he also recognises that all Creations share a common origin. Namely, that in God. Recognising common origins is not

[12] Paul W. Taylor, *Respect for Nature: A Theory of Environmental Ethics*, University Press, Princeton 1986.

[13] John Mizzoni, "St. Francis, Paul Taylor, and Franciscan Biocentrism", *Environmental Ethics*, vol. 26 (1/2004), pp. 41–56.

alien to biocentric theories either. Yet they are founded on science and rational argumentation as opposed to St. Francis's religious inspiration. This is why biocentrism is to be freed from guilt over only partially capturing St. Francis's thought, considering that for rationally guided thought non-living nature can at best attain instrumental value alone.

If we are to keep St. Francis of Assisi within the framework of contemporary ecological discussion, the question is: what does he, in fact, propose? Or in what way could he be read by the contemporary environmental discourse? Discarding theocentrism and anthropocentrism as central constructs and recognising the insufficiencies of alternatively orientated theoretical concepts, such as ecocentrism and biocentrism, lead to the conclusion that the contextualisation of St. Francis's thought in contemporary, ecologically orientated theories falls short since the theoretical groundwork for structuring St. Francis's living practice is insufficiently wide. Biocentrism, which has created almost all the conditions for comfortably placing St. Francis amongst its inspiring forerunners, stumbles at the exclusion of the conditions of life from our moral concern. Appreciation of the value of the conditions of life is naturally implied in the very biocentric concept, but this value is measured in utility. If biocentrism were to take a step forward and explicitly acknowledge that that which it at present qualifies as the *conditions* of life possesses an inherent worth, we would no longer be talking about *bio*centrism. Recent theories within biocentrism may be suggesting a possible solution – adjusting biocentrism to accommodate holistic entities.[14] Motive for such an undertaking is found in the interest of the science of ecology in the whole, the survival of which ensures the survival of all its constituents. Correspondingly, the non-living entities within nature, i.e. all that which circulates through a system, are gradually included.

The thesis on recognising the worth and rights of a given species or collective entity as an individual is not an intellectual novelty for ecological theories. Is it compatible with the all-encompassing system accepted by St. Francis? There are two reasons for an essentially positive answer. First, extended biocentrism also embraces the non-living constituents of nature, which is what St. Francis insists on. Second, Christian theology, which St. Francis draws his inspiration from and finds anchor in, suggests having to be concerned with the survival of species as collective entities already in the story of Noah. Yet, St. Francis of Assisi knew this, but decided to take an individual approach nevertheless, so it seems that he escapes extended biocentrism as well. The inclusion of the non-living constituents of nature in

[14] For example, Sterba (2001) as noted in: John Mizzoni, "Franciscan Biocentrism and the Franciscan Tradition", *Ethics & the Environment*, Vol. 13, No. 1 (Spring 2008), pp. 121-134, here p. 125.

our moral concern and acknowledging their role in sustaining the functioning of the system still keeps them below the standing of living beings which are recognised to possess value as such. St. Francis does not rationalise, he *believes*. And while Noah is building an ark, St. Francis of Assisi allows fire to consume his cloak and a mouse to rove over his ill body. For St. Francis both the mouse and fire possess worth in itself, and their role in the global system is of inferior importance.

The inadequacies of closed theories, including their attempts to compromise and partially extend, point to the fact that "translating" St. Francis's teachings into the language of the contemporary ecological discourse requires a widely integrative approach. Integrativity should here be applied to two levels: with respect to subject and methodology. Methodological integrativity must take account of the source of St. Francis's teachings and the fact that they are deeply rooted in the Christian doctrine, as well as the context of the specific historical moment that St. Francis lived and worked in. Although reputable and respected, his influence has been overpowered by that of Thomas Aquinas; two different views of the world prompted by the same motivation. Thomas Aquinas holds that there is no basis for a correlation between humans and other living beings for the simple reason that there is a clear dividing line between rational and irrational beings. St. Francis of Assisi also grants that humans dominate the rest of nature, but he explains this domination from the position of responsibility – the Christological anchor of St. Francis's thought requires that we serve the rest of nature. Integrativity with respect to subject encompasses the components of St. Francis's thought: egalitarianism of all the constituents of nature within the context of their inclusion in our moral concern, defining one single entity as the smallest unit of our moral concern, sensibility to the fact that all the constituents of nature are profoundly interrelated and well from the same ontic source. St. Francis of Assisi provides insight into a lived sensible interconnection with nature, which also contains the dimension of what will eight centuries later be called 'reverence for life' by Albert Schweitzer. The intensity with which St. Francis lived universal love, coupled with his charisma, have made him become the archetype of a lived sensibility to nature and the prototype of bioethical sensibility. Sadly, although famed and celebrated, his influence on scholastic philosophy and theology was imperceptible. On the other hand, calls for the rehabilitation and reaffirmation of St. Francis's thought have today intensified in both scope and volume – at a time so dramatically different from the time that he lived in, and in a world seeking a powerful symbol of universal sensibility.

KANT'S CATEGORICAL IMPERATIVE AND JAHR'S BIOETHICAL IMPERATIVE*

Igor Eterović[1]

Introduction: some points on the cultural background of the two philosophers

There are at least two similar points underlying lives of two philosophers which should not be overlooked – protestant tradition and Prussian culture: both of them spent their whole lives in their birthplaces[2], towns with significant cultural heritage of the Prussian Kingdom and both were deeply connected with Protestantism.

Immanuel Kant was born in 1724, lived, and died in 1804 in Königsberg, a seaport in East Prussia, located where the River Pregel flows into the Baltic Sea. In Kant's time, the city was an isolated eastern outpost of German culture, and remains as such until the middle of the twentieth century.[3] The city became an university centre and the seat of protestant learning very early. The University of Königsberg was founded by Duke Albrecht of Prussia in 1544 (the so-called Albertina) and became the centre of Protestant teachings. Martin Luther himself assisted Duke Albrecht with advice and some authors say that Wittenberg could be called the intellectual mother of Königsberg.[4] The city has an important place in German history and culture. It "was originally the capital of East Prussia, the base of Prussian power before the acquisition of Brandenburg and the growth of Berlin, and in Kant's time it remained the administrative center of East Prussia and a leading Hanseatic

---

* The paper was presented at 1st International Conference „Fritz Jahr and the Foundations of European Bioethics", Rijeka/Opatija, March 11-12, 2011, and published in Eterovic, I (2011) *Jahr* 2(4) 457-474.

[1] Igor Eterovic, igor.eterovic@medri.hr, University of Rijeka, Faculty of Medicine, Department of Social Sciences and Medical Humanities, Braće Branchetta 20, 51 000 Rijeka, Croatia.

[2] It is very interesting that both philosophers lived their whole lives in their birthplaces. Kant has went outside Königsberg just during his practice as private teacher and even refused a few offers from prestigious universities in other towns (Erlangen, Jena, Halle). Fritz Jahr lived his entire life even on the same address in Halle(!). (Cf. Kuehn, M. (2001) *Kant: A Biography*. Cambridge: Cambridge University Press, passim and Sass, Hans-Martin (2010a) Postscriptum and references, in: Sass, Hans-Martin (ed.), *Selected Essays in Bioethics 1927-1934: Fritz Jahr*. Bochum: Zentrum für Medizinische Ethik, 25. See also in this volume.

[3] Cf. Wood, Allen W. (2005), *Kant*. Oxford: Blackwell Publishing, 3.

[4] Cf. Barnstorff, Hermann (1939), The Rise of the German Universities, in: *The Modern Language Journal*, 23(4), p. 285, and Shennan, Margaret (1995), *The Rise of Brandenburg Prussia*. London–New York: Routledge, 5.

mercantile city, the most important outlet east of Danzig for the vast Polish and Lithuanian hinterlands."[5] The city "was a business, legal, military, and educational center with many connections to the rest of Europe."[6] It was extremely important in the time of the rising of Prussian state, and it was the crowning place of the first Prussian king – Frederick William I.[7]

Fritz Jahr was born in 1895, lived, and died in 1953 in Halle, a city in the south of Saxony-Anhalt, located on the river Saale. As a new university place and protestant centre Halle got in direct rivalry with Königsberg. The University of Halle was founded in 1694, and was very successful: "the free spirit attracted the best teachers, of whom the philosopher and rationalist Christian Wolff[8] became the center."[9] With the tacit approval of the king Frederick himself "the first leader of the Pietists in Germany, Philipp Jacob Spener (1635–1705) (...) and his followers (...) established Pietism in the new university of Halle. Thomasius broke new ground by lecturing in German rather than Latin. In addition, by separating philosophy from theology he was to make Halle the leading centre for the new cultural thought in Protestant Germany."[10]

Königsberg and Halle are undoubtedly the two pillars of German culture, which is especially formed on the basis of Protestantism. Both towns were firm seats of pietism, a movement within Lutheranism (Protestantism), indeed they were rivals in creating the pietism's fundaments of German culture.[11]

Kant was raised in pietism and Jahr himself was a protestant pastor. Having in mind the tradition which connects these two thinkers, these facts are not of little importance. Some authors, despite the historical discontinuities and unsuspected continuities, argue for the idea of the "long story" of German philosophy inside the framework of possibility of a German "national history

---

[5] Guyer, Paul (2006), *Kant*. London-New York: Routledge, 16.

[6] Ibid.

[7] Cf. Shennan, M. (1995), 43, and Gawthrop, Richard L. (1993), *Pietism and the Making of Eighteenth-Century Prussia*. Cambridge: Cambridge University Press, 64-65.

[8] On Leibniz's recommendation Christian Wolff (1679–1754) "was appointed Professor of Mathematics in 1707, a position he held for sixteen years. Wolff was known as an exponent of Rationalism and he became a leading figure in the German Enlightenment." Shennan, M. (1995), 65.

[9] Barnstorff, H. (1939), 287.

[10] Shennan, M. (1995), 63-65.

[11] This rivalry was especially prominent after the founding of the University of Halle, which could be seen "as a continuation of the rivalry between Saxony and Brandenburg-Prussia, initiated by Frederick William, for regional supremacy and for leadership within German Protestantism as a whole." Gawthrop, R. (1993), 61.

of philosophy"[12]. Such framework connects Kant and Jahr especially in terms of "characteristically Prussian ethos", which "first crystallized and gained wide spread acceptance during the reign of Frederick William I (1713-1740)"[13], i.e. during the time of Kant's youth and maturation. This ethos "laid the basis for the subsequent influence of 'Prussianism' on the development of modern Germany"[14], Germany of Fritz Jahr's time. It is important to bear in mind that this ethos "can be best understood in terms of the history of German ascetic Protestantism, especially Lutheran Pietist movement", which "emphasized a 'bornagain' conversion, followed by a highly disciplined life centered around 'doing good for others'."[15]

Kant and Jahr are both deeply enrooted in German culture and protestant tradition. It is important to be aware of such cultural continuum between them if we want to understand Kant's influence present in Jahr's writings. Bearing this in mind, it is not so surprising that Jahr took the fundamental concept of Kant's ethics – the concept of imperative – to ground his vision of bioethics.

Kant's categorical imperative

Kant's work *Groundwork of the Metaphysics of Morals* is "nothing more than the search for and establishment of the *supreme principle of morality*, which constitutes by itself a business that in its purpose is complete and to be kept apart from every other moral investigation." (4:392[16]). Thus, we should focus on the presentation of this "complete business" and our main reference should be the *Groundwork*[17].

At the beginning of the First section of the *Groundwork* Kant asserts: "It is impossible to think of anything at all in the world, or indeed even beyond it, that would be considered good without limitation except a *good will*" (4:393). However, "without the basic principles of the good will they [affects and passions, added: I.E.] can become extremely evil" (4:394), and we need some *a priori* (what means necessary and universal) rule of reason which could guide our will in moral conduct. These thoughts bring us to the concept of

[12] Cf. especially Beck, Lewis White (1969), *Early German Philosophy: Kant and His Predecessors*. Cambridge: The Belknap Press of Harvard University Press, 1-15.

[13] Gawthrop, R. (1993), i.

[14] Ibidem.

[15] Ibidem.

[16] Kant's writings are cited in the body of the text according to volume and page number in *Kants gesammelte Schriften*, edited by Königliche Preußische [now Deutsche] Akademie der Wissenschaften (Berlin, Georg Reimer [now Walter De Gruyter], 1902-). All translations are taken from Guyer, Paul - Wood, Allen W., eds. (1992-), *The Cambridge Edition of the Works of Immanuel Kant*, Cambridge: Cambridge University Press.

[17] I will use abbreviated title further in text.

duty, which "*is the necessity of an action from respect for law*" (4:400). Acting from duty is "acting with inner rational moral constraint, motivated solely by the thought of following a moral principle. The crucial claim is that we think there is something uniquely worthy of esteem about a person who fulfils duty in the absence of (or even in opposition to) all other inducements of inclination or self-interest, solely out of respect for the moral law."[18] "But what kind of law can that be", asks Kant, "the representation of which must determine the will, even without regard for the effect expected from it, in order for the will to be called absolutely and without limitation?" (4:402). His answer is that the purely rational appeal of a universally valid practical principle is the only thing which could motivate us, and he gives the first formulation of what we may call the Formula of Universal Law[19]: "*I ought never to act except in such a way that I could also will that my maxim*[20] *should become a universal law*" (4:402). In this formulation are implicitly contained all main ideas which he goes on to develop and it is the basis for his systematic exposition of the supreme principle of morality which takes place in the Second section of the *Groundwork*. In this section the crucial idea is the idea of categorical imperative as an articulation of the moral law.

Kant begins with the extensive theory of human agency. Shortly, for Kant we are "agents who are self-directing in the sense that we have the capacity to step back from our natural desires, reflect on them, consider whether and how we should satisfy them, and be moved by them only on the basis of such reflections. An inclination (or habitual desire we find in ourselves empirically) moves us to act only when we choose to set its object as an end for ourselves, and this choice then sets us the task of selecting or devising a means to that end."[21] For Kant, setting an end is the basic normative act. Many acts command us to perform some action as a means to some wanted end. Such command is stated in the form of hypothetical imperative. Hypothetical, because it articulates the needed means for some particular end,

---

[18] Wood, Allen W. (2007), The supreme principle of morality, in: Guyer, Paul (ed.), *Cambridge Companion to Kant and Modern Philosophy*. Cambridge: Cambridge University Press, 347.

[19] Kant scholars use this designation widely. There are other abbreviations used in Kantian scholarship which will be introduced bellow. For such usage see for example Wood, Allen W. (2008), *Kantian Ethics*. Cambridge: Cambridge University Press, 66-67. and Guyer, P. (2006) 191-207.

[20] Kant is very clear about the distinction between maxims, as the subjective principles of our volition, and laws, as the objective principles of it: "A *maxim* is the subjective principle of volition; the objective principle (i.e. that which would also serve as subjectively as the practical principle for all rational beings if reason had complete control over the faculty of desire) is the practical *law*." (4:402f).

[21] Wood, A. W. (2007), 348.

and imperative, because it is the command of reason requiring the agent to do something (e.g. If you want to end hunger, take some food).

But the fundamental principle of morality should be unconditionally valid and necessary true, thus the form of hypothetical imperative is inappropriate for moral guidance. Kant claims that only rational beings (including humans) have "the capacity to act *in accordance with the representation* of laws, that is in accordance with principles", while "everything in nature works in accordance with laws" (4:412). With this claim Kant opens the door for his arguments about rational beings as simultaneously free beings. Rational beings could only act according to principles which are not present in nature, as natural laws. Such principles are part of special, moral realm. Thus, the fundamental principle of morality is presented to us in the form of "categorical imperative", because only such imperative is not based on some contingent ends (as is the case with hypothetical imperative) and gives a basis for unconditional and necessary validity. Categorical imperative "tells you what you must do independent of any end you might have."[22] In other words, "a moral imperative is *categorical* because its function is not to advise us how to reach some prior end of ours that is based on what we happen to want but instead to command us how to act irrespective of our wants or our contingent ends. Its rational bindingness is therefore not conditional on our setting any prior end."[23]

Kant then argues that

> When I think of a *categorical* imperative I know at once what it contains. For, since the imperative contains, beyond the law, only the necessity that the maxim be in conformity with this law, while the law contains no condition to which it would be limited, nothing is left with which the maxim of action is to conform but the universality of a law as such; and this conformity alone is what the imperative properly represents as necessary. (4:420-421)

And he further explicates the notions of "maxim" and "law" in a footnote

A *maxim* is the subjective principle of acting and must be distinguished from the *objective* principle, namely the practical law. The former contains the practical rule determined by reason conformably with the conditions of the subject (often his ignorance or also his inclinations), and is therefore the principle in accordance with which the subject *acts*; but the law is objective

[22] Guyer, P. (2006), 184.
[23] Wood, A. W. (2008), 67.

principle valid for every rational being, and the principle in accordance with which he *ought to act*, i.e. an imperative. (4:420-421f)

To put in a few words, Kant argues that human beings, as rational beings, are self-governing beings which should be guided by the moral law, and the only form of such law could be given in the form of categorical imperative. Such imperative articulates and expresses universally valid moral law, on the basis of which we *ought to act*.

"Kant actually formulates the categorical imperative in five different ways[24], although he himself refers to only three."[25] For the purposes of this paper it is enough to follow these three main formulations. Allen Wood differentiates three main formulas of categorical imperative in *Groundwork*:

> **FIRST FORMULA**
> FUL *The Formula of Universal Law*: "**Act only in accordance with that maxim through which you at the same time can will that it become a universal law**" (4:421; cf. 4:402); with its variant,
>
> **SECOND FORMULA**
> FH *The Formula of Humanity as End in Itself*: "**So act that you use humanity, as much in your own person as in the person of every other, always at the same time as an end and never merely as a means**" (4:429; cf. 4:436).
>
> **THIRD FORMULA**
> FRE *The Formula of the Realm of Ends*: "**Act in accordance with maxims of a universally legislative member for a merely possible realm of ends**" (4:439; cf. 4:433, 4:437, 4:438).[26]

Three main formulas could be taken as three standard formulations of categorical imperative. "Kant says that all three are formulations of 'the very same law,' but differ both 'subjectively' (in the way the law is presented to an agent) and also in the aspect of the law that they present."[27] The mentioned

---

[24] Cf. Wood, A. W. (2008), 66-67, Guyer, P. (2006), 191-207., Wood, Allen W. (2006), Kant's Formulations of the Moral Law, in: Bird, Graham (ed.), *A Companion to Kant*, Oxford: Blackwell Publishing, 291-292.

[25] Guyer, P. (2006), 191.

[26] Cf. Wood, A. W. (2008), 66-67; Wood, A. W. (2006.), 291-292; Wood, A. W. (2007), 358. Bold emphasis is mine.

[27] Wood, A. W. (2006.), 294.

aspects of categorical imperative are: "form", "matter" and "complete determination" (cf. 4:436-437).[28]

The first formula of categorical imperative is related to the "form" of practical law. It gives just a formal condition for the possibility of such law. It "was derived from the concept of a maxim that is compatible with this kind of imperative, and the general form that such a maxim would have to have."[29] In Kant's words, "a *form* (...) consists in universality; and in this respect the formula of the moral imperative is expressed thus: that maxims should be chosen as if they were to hold as universal laws of nature". (4:436) For Kant, FUL provides to agent a formal condition for evaluating his maxims. It gives the agent condition for detecting if his maxim is "without contradiction", what means it could be willed as a universal law of every rational beings' conduct with adequate regularity as if it is a law of nature. The first formula provides a test for permissibility of agent's willing of maxims: if the maxim can be universalised it is permissible to follow it.[30]

The second formula is related to the "matter" of this law. It informs us about the conditions of cognitive application of the moral law. It "was derived from the concept of the substantive value (or the end) that could give us a rational ground to follow a categorical imperative."[31] In Kant's words: every maxim has also "a *matter*, namely an end, and in this respect formula says that a rational being, as an end by its nature and hence as an end in itself, must in every maxim serve as the limiting condition of all merely relative or arbitrary ends" (4:436). In FH Kant includes the most important feature of categorical imperative, i.e. the notion of "humanity" as an unconditional value which is the ground of his entire ethical enterprise. Thus, many scholars interpret this formula recently as the most significant one, from which all others derive their strength.[32] This formula is also the best answer to all of the so-called critics who object to Kant that his ethics is too formalistic. A. Wood provides a sharp answer to all such critics and it is important to bring in its fullness, because even today there are still critics who insist on this "formalistic" objection (even F. Jahr is taking this side, about which we will discuss below):

---

[28] Behind Kant's presentation of moral law through the aspects of form, matter and complete determination lies the Leibnizian doctrine of forming the concepts. For a fuller account of Kant's application of these logical assumptions to his concept of the moral law see Wood, A. W. (2006.), pp. 293-294, 303 and Wood, A. W. (2008), 68-69.

[29] Wood, A. W. (2006.), 300.

[30] For a fuller account of the meaning and scope of the First formula see Wood, A. W. (2006.), 293-298 and Wood, A. W. (2008), 69-74.

[31] Wood, A. W. (2006.), 300.

[32] See for example Wood, A. W. (2006) and Guyer, P. (2006), Ch. 5.

For example, Kant's entire approach to ethics has been (and still is) widely described as "formalistic". He has been criticized for not providing (or even for not allowing the possibility of) any substantive value lying behind the moral principle, or providing the rational will with any ground for being able to will one maxim, and not another, to be a universal law (or law of nature). The very concept of a categorical imperative has sometimes been rejected as nonsensical, on the ground that this concept precludes our having any substantive reason for obeying such an imperative. Schopenhauer, for instance, explained the alleged incoherence of Kant's thinking by attributing to him an ethics of divine command but without admitting a divine lawgiver to back up thc command.

Such criticisms are obviated, however, at least in the form they are usually presented, as soon as we turn from Kant's first to his second formulation of the moral principle. For it deals explicitly with the "matter" of the principle, by which Kant means the "end" for the sake of which it is supposed to be rational to follow a categorical imperative. Kant's "formalism" applies only to the first stage of his development of the principle; it is complemented immediately by considering the principle from the opposite, "material" point of view, in which Kant inquires after our rational motive for obeying a categorical imperative, and locates this motive in the distinctive value that grounds morality, which he identifies with a kind of *end*.[33]

The only end which could give the "matter" to the moral law but without jeopardizing the apodictic validity of the moral law is some of substantive value, which is an end in itself (it is not merely means for any other end). For Kant, the rational being is the only candidate for an end which has an "absolute worth" (4:428): "Now I say that the human being, and in general every rational being, *exists* as an end in itself, *not merely as a means* to be used by this or that will at its discretion" (4:428). Kant provides a brief argument for assertion that "only 'humanity', understood in the technical Kantian sense of rational nature regarded as the capacity to set ends, can qualify as an end in itself: we value our own existence as an end in itself, but we do so *rationally* only insofar as we value the existence of other rational beings in precisely the same way."[34]

---

[33] Wood, A. W. (2007), 352. In another place he stresses: "It is deplorably common to regard FUL and FLN (usually not clearly distinguished from each other) as the chief, if not the only, formulation of the moral law. Even some of Kant's most faithful defenders speak of them as 'The Categorical Imperative' (with capital letters) – as if there were no other, and no more adequate, formulations of the moral principle." Wood, A. W. (2008), 69.
[34] Wood, A. W. (2006.), 299.

The third formula is related to "complete determination". "The third formula combines the conception of a law valid universally for all rational beings (in FUL) with the conception of every rational nature as having absolute worth, to get the idea of the will of every rational being as the source of a universally valid legislation."[35] In Kant's words, the one thing left is "*a complete determination* of all maxims by means of that formula, namely that all maxims from one's own lawgiving are to harmonize with a possible kingdom of ends as with a kingdom of nature" (4:436). By "realm of ends" Kant means "a systematic combination of various rational beings through communal laws", or "whole of all ends in systematic connection" (4:433). The third formula brings stronger conclusion and "justifies a correspondingly stronger conclusion about maxims, telling us not merely which ones are permissible and which not, but also which ones we have a positive duty to adopt because they are part of a system of universal moral legislation given by our own rational will."[36] In other words Kant "completely determinates" the meaning and scope of moral law by the third formulation, which articulates universality and necessity of taking all rational beings as interconnected by the obligation of mutual respect of every one of them toward other ones by taking them always (also) as ends in themselves. This special realm of universal moral legislation is exactly the realm of morality, which differentiates all rational beings from other natural living and nonliving beings.

Even if this is just a sketch of Kant's articulation of the moral law, it should be a sufficient demonstration of powerfulness and strength of Kant's grounding of the "supreme principle of morality". This justificatory power should be kept in mind during reflection on other attempts of grounding morality, among which the Jahr's concept of bioethical imperative is a specific and original one.

Jahr's concept of bioethical imperative

Jahr completely abandons Kant's criteria for forming a concept of practical law. He also does not speak in terms of universal and necessary validity. Instead, he brings a few short remarks on the idea of his Bioethical Imperative[37], without extensive theoretical justification (contrary to Kant's practice!).

---

[35] Ibid., 301.
[36] Ibidem. By "realm of ends" Kant means "a systematic combination of various rational beings through communal laws", or "whole of all ends in systematic connection" (4:433).
[37] Writing Bioethical Imperative with capital letters I am trying even on the linguistic level keep the essence of Jahr's idea which will be explained below.

Hans-Martin Sass points out that Fritz Jahr in a few published articles is "in close discussion with Kant, extending the formal Categorical Imperative towards a more encompassing content-based Bioethical Imperative"[38]. Jahr cites Kant in a few articles before the presentation of his Bioethical Imperative, which states:

> ***Respect every living being on principle as an end in itself and treat it, if possible, as such!***[39]

The echo of Kant's second formula is obvious, but Jahr provides at least two very important changes. First, he calls for the extension of the application of practical law on *all living* beings. Second, he points out that following this law is an obligation whenever it is *possible* to follow it. Thus, the question is: can Jahr categorically demand such an imperative, and simultaneously keep the possibility of exceptions in following the law?

Jahr stands in critical relation to Kant, mostly looking on Kant's thoughts in negative light[40], but he also appreciates some parts of his opus, especially his disapproval of unnecessary torturing of animals presented in the *Metaphysics of morals* (cf. 6:443).

Jahr says about it further: [S]enseless cruelty towards animals is an indication of an urefined character becoming dangerous towards the human environment as well. Among other thinkers, philosopher Kant expressively has hinted at this fact of highest importance for social ethics, when in 'Metaphysische Anfangsgründe der Tugendlehre' ['Metaphysical first principles of the doctrine of virtue', added: I.E.] he calls the careful and compassionate treatment of animals a human obligation towards oneself.[41]

But the most direct critical relation to Kant is presented by the extension of his moral imperative. Jahr "extends Kant's moral imperative to all forms of life; but he modifies the inflexible categorical structure of Kant's model into a

---

[38] Sass, Hans-Martin (2009b), European Roots of Bioethics: Fritz Jahr's 1927 Definition and Vision of Bioethics, in: Čović, Ante – Gosić, Nada – Tomašević, Luka (eds.), *Od nove medicinske etike do integrativne bioetike/From New Medical Ethics to Integrative Bioethics*. Zagreb: Pergamena – Hrvatsko bioetičko društvo, 22.

[39] Jahr, Fritz (2010a), Bioethics: Reviewing the Ethical Relations of Humans towards Animals and Plants, in: Sass, Hans-Martin (ed.), *Selected Essays...*, 4; Jahr, Fritz (2010b), The Relationship of Animal Protection and Ethics, in: ibid., 8; Jahr, Fritz, (2010c), Three Studies on the Fifth Commandment, in: ibid., 23. See also in this volume.

[40] See for example Jahr, Fritz (2010d), Egoism and Altruism: Two Basic Moral Problems, their Contradicition and Unification in Social Life, in: ibid., 12, 13; Jahr, F. (2010c), 19. See also in this volume.

[41] Jahr, F. (2010b), 6. See also in this volume.

pragmatic and situational model of balancing moral obligations, rights and visions"[42].

Jahr's thoughts concerning Bioethical Imperative are scattered through a few essays and he did not provide any systematic account of his Imperative. So it is very useful to use Hans-Martin Sass' attempt of providing a coherent reading and interpretation of Bioethical Imperative. He identifies at least six major points in Jahr's enterprise of expanding Kant's imperative:

> (1) *The Bioethical Imperative guides ethical and cultural attitudes andresponsibilities in the life sciences and* ***towards all forms of life****.* (...)
> (2) *The Bioethical Imperative is* ***based on historical and other evidence*** *that 'compassion is an empirical established phenomenon of the human soul'.* (...)
> (3) *The Bioethical Imperative* ***strengthens and complements moral recognition and duties towards fellow humans in the Kantian context*** *and should be followed in respect of human culture and mutual moral obligations among humans.* (...)
> (4) *The Bioethical Imperative has to recognize, to steward, and to cultivate the struggle for life among forms of life and* ***natural and cultural living environments****.* (...)
> (5) *The Bioethical Imperative implements* ***compassion, love, and solidarity*** *with all forms of life as a* ***content-based principle*** *and virtue into the 'golden rule' and into the Kant's Categorical imperative, which are reciprocal and formal only.* (...)
> (6) *The Bioethical Imperative includes* ***obligations towards one's own body and soul as a living being****.*[43]

Bioethical Imperative with such a wide scope gives to Hans-Martin Sass a reason to state that "Jahr develops his vision of bioethics as a discipline, a principle, and a virtue". He stresses that the "Bioethical Imperative is a guide for ethical and cultural attitudes and responsibilities in the life sciences and towards all forms of life."[44]

He further emphasises that Jahr's Bioethical Imperative, as an extension of Kant's formal Categorical Imperative, is a more encompassing content-based imperative. He turns our attention to the fact that "the 'sancticity' of life is the foundation of Jahr's 1927 Bioethical Imperative, while 1788 Kant named

---

[42] Sass, Hans-Martin (2010a), 24.
[43] Sass, Hans-Martin (2009b), 22-25. Bold emphasis is mine.
[44] Sass, Hans-Martin (2009a), Asian and European Roots of Bioethics: Fritz Jahr's 1927 Definition and Vision of Bioethics, in: *Asian Bioethics Review* 1(3), 188.

'sanctity of moral law'[45] as the foundation of the Categorical Imperative."[46] Taking in consideration Jahr's critical stand towards Kant he concludes that "Jahr's imperative is content-rich; Kant's imperative allows himself the luxury of formality only."[47] In another place Sass states more sharply: "While Kant reserved his moral imperative to humans only kept it formal, Jahr's imperative encompasses all living beings and their interactions in order to present a flexible and pragmatic imperative, which takes the struggle for life as an essential part of all life into account."[48]

Even though Sass points out that Kant's imperative is unconditional and that Jahr's one is conditional, he is convinced that they have similar motivation:

But Jahr and Kant share the concept that ethics and bioethics are categorical in as far as they do not depend on reciprocity. For Kant the moral imperative is unconditional regardless of whether fellow persons reciprocate. For Jahr reciprocity cannot be expected from nonhuman forms of life, and unfortunately also not from every human person. For Kant and Jahr, it is human dignity, expressing itself in the dignity of the individual conscience, that calls for moral action and attitude, for compassion, and for integrating ethics and expertise.[49]

If these thoughts are even partially true, the question is: how could a Kantian take into consideration Jahr's views seriously, but at the same time not to jeopardize the stability and coherence of his ethical system?

Taking all these points into consideration, we could now ask: do these pose some challenges for Kant's ethical thought? We can ask further: if they do, which challenges are in question?

Is there a challenge for Kant?

Kant would have straightforward answers to all six features of Bioethical Imperative listed above.

We can start with the last – sixth – consideration. Kant would agree with Jahr that there are obligations of beings toward themselves, but exclusively on the

---

[45] Kant speaks about "holiness of the moral law" in his *Critique of practical reason*. (Cf. 5:87)

[46] Sass, Hans-Martin (2009a), 188.

[47] Ibidem.

[48] Sass, Hans-Martin (2010a), 27.

[49] Sass, Hans-Martin (2008), Fritz Jahr's 1927 Concept of Bioethics, in: *Kennedy Institute for Ethics Journal* 17(4), 287.

ground of rationality. For Kant, rational beings exclusively have obligations toward themselves, because they only can be moral agents. The ground for this claim is located in Kant's theory of rationality and freedom, where those two concepts are taken as two faces of the same coin.

This is closely related to the first point. Jahr demands extension of moral obligations on every living being. As said, Kant strongly argues for the thesis that just rational beings could be moral agents. Jahr's advocates should give a firm justification of this demand, which not necessarily should be in contradiction with Kant's imperative, but can substantially upgrade it. The justification of non-anthropocentric, more precisely bio-centric ethics is still needed.[50]

Jahr is going further in the fourth point: he advocates for the extension of moral obligations on all natural and cultural environment. This point is even harder to justify because although we can find some basis in arguing for obligations toward all living beings, there is much harder to find such basis for a nonliving matter.

Jahr's guidelines for the justification of the mentioned extension of the ethical imperative are contained in the second and the fifth point. The second point suggests that empirical facts in various cultures (customs and practices) provide solid bases for the bioethical imperative. The fifth point indicates that many values should be included in the ethical imperative. Kant is clear about these two points. Against the first, he argues that there are none empirical sources for practical law, because it has its seat exclusively in reason and *a priori* reasoning. Against the second, Kant argues that there are only two absolute values, which can be the basis for the categorical nature of practical law: "The only things which are good in themselves and thus of absolute worth are a good will and a person."[51]

The third point is the most interesting. It suggests that Jahr does not advocate for the cancellation of Kant's imperative, but just for its upgrading and extension of his scope. This claim, as a postulate, has legitimate position in calling every Kantian ethics to provide firm grounds of Bioethical Imperative in Jahr's sense.

At the end three very important points should be noted. First, Kant's categorical imperative, as we have seen above, is not at all formal as Jahr and

[50] For a one way of criticism of non-anthropocentric ethics, as the ones which lead in absurd positions see for example Čović, Ante (2009), Biotička zajednica kao temelj odgovornosti za ne-ljudska bića, in: Čović, A. – Gosić, N. – Tomašević, L., op. cit., 33-46.
[51] Caygill, Howard (2005), *A Kant Dictionary*. Oxford: Blackwell Publishing, 411.

other critics assume. Kant provides a firm material basis for his moral law in the form of humanity as the substantive value. It is misleading to read Kant as a formalistic in the articulation of his moral imperative.

Second, Jahr is starting from the results of (empirical) sciences of his days[52] to give a basis for his arguments in favour of the Bioethical Imperative. Kant would never accept such a methodology. For him, the "supreme principle of morality" can be determined exclusively *a priori*, because the practical law could not be based on the contingent empirical truths. Moral imperative should have its seat in reason alone, because this is the only way of securing its universality and necessity.

Third, Kant's categorical imperative is, as its name says, a categorical demand. When the practical law is in question, the only acceptable form is categorical imperative. Jahr's Bioethical Imperative leaves space for exceptions. It is hypothetical in its form.

These three final remarks are crucial in comparison of Kant's and Jahr's thought. They form sufficient reason to rule out, from Kantian standpoint, the Bioethical Imperative as a possible constitutive principle[53] of our morality. There is eventually some possibility for its justification as a regulative principle of our moral conduct.

Jahr does not provide thorough theoretical justification for his Bioethical Imperative, but he poses, on the basis of new cognitions about living world, reference to Eastern wisdom and some intuitions of his contemporaries, a challenging question for Kantian ethics: could Kantian ethics provide a justification for our responsible behaviour to other living beings? This challenge could be formulated differently: can we find some guidelines inside the Kantian philosophy to justify the bioethical imperative?

Conclusion: possible Kantian answers to Jahr's objections

Kantian scholars could reject Jahr's Bioethical imperative on the basis of insufficient justification. But, Jahr's thought, although not precise and thorough as Kant's, could be a valuable guiding thread for a kind of bioethical rereading of Kant's philosophy. Such bioethical rereading of Kant is simply

---

[52] Cf. Sass, Hans-Martin (2010a), 25, 26.

[53] Kant's categorical imperative is constitutive in the sense that it defines moral agents as rational/free beings, and *vice versa*: rationality and freedom are conditions for morality. Thus, categorical imperative is in the essence of rational being as practical being.

necessary today in the "epoch of bioethics"[54] and it is a challenge for every Kantian who tries to find a true place of Kant's thought in struggling with and answering to current bioethical issues. Having in mind the global environmental deterioration and endangering all life on earth[55] Jahr's ideas could be really seen as a "cultural and moral innovation for the 3rd millennium"[56].

I can identify one possible way of trying to accomplish such project of rereading of Kant. It starts with Kant's philosophy of biology and on the grounds of teleological power of judgment tries to give *a priori* ground for establishing a kind of Kantian ecology[57], which could be guided with bioethical imperative as its main regulative principle. In such reading, Bioethical Imperative maybe would never accomplish the strength of a constitutive principle of reason, but could take place of a regulative principle of reason which could be an excellent supplement to the constitutive force of the categorical imperative in determination of our moral conduct.

To conclude, I see the central strength of Jahr's thought, inside Kantian horizon, as a stimulator of a new bioethical rereading of Kant's, not just ethical, but whole philosophical heritage.

---

[54] Cf. Čović, Ante (2007), Integrativna bioetika i pluriperspektivizam, in: Valjan, Velimir (ed.), *Integrativna bioetika i izazovi suvremene civilizacije*, Bioetičko društvo u BiH, Sarajevo, 71-72; Čović, Ante (2004), *Etika i bioetika: Razmišljanja na pragu bioetičke epohe*, Zagreb: Pergamena, passim.

[55] Cf. for example Cifrić, Ivan (2007), *Bioetička ekumena: Odgovornost za život susvijeta*, Zagreb: Pergamena, passim (especially pp. 30-31).

[56] Sass, Hans-Martin (2010b), Bioethics as a European Innovation. Fritz Jahr's 1927 Concept of Bioethics, in: Gimmler, Antje - Holzinger, Markus - Knopp, Lothar (eds.), *Vernunft und Innovation: Über das alte Vorurteil für das Neue. Festschrift für Walther Ch. Zimmerli zum 65. Geburtstag*, Paderborn: Wilhelm Fink, 369.

[57] An attempt of providing such rereading of Kant is given in: Eterović, Igor (2011), Kantova teleologija kao podloga orijentiranju u ekologiji, *Filozofska istraživanja* 31(2), 299-309.

THE IMPORTANCE OF CHARLES DARWIN'S THEORY FOR FRITZ JAHR'S CONCEPTION OF BIOETHICS*

Eve-Marie Engels[1]

> "Man in his arrogance thinks himself a great work. worthy the interposition of a deity, more humble & I believe true to consider him created from animals."[2]
>
> Charles Darwin, 1838

Introduction

Fritz Jahr's short article „Bioethics – A Survey of the Ethical Relations of the Human Being towards Animals and Plants"[3] of 1927 does not only predate the "birth" of that branch of bioethics called "medical ethics" but also the discussions on a "global bioethics" including our ethical concern for the preservation of animals and plants, as it was particularly stimulated by Van Rensselaer Potter who introduced the term "bioethics" in the 1970s (Potter 1970, 1971, 1988) and who built on the legacy of Aldo Leopold, a famous founder of environmental ethics.[4] To date we can consider Jahr as the first one who used and thus coined the term "Bio=Ethik"[5].

---

* The paper was presented at 1st International Conference „Fritz Jahr and the Foundations of European Bioethics", Rijeka/Opatija, March 11-12, 2011. This article is a shortened and slightly modified version of my contribution in Engels, E-M (2011) *Jahr* 2(4) 475-504.

[1] Eve-Marie Engels, eve-marie.engels@uni-tuebingen.de, Chair for Ethics in the Biosciences, Eberhard Karls Universität Tübingen, Dept. of Biology, Faculty of Science, Wilhelmstraße 19, 72076 Tübingen, Germany.

[2] Barrett P H, Gautrey P J, Herbert S, Kohn D, Smith S (eds.) (1987): Charles Darwin's Notebooks, 1836-1844. Ithaca, New York: Cornell University Press 1987, 300, C 196f. The unusual punctuation is used by Darwin in his private notes.

[3] Jahr F (1927) Bio=Ethik – Eine Umschau über die ethischen Beziehungen des Menschen zu Tier und Pflanze, in: Kosmos. Handweiser für Naturfreunde 24 (1)2-4. If not stated otherwise the English translations of the quotations from Jahr are mine. A helpful support were the English translations by Irene M. Miller and Hans-Martin Sass: Jahr F (2011) Essays in Bioethics and Ethics 1927-1947. Bochum: Zentrum für Medizinische Ethik (Medizinethische Materialien 188). See also in this volume.

There are similarities but also differences between these translations and mine. As a good introduction into Jahr see Sass HM (2007) Fritz Jahr's bioethischer Imperativ. Bochum: Zentrum für Medizinische Ethik (Medizinethische Materialien 175).

[4] Potter V R (1970) Bioethics, the Science of Survival, in: Perspectives in Biology and Medicine. 14, 127-153; Potter (1971) Bioethics. *Bridge to the Future.* Englewood Cliffs, New Jersey: Prentice-Hall; Potter (1988) Global Bioethics: Building on the Leopold Legacy. East Lansing, Michigan: Michigan State University Press. Potters pioneer work was overshadowed by the foundation of the "Joseph and Rose Kennedy Institute for the Study of Human Reproduction and Bioethics" at Georgetown University Washington where the term "bioethics" was reduced to the meaning of biomedical ethics. For an

As a guiding principle for our action Jahr formulates the "bioethical postulate": "*Respect every living being in principle as an end in itself and treat it, whenever possible, as such.*"[6] One year later he formulates this postulate again as a "bioethical imperative", specifying it: "*Respect every living being, therefore also the animals, as an end in itself, and treat it whenever possible as such.*"[7] Jahr includes in his imperative explicitly the protection of animals as well as plants. And he also uses the terms "Tierethik" and "Pflanzenethik".

Bioethics as it has been practiced since the 1970s until today is an interdisciplinary and application oriented discipline which aims at a normative understanding about the range and limits of human action towards living nature including the human being. Thus, bioethics is not anthropocentric concerning its objects or subject-matter. Human action encompasses our conduct towards nature in everyday life as well as in the theoretical and practical contexts of science and research. Two developments triggered the emergence of bioethics in the second part of the twentieth century: Firstly increasing environmental problems, i.e. particularly the damage and destruction of the natural environment, of animals and plants, in many countries up to a global problem which we all are facing now, and secondly scientific and biotechnological developments in quite different branches of the life sciences, which confronted us with completely new questions and prompted discussions about the nature of humans, animals and plants, the meaning of life and death, the beginning and end of individual life. These developments challenged our traditional intuitions about the human being and nature, causing fear of unintended consequences and risks as well as hopes for overcoming severe diseases and extending the span of life.

---

overview see Callahan D (1995) Article "Bioethics", in: Reich W Th (ed.): Encyclopedia of Bioethics. Revised Edition, Vol. I, 247-256; Reich W Th (1995) The Word ‚Bioethics': The Struggle over its Earliest Meanings, in: Kennedy Institute of Ethics Journal 5 (1)19-34.

[5] I came to know Fritz Jahr through a paper given by Prof. Dr. Rolf Löther in June 1997 at the 6th annual conference of the German Society for the History and Theory of Biology (DGGTB) organized by the Chair for Ethics in the Biosciences at the University of Tübingen. Löther mentioned Jahr in his paper very shortly. See Löther R (1998) Evolution der Biosphäre und Ethik, in: Engels EM, Junker Th, Weingarten M (eds.): Ethik der Biowissenschaften, Geschichte und Theorie. Verhandlungen zur Geschichte und Theorie der Biologie. Bd. 1. Berlin: Verlag für Wissenschaft und Bildung, 61-68.

[6] Jahrs „bioethische Forderung": „*Achte jedes Lebewesen grundsätzlich als einen Selbstzweck, und behandle es nach Möglichkeit als solchen*!" (Jahr Bio=Ethik, 4).

[7] Jahrs „bio=ethischer Imperativ": „*Achte jedes Lebewesen, also auch die Tiere, als einen Selbstzweck, und behandle es nach Möglichkeit als solchen*!" p. 102 in Jahr F (1928a) Tierschutz und Ethik in ihren Beziehungen zueinander, in: Ethik. Sexual= und Gesellschafts=Ethik. 4 (6/7)100-102.

Confronted with and often surprised by these new options, which called for ethical as well as legal decisions, bioethics became an inescapable necessity for scientists, physicians and scholars of quite different disciplines as well as for politicians and the public.

Although bioethics is application oriented, it is not simply an "applied" ethics like a recipe or an instruction manual is applied[8]. Bioethical problems are too complex to be solvable by a simple application of ethical principles and norms to cases. Many times one has first to clarify the objects and cases which are at stake, before one can apply principles and norms. As mentioned above, the described developments have challenged our traditional view of humans, animals and plants. Therefore *natural philosophy* as well as *philosophical anthropology* and *other philosophical disciplines* are essential elements of bioethics. Another indispensable element of bioethics is the *empirical and scientific knowledge* about the objects dealt with in our ethical considerations. It is crucial to know something about the specific constitution of plants and animals when we have to decide about the best way to treat and protect them. We have to know the biological status of an entity before we can judge what their moral status is, that is if they deserve moral consideration and which one or in which way. Last but not least, depending on the basic normative ethics one chooses, bioethical judgments and decisions can be founded on deontological, utilitarian, virtue ethical or other normative theories.

Fritz Jahr anticipates these ideas in his short articles in an impressive way and pays tribute to the role of the thriving biosciences for bioethics. He opens his article of 1927 with the crucial statement that „the strict separation between animal and human, predominant since the beginning of our European culture until the end of the 18th century, today cannot be maintained any more."[9] Jahr refers to the changing relationship between science, philosophy and religion and highlights the importance of the *natural sciences* for our understanding of nature. He describes the *dialectical impact* of the new scientific knowledge. On the one hand "it will always be the merit of modern natural sciences of having made possible in the first place an unbiased view of the events of nature [Weltgeschehen]." (Jahr 1927, 2) Jahr mentions successes of animal experiments, blood research etc. as examples of our search for truth.

---

[8] Engels EM (2005) Ethik *in* den Biowissenschaften, in: Maring M (ed.): Ethisch-Philosophisches Grundlagenstudium 2. Ein Projektbuch. Münster: LIT, 135-166, 163f.

[9] „Die scharfe Scheidung zwischen Tier und Mensch, die seit Beginn unserer europäischen Kultur bis zum Ende des 18. Jahrhunderts herrschend war, kann heute nicht mehr aufrecht erhalten werden." (Jahr Bio=Ethik, 2).

> "On the other hand we must not underrate the fact that exactly these scientific triumphs of the human mind have deprived the human being himself/herself of his/her dominant position in the cosmos. Philosophy which formerly used to prescribe its leading ideas to the natural sciences now had itself to found its systems on detailed scientific knowledge, and it was only a poetic-philosophical formulation of Darwin's insight, when Nietzsche considered the human being as a rather inferior transitory stage to a higher evolution, as a ‚rope drawn between animal and superman [Übermensch]'."[10]

I cannot discuss the question here if Jahr's interpretation of Nietzsche as recipient of Charles Darwin's idea of evolution is adequate. It may at least be said that Darwin did not advocate the idea of an evolutionary progress from animal through man to "superman" in Nietzsche's sense. What is important for our context is the fact that Jahr acknowledges Darwin's great achievement. Darwin laid the scientific foundation for bridging the presupposed gap between animals and humans. By his theory of descent and the evidence supplied for it, he provided convincing arguments for the existence of a *real relationship* between animals and humans.

Thus these "scientific triumphs of the human mind" by which we have gained an "unbiased view of the events of nature", have deprived us of our "dominant position in the cosmos". They include the disillusioning insight that we humans have descended from animals. Only a highly developed mind like ours, which raises us above the rest of nature, is capable of gaining this knowledge. But the outcome of our search for truth is the discovery that we are not the crown of creation, but just animals, even though special animals. Jahr justly describes Darwin's insight as a "revolution" [Umwälzung]. What followed from this revolution, he asks.

First of all, a consequence of this revolution is the "fundamental equality of man and animal ["grundsätzliche Gleichstellung von Mensch und Tier"] as test object of psychology. Psychology is not restricted any more to the human

---

[10] „Es wird stets das Verdienst der modernen Naturwissenschaft bleiben, daß sie eine vorurteilslose Betrachtung des Weltgeschehens erst möglich gemacht hat. Wir würden uns heute als Wahrheitssucher aufgeben, wenn wir die Erfolge der Tierexperimente, Blutversuche, Serumforschung u.v.a. ablehnen wollten. Andererseits dürfen wir nicht verkennen, daß gerade diese wissenschaftlichen Triumphe des Menschengeistes dem Menschen selbst seine beherrschende Stellung im Weltganzen genommen haben. Die Philosophie, die früher der Naturwissenschaft ihre Leitgedanken vorschrieb, mußte nun selbst ihre Systeme auf naturwissenschaftlichen Einzelerkenntnissen aufbauen, und es war nur eine dichterphilosophische Formulierung der Erkenntnis Darwins, wenn Nietzsche den Menschen als ein recht minderwertiges Übergangsstadium zu einer höheren Entwicklung, als ein ‚Seil, gespannt zwischen Tier und Übermensch' ansah." (Jahr Bio=Ethik, 2).

being, but applies the same methods also to animals, and as in comparative anatomical-zoological research, most instructive comparisons are also drawn between human and animal soul. Indeed, even the beginnings of a plant psychology become noticeable."[11] Jahr mentions G. Th. Fechner, R. H. Francé, A. Wagner and the Indian Bose as the most prominent representatives. Therefore Jahr considers it only as consistent that Rudolf Eisler uses the term "Bio-Psychik" ("soul study of all that lives").

In his article "Death and the Animals. Contemplating the 5th Commandment"[12] Jahr again refers to Darwin and his influence on "biology, the science of life", which particularly since Darwin has discovered many related characteristics between humans and animals and which are now utilized in medicine. Besides the applications already mentioned, Jahr cites as an example the transplantation of animal tissues into humans, nowadays called "xenotransplantation" ["Ueberpflanzung von tierischem Gewebe auf den Menschen"]. Again he points to the physiological as well as psychic affinities of humans and animals. In a certain way, humans and animals are "close to each other". Indeed, only few years after the publication of Darwin's *Origin of Species* (1859)[13] as well as its German translations several books with comparative reflections on animals and humans were published, as well as innumerable articles on the impact of Darwin's new theory on our general view of the human being and nature.[14]

---

[11] „Zunächst die grundsätzliche Gleichstellung von Mensch und Tier als Versuchsobjekt der Psychologie. Diese beschränkt sich heute nicht mehr auf den Menschen, sondern arbeitet mit denselben Methoden auch auf dem Gebiet des Tierischen, und wie es eine vergleichende anatomisch-zoologische Forschung gibt, so werden auch höchst lehrreiche Vergleiche zwischen Menschen= und Tierseele angestellt. Ja, sogar die Anfänge einer Pflanzenpsychologie machen sich bemerkbar." (Jahr Bio=Ethik, 2).

[12] Jahr F (1928b) Der Tod und die Tiere. Eine Betrachtung über das 5. Gebot, in: Mut und Kraft. 5 (1), 5-6. Translation by Miller I M: Death and the Animals. Contemplating the 5th Commandment, in: Miller I M, Sass, HM (eds.): Essays in Bioethics and Ethics 1927-1947 Fritz Jahr. Translated by I M Miller and HM Sass. Bochum: Zentrum für Medizinische Ethik (Medizinethische Materialien 188). See also in this volume.

[13] Darwin Ch (1964): On the Origin of Species by Means of Natural Selection, or the Preservation of Favoured Races in the Struggle for Life. Facsimile of the 1st ed. London 1859, Introduction by E Mayr. Cambridge, Mass, London, England: The Balknap Press of Harvard University Press.

[14] Büchner L (1868): Die Stellung des Menschen in der Natur in Vergangenheit, Gegenwart und Zukunft. Leipzig: Theodor Thomas; Haeckel E (1863) Über die Entwicklungstheorie Darwins, in: Haeckel E (1924): Gemeinverständliche Werke. Vol. V. Leipzig: Alfred Kröner, Berlin: Carl Henschel, 3-32; Haeckel E (1868): Natürliche Schöpfungsgeschichte. Berlin: Georg Reimer; Huxley T H (1863) Evidence as to Man's Place in Nature. London: Williams & Norgate, Reprint Cambridge 2009; Rolle F (1866): Der Mensch, seine Abstammung und Gesittung, im Lichte der Darwin'schen Lehre von der Art-Entstehung und auf Grundlage der neuern geologischen Entdeckung dargestellt. Frankfurt am Main: Joh. Christ. Hermann'sche Verlagsbuchhandlung; Vogt C (1863): Vorlesungen über den

A second consequence of Darwin's revolution is the extension of our moral obligations beyond the human being: From "Bio-Psychik there is only a step to *Bio=Ethik*, i.e. to the assumption of moral obligations not only towards the human being but towards all living beings."[15] Thus for Jahr science and ethics are closely linked to each other. Comparative scientific research on humans, animals and plants (comparative anatomy, zoology, psychology) which reveal the similarities between humans and the rest of living nature leads to an ethical insight, that of our moral obligations towards all living beings.

Two aspects have to be highlighted in this context. This is firstly Jahr's conviction that scientific knowledge plays a crucial role for the protection of nature and thus for bioethics: "Useful and efficient animal protection can only be practised well if enough knowledge of nature and at least some understanding of nature is available. For, in fact, we will only be able to really protect animals, if we know to some degree their physiological and psychological traits and conditions of life. Therefore it is one of the main goals of the animal protection movement to arouse, spread and deepen such knowledge and such understanding of nature as far as possible. This interest in nature then will quite on its own not be restricted to animals but will have to include on the one hand plants, on the other hand (and for us this is more important in this context) the human being."[16]

According to Jahr bioethics has not been discovered in the present. In his articles he mentions several important predecessors of animal ethics, beginning with "a particularly attractive example from the past", Francis of Assisi (1182-1226).[17] For the protestant pastor also the bible was a rich

---

Menschen seine Stellung in der Schöpfung und in der Geschichte der Erde. 2 vols. Gießen: Ricker'sche Buchhandlung; Wundt W (1863): Vorlesungen über die Menschen- und Thierseele. 2 vols. Ed. W Nitzsche. Reprint of the 1st ed. Berlin et al: Springer 1990. See also the bibliography in Engels EM (ed.) (1995) Die Rezeption von Evolutionstheorien im 19. Jahrhundert. Frankfurt am Main: Suhrkamp, 395-414.

15 Jahr Bio=Ethik, 2.

16 „Nun ist jedoch ein zweckmäßiger, leistungsfähiger Tierschutz nur dann gut möglich, wenn genügende Naturerkenntnis und wenigstens einiges Naturverständnis vorhanden ist. Denn tatsächlich kann man die Tiere nur dann wirklich schützen, wenn man ihre physiologischen und psychologischen Eigenschaften und Lebensbedingungen einigermaßen kennt. Daher ist es mit ein Hauptziel der Tierschutzbewegung, solche Kenntnis und solches Verständnis der Natur nach Möglichkeit zu wecken, zu verbreiten und zu vertiefen. Solches Naturinteresse wird sich dann ganz von selbst nicht auf die Tiere beschränken, sondern nach der einen Seite die Pflanzen, nach der anderen Seite (und das ist für uns in diesem Zusammenhange das Wichtigere) den Menschen mit einbeziehen müssen." (Jahr Tierschutz und Ethik, 101).

17 He also mentions Michel de Montaigne (1533-1592), Jean-Jacques Rousseau (1712-1778), Johann Gottfried Herder (1744-1803), Immanuel Kant (1724-1804), Friedrich

source of bioethical insights. Jahr thus draws on secular as well as biblical sources as bioethical references. And he can already point to a growing sensitivity towards animals and plants which manifests itself also at the legal level.

It is however surprising that Jahr did not know the writings of Albert Schweitzer and his "Ethics of Respect for Life" published in 1923 in his work *Philosophy of Culture*[18]. Neither was Schweitzer familiar with Jahr's writings.

Jahr does not go into detail about Darwin's theory. Therefore I will describe the essential core of Darwin's theory and his scientific revolution which was also a philosophical revolution, because it changed our view of living nature and the relations between humans and all other living beings.

Charles Darwin's Scientific and Philosophical Revolution

What does the "Umwälzung", the *revolution*, which Jahr mentions in his essay, consist in? Darwin's *scientific revolution* consists in explaining the origin of species within the framework of natural science, i.e. by drawing on *natural laws*.[19] Species come into being by the *transformation of other species*. Darwin does not claim to be able to explain the origin of life as such. Rather he wants to describe the mechanisms by which new species of plants and animals come into being when we presuppose the existence of a few or only one simple form of life. Thus Darwin rejects the idea of a special or separate creation of each species by the Creator as well as the idea of the fixity of species. He only allows for natural laws and causes. Thus he wants to raise natural history, botany and zoology to the same scientific level already reached in other natural sciences, in physics and astronomy. Darwin

---

Schleiermacher (1768-1834), Karl Christian Friedrich Krause (1781-1832) and Arthur Schopenhauer (1788-1860) who influenced Richard Wagner (1813-1883). An influential source of animal ethics is Indian philosophy, particularly the idea of reincarnation. As forerunners of plant ethics he mentions Richard Wagner as well as the philosopher Eduard von Hartmann (1842-1906). My list is not complete.

[18] Schweitzer A (2007): Kulturphilosophie. Bd. 1 Verfall und Wiederaufbau der Kultur. Bd. 2 Kultur und Ethik. München: C. H. Beck (1st ed. 1923).

[19] As a more detailed presentation of Darwin's theory of descent, his evolutionary anthropology and his theory of the moral sense see my monograph Engels EM (2007) Charles Darwin. München: C.H. Beck and e.g. Engels EM (2011b) Der Mensch, das moralfähige Tier- Zur Anthropologie und Ethik von Charles Darwin, in: Engels EM, Betz O, Köhler H-R, Potthast Th. (eds.) Charles Darwin und seine Bedeutung für die Wissenschaften. Tübingen: Attempto Narr Francke, 145-179.

however does not claim to be able to refute the existence of God by his theory. Rather he claims to formulate a theory with greater explanatory power than the traditional doctrine of special creation and which avoids the difficulties of this doctrine.

Darwin names his theory "theory of descent with modification through natural selection"[20]. He proceeds from the observation that two organisms of the same species are never completely identical. There are always *variations*, however small, and thus also differences in adaptation to an environment. Those organisms of a species whose traits are better adapted to their environment, that is, more purposively outfitted than their conspecifics, have a higher chance of survival and thus can more successfully reproduce than the others. This means that a *natural selection* of the better adapted takes place. Those traits which are advantageous for survival can accumulate during generations by inheritance and thus increasingly change, compared to the traits of the aboriginal stock. Thus in the course of long periods of time, from individual variants hereditary varieties, subspecies and finally new species evolve. Darwin advocates a *gradualism* and draws on the *principle of continuity* of natural philosophy. The old principle "Natura non facit saltum" (nature makes no leaps) "is on this theory simply intelligible." (ibid. 471). However, the principle of continuity is not static any more, it becomes *dynamic* and it stands for a *real relationship* between species.

Thus natural selection not only leads to the dying out of species but also fulfils the *constructive* function of bringing forth new species. It is important to note that not every origin of a new species leads to the extinction of an already existing species. Darwin points to the principle of "divergence of character" which means that under the pressure of natural selection not only one, but several species can evolve from one and the same stock in adaptation to different ecological niches. For the evolution of such divergence, which presupposes the possibility of reproductive isolation, the Galapagos Archipelago was an exemplary laboratory.

The mechanism which exerts selection in nature is what Darwin calls the "struggle for life" or "struggle for existence", drawing on Malthus' principle of population. Particularly his metaphors "natural selection" and "struggle for existence" were subject to much misunderstanding. The term "struggle for life" or "struggle for existence" can have quite different meanings:

1. competition among individuals of the same species (intraspecific competition), 2. competition among individuals of different species (interspecific competition), 3. struggle for existence of an individual against

[20] Darwin Origin of Species, 459.

environmental dangers (drought, coldness, wetness etc.). Darwin moreover emphasises another aspect and mentions more meanings of the term. He uses the term "Struggle for Existence in a large and metaphorical sense, including dependence of one being on another, and including (which is more important) not only the life of the individual, but success in leaving progeny." (ibid. 62). The phrase "struggle for existence" has often been interpreted as a bloody or deadly fight between individuals, races or species. Depending on the situation the *struggle for existence* can however be coped with by competition or cooperation. "Mutual aid" is a line of Darwin-reception which was pursued particularly in the Russian reception by Peter Kropotkin and others[21].

Darwin adduces evidence from a variety of disciplines and compiles facts like homologies, examples from embryology, rudiments, the fossil record etc. Homologies are similarities between the bodily structure of animals of the same class, like dogs, cats and humans (mammals), as well as between animals belonging to different classes of the same phylum or clade, like fish, amphibians, reptiles, birds and mammals. The facts of embryology adduced by Darwin are the similarities between embryos of different species and even of different classes. They reflect a common descent. Rudiments are traits of organisms which have no function in these organisms. They can be explained as remains of former organisms in which they had functions. These facts as such have already been discovered by well-respected experts in different fields before Darwin and independently of his *theory of descent*. But the philosopher of science and scientist Darwin can make use of them to back up his theory by showing that they can be best explained by a *common descent* of the human being and other animals whereas the belief of the separate creation of each species cannot provide such an explanation and leaves many questions open.

Already at the end of the first edition of *Origin of Species* Darwin alludes to the importance of his theory of descent for our understanding of the human being. "In the distant future I see open fields for far more important researches. Psychology will be based on a new foundation, that of the necessary acquirement of each mental power and capacity by gradation. Light will be thrown on the origin of man and his history." (ibid. 488). Although Darwin's conviction of the relationship between the human being and other animals can be traced back to his early *Notebooks* of 1837[22], Darwin did not publish his

[21] See Todes D (1989): Darwin without Malthus. The Struggle for Existence in Russian Evolutionary Thought. New York, Oxford: Oxford University Press; (2009): Darwins malthusische Metapher und russische Evolutionsvorstellungen, in: Engels EM (ed.): Charles Darwin und seine Wirkung. Frankfurt am Main: Suhrkamp, 203-230.

[22] Barrett P H, Gautrey P J, Herbert S, Kohn D, Smith S (eds.) (1987): Charles Darwin's Notebooks, 1836-1844. Ithaca, New York: Cornell University Press 1987.

Descent of Man and *The Expression of the Emotions in Man and Animals* before 1871 and 1872 for fear of more prejudices against his views. These books are milestones in comparative ethology and comparative psychology.

Darwin's theory contains four theorems which have consequences for our understanding of the relationship between the human being and other living beings: 1. The theorem of descent means that the human species has ultimately descended from another, a nonhuman species. 2. Gradualism means that the human being has evolved from nonhuman animals by a process of gradual evolution. 3. Natural selection was the important mechanism of the evolution of the human being. 4. The principle of divergence means that the human species and other species have evolved from common progenitor species.

Children share their characters with their parents and their brothers and sisters. Due to the principle of descent and to that of divergence there is a *real kinship* between the evolutionary line of humans and our apelike progenitors (our parent species) as well as with those apes that evolved as our sibling species from this common progenitor. Apes and humans have common ancestors. Today we know that bonobos are our closest living nonhuman relatives.

But the human being is not only related to its next relatives but also to remote relatives. *The whole of living nature is a unity and there is a real kinship between human beings and the rest of living nature.*

Already before Darwin there had been developments in medicine and biology showing the similarities between the plant and the animal kingdom. Remarkable similarities in structure, growth and development were discovered within as well as between the animal and plant kingdoms by comparative anatomy, embryology, morphology and cell theory. Although scientific progress in these disciplines was impressive in the 19th century, the similarities within and between the animal and plant kingdom could not be explained scientifically without a unifying theory. This *unifying framework* was provided by *Darwin's theory of descent*. It were particularly this aspect of *unification* and the *explanatory power* of Darwin's theory which fascinated many scientists and philosophers[23], among them the famous physicist and philosopher Hermann von Helmholtz (1821-1894).

---

[23] Engels EM (2011a) Darwin, der, bedeutendste Pfadfinder der Wissenschaft' des 19. Jahrhunderts, in: Samida S (ed.): Inszenierte Wissenschaft. Zur Popularisierung von Wissen im 19. Jahrhundert. Bielefeld: transcript, 213-243.

He wrote: "Besides we do not want to forget which clear understanding *Darwin's* grand idea brought into the until then so mysterious notions of natural relatedness, of the natural system and homology of organs of different animals […] Formerly affinity appeared to be only a mysterious, but completely groundless similarity of forms; now it has become a *real blood relationship*. The natural system […] now gains the meaning of a real phylogenetic tree of organisms. *Darwin* has raised all these isolated areas from a state of mysterious quaintness into the connection of a great evolution […]."[24]

Darwin's theory also promoted the initiation of new research programmes, like comparative psychology (Wundt 1863)[25] and comparative ethology.

The Descent of the Human Being from Other Animals

In his *Descent of Man* (1871)[26] Darwin outlines his evolutionary anthropology, thus pursuing the programme which he already hinted at in his *Origin of Species*. Darwin adduces "three great classes of facts" in order to substantiate his assumption that the human being has evolved from other animals. These are the already mentioned *homologies* as well as facts from *embryology* and thirdly *rudiments*. They reflect the common descent of the human being and other animals. Darwin describes features which are unique for the human being but which can be explained by having gradually evolved from other animals.

He shows many similarities between the human being and other animals in their bodily structure by which he supports his theory of a common descent of

---

24 "Daneben wollen wir nicht vergessen, welch' klares Verständniss *Darwin*'s grosser Gedanke in die bis dahin so mysteriösen Begriffe der natürlichen Verwandtschaft, des natürlichen Systems und der Homologie der Organe bei verschiedenen Thieren gebracht hat; […] Die natürliche Verwandtschaft erschien sonst nur als eine räthselhafte aber vollkommen grundlose Aehnlichkeit der Formen; jetzt ist sie zur *wirklichen Blutsverwandtschaft* geworden. […] jetzt erhält es [das natürliche System] die Bedeutung eines wirklichen Stammbaums der Organismen. […] *Darwin* hat alle diese vereinzelten Gebiete aus dem Zustande einer Anhäufung räthselhafter Wunderlichkeiten in den Zusammenhang einer grossen Entwickelung erhoben […]."(Emphasis on „wirklichen Blutsverwandtschaft", engl. "real blood relationship" by EM E). Helmholtz H von (1869) Über das Ziel und die Fortschritte der Naturwissenschaft. Eröffnungsrede für die Naturforscherversammlung zu Innsbruck, in: Helmholtz H von (1968) Das Denken in der Naturwissenschaft. Darmstadt: Wissenschaftliche Buchgesellschaft, 31-61, 53f.

25 Wundt W (1863): Vorlesungen über die Menschen- und Thierseele. 2 vols. Ed. W Nitzsche. Reprint of the 1st. ed. Berlin et al: Springer 1990.

26 Darwin Ch (1989) The Descent of Man, and Selection in Relation to Sex, 2nd. ed. 1874, rev. augm. 1877. Barrett P H, Freeman R B (eds.) The Works of Charles Darwin, Vols. 21, 22. London: William Pickering (1st ed. London 1871).

the human being and other animals. This may be true for bodily structures and traits, but does this also hold for the emotional and cognitive faculties? Yes: "Nevertheless the difference in mind between man and the higher animals, great as it is, certainly is one of degree and not of kind."[27]

In the chapters on the mental powers of man and animals Darwin describes a broad range of emotional as well as cognitive faculties which can be found in humans as well as other animals. Darwin's elaborations on animals are remarkably up to date, he discusses all those faculties which are discussed today in the context of animal intelligence and feelings. There is a large scale of mental powers among animals, there is no all or nothing. In *The Expression of the Emotions in Man and Animals* (1872)[28] Darwin describes a rich variety of facial and bodily expressions of emotions in animals as well as the human being. Animals feel pleasure and pain, happiness and misery, joy and affection, jealousy, love as well as the desire to be loved, shame, fear, modesty, anger, rage. Also more intellectual emotions and faculties can be found in animals, like excitement and suffer from ennui, wonder, curiosity, suspicion, imitation. There is also a broad range of cognitive faculties. Depending on the species, animals display memory, attention, association, imagination, abstraction and reason, mental individuality and perhaps even self-consciousness. Animals use tools, they also communicate with each other, although articulate language is a specific faculty of the human being. Darwin also describes the social life of animals. Animals of many kinds are social and feel miserable when separated from their companions. They render important services to each other, utter cries which express both danger as well as safety, and mutually defend each other.

Inspite of all these similarities between the human being and other animals for Darwin the human being takes a special position in nature. "Man in the rudest state in which he now exists is the most dominant animal that has ever appeared on this earth. He has spread more widely than any other highly organized form: and all others have yielded before him. He manifestly owes this immense superiority to his intellectual faculties, to his social habits, which lead him to aid and defend his fellows, and to his corporeal structure. The supreme importance of these characters has been proved by the final arbitrament of the battle of life."[29]

---

[27] Darwin Descent of Man vol. 21, 130.

[28] Darwin Ch (1998) The Expression of the Emotions in Man and Animals. Ed., Introd., Afterword, Commentaries Ekman P, 3rd ed. London: ParperCollins Publishers (1st ed. London 1872).

[29] Darwin Descent of Man vol. 21, 52.

Moreover the human being has a unique character which no other animal possesses, a *moral sense* or *conscience* and thus the ability of being moral.[30] Humans are able to *evaluate* their thoughts and actions according to moral and ethical principles. Darwin also has a clear concept of moral progress in the course of history. Moral progress is the increasing enlargement of our sympathy towards members of other groups, nations and races, to the imbecile and maimed and finally to members of other species, to the animals.[31]. The wellbeing of animals was near and dear to Darwin and became for him sort of a yardstick for progress in humanity. Darwin did not only advocate animal protection in the theoretical context of his *Descent of Man*, but also in practical life. As Darwin's son Francis Darwin writes, his father's "humanity to animals was well known in his own neighbourhood."[32]

Darwin's theoretical as well as practical attitude towards animals is interesting for our bioethical context: Although animals are no moral beings because they are not capable of morality, they have to be treated with sympathy and moral consideration. They are "moral patients" and belong to the "moral community", although they cannot be "moral agents".[33]

From this presentation it should have become clear that Jahr has good reasons to refer to Darwin. The Darwinian revolution was well known at his time, and particularly Germany had been a stronghold of Darwin-reception since the 19th century. So Jahr may have known more about Darwin than becomes apparent in his texts. Jahr's articles are generally very short, and also the other thinkers and authorities whom he mentions, are hardly dealt with more elaborately.

The Foundations of Fritz Jahr's Bioethics

"*Respect every living being, therefore also animals, as an end in itself, and treat it whenever possible as such.*"[34] is Jahr's "bioethical imperative".

Considering an entity as an "end in itself" usually means that we have *direct duties* towards this entity. Thus Jahr's bioethical imperative implies that we

[30] Darwin Descent of Man vol. 21, 115f.

[31] Darwin Descent of Man vol. 21, 127-129.

[32] Darwin F (ed.) (1887) The Life and Letters of Charles Darwin. 3 Vols. London: John Murray, Vol. 3, 200.

[33] This is Tom Regan's terminology in his classic book (2004) The Case for Animal Rights. Updated with a new Preface. Berkeley, Los Angeles: University of California Press (1st ed. Berkeley, Los Angeles 1983).

[34] Jahrs „bio=ethischer Imperativ": „*Achte jedes Lebewesen, also auch die Tiere, als einen Selbstzweck, und behandle es nach Möglichkeit als solchen*!" (Jahr Tierschutz und Ethik, 102).

have direct duties not only towards humans but also towards animals and plants. They are not just instruments for us humans, not just means for our manifold purposes, but have to be respected as beings that possess an inherent value, a "Selbstwert". This means that also in those situations, when we use animals and plants for our purposes, we must not lose their inherent value out of sight. That's what we postulate with regard to the treatment of humans. We all are means for others in various situations and roles. But we have to be respected *at the same time* as an end in itself, never as mere means for this or that will. This is the meaning of Kant's categorical imperative which he formulates in different versions with respect to persons.[35]

What are Jahr's arguments for his bioethical imperative? Four elements are important here: *Firstly* empirical scientific knowledge about the relationship between the human being and the rest of living nature, *secondly* a certain standpoint of normative ethics, supported by a variety of philosophers and theologians, *thirdly* a growing sensitivity towards animals which manifests itself in the existence of animal protection paragraphs, i.e. a legal practice which is improved in the course of time and *fourthly* a certain kind of realism, recognizing the fact that we humans are also living beings who have the right to life and to means of subsistence. I will elaborate on this in my following remarks.

The importance of scientific knowledge for animal ethics and protection was already demonstrated. Helmholtz summarised Darwin's position to the point by describing the relationship between humans and animals as a "real blood relationship". Although it is not possible to deduce directly and merely from the existence of a blood relationship and its scientific explanation moral obligations towards animals and plants – this would be a *naturalistic fallacy* –, such a relationship may well be ethically relevant when its statement is combined with other premises, with ethical premises. We will later see what is meant by this.

What is Jahr's foundation of bioethics and particularly of this kind of normative bioethics which claims that we have moral obligations also to animals and plants, not only to humans?

A prominent representative of duty ethics (deontology) is Immanuel Kant. He makes a distinction between *direct* and *indirect duties*. According to Kant only the human being and every being endowed with reason, self-consciousness, i.e. a *person*, has an "*absolute value*" [absoluter Wert] and thus exists as an *end in itself* [Zweck an sich selbst]. Beings without reason

[35] Kant I (1999): Grundlegung zur Metaphysik der Sitten. Kraft B, Schönecker D (eds.). Hamburg: Felix Meiner (1st ed. Riga 1785).

have only a "relative value" as means and are therefore called "things" ("*Sachen*"). For Kant we therefore have direct duties only towards persons. Only persons are able to reciprocally obligate themselves. Since animals and plants are no persons, they are no end in itself and therefore we have no direct duties towards them according to Kant. But animals are "an analogon of humanity" ["ein Analogon der Menschheit"][36]. Our meekness and mercy towards animals as well as our violence and cruelty towards them are conveyed to our attitude and behaviour towards humans. Therefore violence and cruelty towards animals infringe on the duty of the human being towards humanity, because our compassion with other humans is weakened and finally destroyed by this cruelty towards animals. Compassion, however, is a predisposition [Anlage] in the human being which is very conducive to our moral behaviour towards our fellow humans. Therefore Kant argues: "Even gratitude towards an old horse or dog for having provided their service for a long time (as if they were household members) belongs *indirectly* to the human being's duty, namely in relation to these animals [*in Ansehung* dieser Tiere], considered *directly* however it is always only a duty of man towards himself."[37]

Kant also argues for the protection of "beautiful crystals and the indescribable beauty of the plant kingdom"[38]. The propensity for mere destruction ("spiritus destructionis") violates the duty of the human being towards himself/herself, because it weakens and destroys a feeling in us which is not by itself moral but which has an important function for morality, because it prepares us for loving something without having its utility in mind.

Arthur Schopenhauer who advocates an ethics of compassion, of pity [Mitleidsethik], formulates a sharp criticism of Kant's version of animal ethics. "Thus, one should only feel pity for animals for exercise, and they are so to speak the pathological phantom for the exercise of pity for humans."[39]

---

[36] Kant I (1990): Eine Vorlesung über Ethik. Gerhardt G (ed.). Frankfurt am Main: Fischer, 256f.

[37] "Selbst Dankbarkeit für lang geleistete Dienste eines alten Pferdes oder Hundes (gleich als ob sie Hausgenossen wären) gehört *indirekt* zur Pflicht des Menschen, nämlich *in Ansehung* dieser Tiere, *direkt* aber betrachtet ist sie immer nur Pflicht des Menschen *gegen* sich selbst." (Kant 1993 [1797, A 108, § 17] 579. Kant I (1993): Die Metaphysik der Sitten. Werkausgabe Band VIII. Weischedel W (ed.). 10th ed. Frankfurt am Main: Suhrkamp (1st ed. Königsberg 1797).

[38] "schöne Kristallisationen, das unbeschreiblich Schöne des Gewächsreichs". (Kant 1993 [1797, A 107] § 17, 578)

[39] "Also bloß zur Uebung soll man mit Thieren Mitleid haben, und sie sind gleichsam das pathologische Phantom zur Uebung des Mitleids mit Menschen." Schopenhauer A (1977): Über die Freiheit des menschlichen Willens (1839). Über die Grundlage der Moral (1840). Kleinere Schriften II. Zürcher Ausgabe. Werke in zehn Bänden. Bd. VI. Zürich: Diogenes, 202.

For Schopenhauer Kant's position means that our compassion towards animals and our humane treatment of them is only a means of training our moral sense towards humans. Schopenhauer confronts Kant's position with his "formula of morals" "*Neminem laede, imo omnes, quantum potes, juva*" [Don't hurt anyone, but help everybody as far as you can.][40].

Jahr takes the part of Schopenhauer then, not that of Kant. For Jahr the Golden Rule as well as Kant's categorical imperative only describe a formal criterion of a "good" way of action. "In spite of this criterion the motive could even be blatant egoism, a kind of reciprocal contract: Don't harm me, then I won't harm you either. (that's what Schopenhauer shows in his, 'Grundlage der Moral')."[41] Jahr refers to Schopenhauer who "here provides help. He quotes as the best concrete description of a moral way of conduct the sentence 'Neminem laede, imo omnes, quantum potes, juva!'"[42]

For Jahr this insight is already formulated in the Fifth Commandment under the aspect of the *sanctity of life* and of its manifestations. The Fifth Commandment "does not only forbid the malicious or careless destruction of a life, but everything which can somehow interfere with life or inhibit it."[43]

This means that Jahr's bioethical imperative is not founded on a denontological position like that of Kant. Since it includes all living beings, humans, animals, and plants as an end in itself, reason, self-consciousness or personhood obviously are not the only criteria or presupposition for deserving respect as "end in itself". Jahr's ethics is based on compassion regarding all living beings who are capable of feeling pleasure and pain and who can be an object of welfare and harm.[44] Moreover it is based on the Christian religion

---

[40] „Verletze niemanden, vielmehr hilf allen, soweit du kannst." Schopenhauer Grundlage der Moral, 199. Heike Baranzke takes a more generous interpretation of Kant's animal ethics than Schopenhauer. In short, she claims that Kant's ethics allows for a very demanding protection of animals because the ultimate goal, respect for humanity, is a very demanding imperative (see Baranzke H (2002) Würde der Kreatur? Die Idee der Würde im Horizont der Bioethik. Würzburg: Königshausen & Neumann; Baranzke H (2005) Tierethik, Tiernatur und Moralanthropologie im Kontext von § 17 Tugendlehre, in: Kant-Studien 96, 336-363.

[41] „Das Motiv könnte trotz dieses Kennzeichens sogar krasser Eigennutz sein, nämlich eine Art Vertrag auf Gegenseitigkeit: Tue mir nichts, dann tue ich dir auch nichts. (Das zeigt Schopenhauer in seiner ‚Grundlage der Moral'.) Jahr F (1934) Drei Studien zum 5. Gebot, in: Ethik. Sexual= und Gesellschafts=Ethik. 11, 183-187, 183f.

[42] Jahr Drei Studien zum 5. Gebot, 184.

[43] „Es verbietet nicht nur die böswillige oder fahrlässige Vernichtung eines Lebens, sondern alles, was irgendwie störend oder hemmend auf ein Leben einzuwirken geeignet ist." ibid.

[44] In today's animal ethics the members of certain species of animals are also called "persons". An example is the practical ethics of Peter Singer (Singer 1993). For Singer

and its Fifth Commandment which, as Jahr points out, is not explicitly restricted to the prohibition of killing humans. Should it not be applicable correspondingly also to animals and plants, is Jahr's question.

"But are the animals really so close to us that we have to regard and treat them quasi as our neighbours? – Without doubt there are quite huge differences between the human being and animals, and also modern natural science only confirms this fact."[45] This by no means excludes many similarities between animals and humans which have been discovered particularly since Darwin, as Jahr emphasises.

Here we can come back to the fact that animals are our "blood relatives" and expand on this idea. We may ask on what basis we humans claim for ourselves to be treated as an end in itself, not to be tortured and killed but to be treated in a way which is beneficial for our well being. The answer is that we are living beings who have needs and desires regarding the presence and the future and who can feel pleasure and pain. Veracity and consistency call for an equally respectful treatment of humans and animals. Even if perhaps most animals have no self-consciousness like we have they are nevertheless aware of their pain and pleasure, they have a *self-awareness*. If animals have a choice they chose the environment which is conducive to their wellbeing and they avoid less comfortable and harmful situations. Drawing upon the relationship between animals and us for claiming animal protection does not imply a naturalistic fallacy if we combine the statement of kinship with normative premises about our own right to protection and the demand for consistency and veracity.

But can this conception of bioethics and its imperative "Respect every living being principally as an end in itself and treat it, whenever possible, as such!" be realized in practice? Is it not utopian, asks Jahr. He presents several arguments for refuting this possible objection, drawing on Herder, Schleiermacher and Krause.

---

some nonhuman animals like the great apes are persons, because they are endowed with reason, self consciousness and a sense of time. And if other animals are also endowed with these traits they are as well persons. However this does not at all mean that in Singer's system all other animals are regarded as instruments or means and only persons included certain animals have an intrinsic value. For Singer it means however that the prohibition of killing persons is stricter than that of killing non-personal animals and non-personal humans. Singer P (1993): Practical Ethics. Cambridge: Cambridge University Press (1st ed. 1979).

[45] „Stehen uns die Tiere aber wirklich so nahe, dass wir sie gleichsam als unsere ‚Nächsten' einschätzen und behandeln müßten? – Ohne Zweifel sind ganz gewaltige Unterschiede zwischen dem Menschen und den Tieren vorhanden, und auch die moderne Naturwissenschaft bestätigt diese Tatsache nur." (Jahr, Tod und Tiere, 5)

First of all, our ethical obligations towards a living being practically conform to the "needs" (Herder) or to the "destiny" ["Bestimmung"] (Krause) of that living being. Plants, animals and humans have equal rights ["gleichberechtigt"], however not to "equal things but each of them only to that which is a necessary prerequisite for the attainment of its/his/her destiny."[46] This means that the specific features of a living being have to be taken into account in our treatment of this being, features which are species specific but also dependent on the individual situation of an organism. The claim of equal rights for plants, animals and humans means an equal consideration of their specific kind and needs and does not mean an equal treatment in every sense and way. Peter Singer expresses something similar by using the term "equal consideration of interests" as distinct from an equal treatment.[47] Gotthard M. Teutsch refers to the principle of equality which is a principle of justice. It means that we have to treat equal entities equally and different entities differently.[48] Thus justice may call for a different treatment of entities for the purpose of the equal consideration of their interests.

Jahr thinks that animals have fewer and less complicated needs than humans and that we have therefore fewer practical moral obligations towards animals. This holds the more with respect to plants. These differences of needs between humans, animals and plants facilitate our compliance with the bioethical imperative.[49] It may well be true that we humans can appropriately satisfy the needs of animals and plants. But in the light of our immense increase of knowledge about the emotional, cognitive and behavioural faculties of nonhuman animals we have to be careful not to underestimate the complexity of animals' needs. Here we have to remember Jahr's own statement quoted at the beginning: "Useful and efficient animal protection can only be practised well if enough knowledge of nature and at least some understanding of nature is available. For, in fact, we will only come to a real protection of animals, if we know to some degree their physiological and psychological traits and conditions of life."[50] Particularly in the tradition of Charles Darwin cognitive ethology has made much progress.[51]

Does our bioethical imperative also imply the prohibition of killing animals? The 5th Commandment, "Thou shalt not kill!" does not explicitly forbid only

[46] „allerdings nicht zu gleichem, sondern ein jedes nur zu dem, was ein notwendiges Erfordernis zur Erreichung seiner Bestimmung ist." (Jahr Bio=Ethik,3; Jahr Tierschutz und Ethik,101)
[47] Singer Practical Ethics, Chapters 2 and 3.
[48] Teutsch G M (1987): Lexikon der Tierschutzethik. Göttingen: Vandenhoek & Ruprecht.
[49] Jahr Drei Studien zum 5. Gebot, 187.
[50] Jahr Tierschutz und Ethik, 101. German quotation see note 15.
[51] See the new *Journal of Animal Ethics* 2011.

the killing of humans. Should it not be valid for our behaviour towards animals and plants as well? The extension of the 5th Commandment to our treatment of animals at first sight seems to be utopian, "for slaughter and killing of animals are virtually unavoidable, even if the last one only happens collaterally by depriving animals of their necessary conditions of life, which is a consequence of the distribution [Ausbreitung] of the human species. It is the struggle for life which inflicts this necessity on us."[52] There is a struggle for life between humans and animals, which seems to render the extension of the 5th Commandment to our treatment of animals and thus the application of the bioethical imperative impossible.

Jahr invalidates this critique by pointing to a similar situation in the relation among humans themselves. The principle of the struggle for life ["Prinzip des Kampfes ums Dasein"] is influencial in everyday life and influences as well as "modifies our moral obligations towards our fellow humans in a certain way, much as we may regret this."[53] According to Jahr, in all branches of our life and in all professions (politics, business, laboratory, workshop, in the field etc.) our entire life and activity is in the first place not focused on love, frequently however on struggle with our competitors.

"Mostly we are not quite aware of this, as long as the struggle does not breach the limits of the law. In such struggle for life we humans deliberately and consciously use human power, human health, human life, not only in times of war, but also in 'peaceful' life such as in cultural development, especially in some of the industries. In spite of all this no one considers the 5th commandment a utopian charge. As our attitude towards animals – as determined by struggle for life – basically does not fall outside our attitude towards man, the commandment can and must be valid here as well, an ideal and a point of reference for our moral strife." (Jahr 1928b, 6; transl into English by I. M. Miller; Miller, Sass 2011, 5f.)[54]

---

[52] „[…] denn das Schlachten und Töten der Tiere, möge dieses letztere auch nur mittelbar geschehen durch Entziehung der notwendigen Lebensbedingungen infolge der Ausbreitung des Menschengeschlechtes, ist schlechterdings unvermeidlich. Der Kampf ums Dasein ist es, der uns diese Notwendigkeit auferlegt." (Jahr Tod und Tiere, 5)

[53] „Des weiteren ist hier noch das Prinzip des Kampfes ums Dasein von Einfluß, ein Grundsatz, der auch unsere ethischen Pflichten gegen unsere Mitmenschen in gewisser Weise modifiziert, so sehr wir dies auch bedauern mögen." (Jahr Tierschutz und Ethik, 101)

[54] „Wir werden uns dessen meist nur nicht bewußt, solange dieser Kampf in gesetzlich erlaubter Weise geführt wird. In diesem Kampfe der Menschen ums Dasein wird auch mit vollem Bewußtsein Menschenkraft, Menschengesundheit und Menschenleben verbraucht, und das gilt nicht etwa nur für Kriegszeiten, sondern auch für das ‚friedliche' Leben der fortschreitenden Kulturentwicklung, besonders in manchen Industriezweigen. Trotz alledem wird niemand das 5. Gebot als eine utopische Forderung ansehen. Und da das Verhalten gegen die Tiere, wenn es durch den Kampf ums Dasein bestimmt wird,

Jahr calls for consistency of argumentation: The discrepancy between ideal norms and practice which does not keep us from maintaining moral obligations among humans as well is no reason to discard the bioethical imperative regarding animals.

Although Jahr considers "slaughter and killing of animals as virtually unavoidable", it would be nevertheless in accordance with his line of argumentation to plead for vegetarianism. For it is not necessary for humans to live on meat, the slaughter of animals *is* avoidable. Particularly in numerous highly industrialized countries since the 1960s Jahr's principle to respect animals as an end in itself is violated by mass (intensive) animal keeping, by the breeding of high-performance animals to the point of what may be called "tormentive breeding", by cruel animal transport, and by feeding in ways inappropriate to a given species. All such practices derive from the desire to maximize profit and to accommodate the excessive consumption of animals and their products. For Jahr "the moral postulate also to protect animal-life is absolutely valid, without any consideration, if it is advantageous for us, as ethics in general does not and must not ask these questions."[55] This however means that people in the industrialized countries have at least to reduce their meat consumption, which would not only be in accordance with our obligation towards animals but also towards humans as well as the environment. For only then we could break the cycle of producing food for feeding animals, destroying the environment, causing hunger in third world countries and torturing animals.

Nevertheless we live in many ways at the expense of other living beings, of animals and plants. Which practical influence can the application of the 5th Commandment for the protection of animals and plants have, considering these restrictions? We must not kill or destroy animals and plants without a "reasonable purpose" ["vernünftiger Zweck"] for doing so (Schleiermacher, Krause).[56] In the first paragraph of the current German Animal Protection Law we find a similar formulation: "The purpose of this law is the protection of the life and wellbeing of the animal, based on the human being's responsibility for the animal as fellow creature. Nobody must inflict pain,

---

grundsätzlich nicht aus dem Rahmen unseres Verhaltens gegen die Menschen herausfällt, so kann und muß das Gebot als Ideal, als Richtungspunkt unseres sittlichen Vorwärtsstrebens, auch hier seine Geltung behalten." (Jahr Tod und Tiere, 6; transl into English by I. M. Miller; Miller, Sass 2011, 5f. see in this volume.

[55] „Die Forderung, auch das tierische Leben zu schonen, hat absolute Geltung, ohne jede Rücksicht darauf, ob uns ein äußerer Vorteil daraus erwächst, wie denn überhaupt die Ethik nach solchen Dingen nicht fragt und nicht fragen darf." (Jahr Tod und Tiere, 6).

[56] Jahr Bio=Ethik, 3; Tierschutz und Ethik, 101.

suffering or damage to an animal without reasonable purpose [ohne vernünftigen Grund]."[57]

Jahr also points to animal-protection paragraphs for further orientation how we can take pity on animals. It is however an empirical and ethical challenge to decide, which purposes for inflicting pain, suffering or damage to animals are "reasonable".

Jahr does not defend exaggerated practices of protection of nature like the "fanatic self-harm of the school of yoga" which allows the eating of plants only under certain circumstances. Jahr's wording is interesting here. "The addiction not to harm any living being in its self-preservation even today leads certain Indian penitents to live off horse manure."[58] Not only animals and plants have the right to live but also humans. It would be inconsistent to defend animal and plant protection at the expense of humans, because we too are living beings. And we have the duties towards ourselves, particularly the duty of self maintenance.[59]

According to Jahr the bioethical imperative has become self-evident as far as the protection of animals is concerned. Although Jahr formulates a bioethical imperative and not just one for the protection of animals and although he explicitly includes plants in his claim, in his examples he more often refers to animals than to plants. He rejects for instance the malicious destruction of flowers and the wilful damage of trees. Intuitively we agree with him. Watching this kind of vandalism causes outrage and anger in us. But what is the argumentative basis for considering a plant or species of plants as an *end in itself* and for protecting them not just for esthetical, environmental and ecological reasons but in the first place with respect for themselves? Regarding "plant ethics" Jahr refers to our intuitions [das Gefühl] which during a walk keep us from "beheading" plants by our cane or from picking flowers and discard them carelessly after a short time.[60] It is however more

---

[57] Tierschutzgesetz (TierSchG) „§ 1 Zweck dieses Gesetzes ist es, aus der Verantwortung des Menschen für das Tier als Mitgeschöpf dessen Leben und Wohlbefinden zu schützen. Niemand darf einem Tier ohne vernünftigen Grund Schmerzen, Leiden oder Schäden zufügen." (Emphasis by E.-M.E.) *Tierschutzgesetz (TierSchG)* (German Animal Protection Law) 24.07.1972, last change by Art. 20 G v. 9.12.2010 I 1934. http://bundesrecht.juris.de/bundesrecht/tierschg/gesamt.pdf (last check by EM E 14th November 2011).

[58] „Die Sucht, keinem Lebewesen bei der Selbsterhaltung zu schaden, führt auch noch heute gewisse indische Büßer dazu, sich von Pferdemist zu nähren." (Jahr Bio=Ethik, 2).

[59] Jahr Drei Studien zum 5. Gebot, 184f.

[60] „Aber wir sind doch auf dem Wege des Fortschritts, und der Tierschutz gewinnt in immer weiteren Kreisen Raum, so wie es ja auch kein anständiger Mensch widerspruchslos ansieht, wenn irgendein Flegel gedankenlos Blumen am Wege mit dem

difficult to give an ethical foundation for *biocentrism*, a bioethical position that claims an inherent value of all living beings including plants and thus postulates direct duties towards plants, than to give an ethical foundation of sentientism or zoocentrism, of direct duties towards all sentient animals. The question of how biocentrism can be founded is a central issue in today's bioethical debates.

We might ask if it is not also a question of consistency and veracity to respect plants for their own sake and not just as means for our or other animals' ends. Our species is a very late product of evolution, whereas the first life forms came into being to our estimation about 3.5 billions [Milliarden] years ago. Humans are only one single species of thousands and thousands other species, and we owe our existence a long line of ancestors. Without them and without the very first living organisms on this planet we wouldn't exist. Plants and animals have a common origin and are siblings. We are not only the relatives of other animals but also of plants. And regarding ourselves we do not only claim protection for our self-consciousness and the so called higher functions of our body but for our whole body including its vegetative functions. Ethical anthropocentrism is based on a reduced image of the human being. Not only does it draw a demarcation line between humans and the rest of nature but it also segregates essential parts of the human being himself/herself. Instead of arguing for an exclusive bioethics which claims an exquisite position for humans we should make an effort to work out an inclusive bioethics which views the human being as part of nature and emphasises the mutual relationships between humans and other living beings.[61] What an inclusive bioethics means in detail and how it can be founded still has to be worked out.

Jahr finally faces the question: "What is the impact of the extension of our moral obligations beyond our fellow humans to animals on our relationship towards other humans? Don't we have to fear that our attention is diverted from our fellow humans' misery when we focus on animals?"[62] For Jahr the opposite is the case, and here he draws on Immanuel Kant's claim that the protective and merciful treatment of animals is a duty of the human being

Spazierstock köpft oder wenn Kinder Blumen abreißen, um sie nach wenigen Schritten achtlos wegzuwerfen." Jahr Bio=Ethik, 4.

[61] See the short description of „inclusive ethics" in Eser U, Potthast Th (1999) Naturschutzethik. Eine Einführung in die Praxis. Baden-Baden: Nomos, 48.

[62] „Welche Wirkung hat die Ausdehnung unserer sittlichen Verpflichtungen über den Menschen hinaus auf die Tiere auf unser Verhältnis zu unseren Mitmenschen? Ist nicht zu befürchten, daß unsere Aufmerksamkeit von der Not der letzteren abgelenkt wird, wenn wir unser Augenmerk auf die ersteren richten?" (Jahr Tod und Tiere, 6)

towards himself/herself. It can be of highest importance for an "ethics of society" ["Gesellschaftsethik"].[63]

"*Respect every living being, therefore also the animals, as an end in itself and treat it whenever possible, as such*! And if one cannot recognise the *absolute validity* of this principle insofar as it refers to the animals and plants, one may, as I already said, nevertheless follow it out of consideration for the moral obligations towards the whole human society."[64]

Thus Jahr offers a double normative foundation of his bioethical theory, one which includes direct duties not only towards humans but also towards animals and plants – this is his conviction – and another one which accepts direct duties only for the treatment of humans but which nevertheless can lead to a strict and effective animal and plant protection.[65] The imperative of the protection of living nature has entered legal and regulatory frameworks at the national and international level. The protection of the natural basis of existence and of the animals [Schutz der natürlichen Lebensgrundlagen und der Tiere] has become a special article in the German Basic Law (Art. 20a), and "the dignity of the creature" [die "Würde der Kreatur"] is protected by the Swiss constitution since 1992.

I will finish with a quotation from Hans Jonas. Long before he published his famous book *The Imperative of Responsibility* (*Das Prinzip Verantwortung*) he presented results of his philosophical biology. This leads us back to the importance of Charles Darwin for our understanding of nature and for bioethics:„In the hue and cry over the indignity done to man's metaphysical status in the doctrine of his animal descent, it was overlooked that by the same token some dignity had been restored to the realm of life as a whole. If man was the relative of animals, then animals were the relatives of man and in degrees bearers of that inwardness of which man, the most advanced of their kin, is conscious in himself [...] So it happened that in the hour of the

---

[63] Jahr Tierschutz und Ethik, 101.

[64] Emphasis on „absolute validity" by E.-M.E. „*Achte jedes Lebewesen, also auch die Tiere, als einen Selbstzweck, und behandle es nach Möglichkeit als solchen*! Und wenn man die absolute Geltung dieses Grundsatzes, soweit er sich eben auf die Tiere und Pflanzen bezieht, nicht anerkennen will, so möge man ihn, um schon Gesagtes zu wiederholen, mit Rücksicht auf die sittlichen Verpflichtungen gegen die gesamte menschliche Gesellschaft dennoch befolgen." (Jahr Tierschutz und Ethik, 102)

[65] This reminds us of Bryan Norton's "convergence hypothesis", meaning, that when we pursue practical goals of environmental protection policy, anthropocentrists and nonanthropocentrists can act in concert in spite of the differences in their basic premises. Nevertheless the question remains which position provides a more stable protection of nature throughout all the changing practical and political situations. Norton B G (1991): Toward Unity among Environmentalists. New York, Oxford: Oxford University Press.

final triumph of materialism, the very instrument of it, ‚evolution', implicitly transcended the terms of materialism and posed the ontological question anew – when it just seemed settled. And Darwinism, more than any other doctrine responsible for the now dominant evolutionary vision of all reality, turns out to have been a thoroughly dialectical event."[66]

[66] Jonas H (2001) The Phenomenon of Life. Toward a Philosophical Biology. Evanston, Illinois: Northwestern University Press. (1st ed. New York 1966), 57f.
„In der lauten Entrüstung über den Schimpf, den die Lehre von der tierischen Abstammung der metaphysischen Würde des Menschen angetan habe, wurde übersehen, daß nach dem gleichen Prinzip dem Gesamtreich des Lebens etwas von seiner Würde zurückgegeben wurde. Ist der Mensch mit den Tieren verwandt, dann sind auch die Tiere mit dem Menschen verwandt und in Graden Träger jener Innerlichkeit, deren sich der Mensch, der vorgeschrittenste ihrer Gattung, in sich selbst bewußt ist. […] So geschah es, daß in dem Augenblick, da der Materialismus seinen vollen Sieg gewann, das eigentliche Mittel dieses Sieges, die ‚Evolution', nach seiner inneren Konsequenz die Grenzen des Materialismus sprengte und die ontologische Frage neu aufwarf – als sie gerade entschieden schien. Und es stellt sich heraus, daß der Darwinismus, der mehr als jede andere Lehre für die nunmehr dominierende evolutionäre Schau aller Wirklichkeit verantwortlich ist, ein von Grund auf dialektisches Ereignis war." Jonas H (1973): Organismus und Freiheit. Ansätze zu einer philosophischen Biologie. Göttingen: Vandenhoeck & Ruprecht, 84f.

VIKTOR VON WEIZSÄCKER AND FRITZ JAHR. A CHALLENGE FOR CULTURAL ANALYSIS

Fernando Lolas[1]

The 20`s of the XXth Century were a time of intense political and ideological debate in Germany and Europe. The aftermath of the First World War, the tensions between the states, and the period of feeble peace created a space where conceptual and esthetic innovations took place. The sciences experienced several crises. Some of them had to do with the reliability of their findings, others with the uses of science and technology at the service of mankind. The *Zeitgeist* is one of effervescence, confusion, hope, and despair, all mixed with the effects of philosophical reflection. It is a time when the Vienna Circle eliminates all speculation not based on positive fact. It is also a time when the gap between philosophy and life moves the spirits in search of the value of life and the need to experience "wholeness" becomes prevalent in biology and medicine[2].

No wonder that a parallel can be established between the pioneer work of Fritz Jahr and the equally pioneering work of Viktor von Weizsäcker and the founders of the German anthropological medicine[3]. In point of fact, the seminal article by Jahr appeared in 1927 and established the foundations for a line of thought that later would be reconstructed and expanded in different parts of the world[4]. Around the same time, Viktor von Weizsäcker, in the journal which he established and directed with Joseph Wittig and Martin Buber, "*Die Kreatur*", formulated what at the time appeared as a new direction in medical thinking, which he christened medical anthropology. In this and other writings, Weizsäcker tries to overcome the irreparable separation between philosophy, as the working of reason alone, and the sciences, as the expression of the need to find causal relations in nature. His was an attempt to vindicate life as the unreflected source of meaning for humans, animals, and nature. Although his work was mainly devoted to

---

[1] Fernando Lolas Stepke, flolas@u.uchile.cl, Interdisciplinary Centre of Studies in Bioethics, University of Chile, Chile. - During the preparation of this paper the author received support from NIH-Fogarty, Grant 5R25TW006056 and the Alexander von Humboldt Foundation (Humboldt Alumni-Preis).

[2] Neumann, J. Viktor von Weizsäcker im Kontext medizinischer Theorieansätze zur Zeit der Weimarer Republik. In Gahl, K., Achilles, P., Jacobi, R. (editors) *Gegenseitigkeit. Grundfragen medizinischer Ethik*. Königshausen & Neumann, Würzburg, 2008. (pp. 87-102)

[3] Weizsäcker, Viktor von, *Gesammelte Schriften*. 10 volumes Suhrkamp, Frankfurt/Main, 1987/1996.

[4] Jahr, F. Bio-Ethik. Eine Umschau über die ethischen Beziehungen des Menschen zu Tier und Pflanze. *Kosmos. Handweiser für Naturfreunde* 24(1): 2-4, 1927. See in this volume.

medicine as "*Umgang*", and the need to enrich the positivistic outlook with an empathic participation in life in all its forms, his proposals about reintroducing the subject in biology and medicine, and the recognition of the respect due to all living beings bring his ideas close to those of Fritz Jahr[5].

The notion of life acquires, in the work of von Weizsäcker, a value-laden character and is logically akin to an all embracing sense and meaning of actions in the medical and life research areas. "*Um Lebendes zu erforschen muss man sich am Leben beteiligen*" (In order to research the living one has to take part in life). This opening sentence of his work "*Der Gestaltkreis*" suggests that life is there before any attempt is made at its analysis or comprehension[6]. It is an unreflected given, common to humans, animals, and plants. Its primacy derives precisely from this aprioristic nature. We are immerse in life and cannot escape from having to live. Such essential character of life is also present in Jahr's writings, with the addition that he considers compassion and altruism essential features of human nature, and hence, naturally given[7].

A curious fate is common to both endeavors. Although much must be done on the way of scholarship to ascertain the real influence of Jahr`s ideas on the scenario of European humanities and sciences, it may be said that aside from labeling *Bio-Ethik* no institutional impact was evident afterwards. For instance, it is not visible in American neo-bioethics of the seventies, however familiar the formulations of Potter's global ethics seem to be after reading Jahr articles. It did not express itself in the writings of scholars or academics. It was not the source of new institutions. By the same token, the main tenets and ideas of the anthropological medicine developed at Heidelberg did not have a persistent influence on medical research and practice. Efforts were made to translate the main intuitions into viable institutions but in the end the crisis of medicine was not solved by the contributions of von Weizsäcker, nor were the institutions he promoted the sites where the conceptual revolution took place. Both pioneers shared the fate of many cultural products of Germany during the first half of the XXth century. Like psychoanalysis, bioethics and anthropological medicine were exported to other parts of the world, mostly North America, re-discovered after some time and then given diffusion. This fact merits further consideration for students of culture.

---

[5] Lolas Stepke, F. *La medicina antropológica y el juicio de Nürnberg. El aporte de Viktor von Weizsäcker*. Monografías de Acta Bioethica N°5, CIEB, Universidad de Chile, 2010.

[6] Weizsäcker, Viktor von, *Der Gestaltkreis*. Thieme, Stuttgart, 1940.

[7] Jahr, F. Zwei ethische Grundprobleme in ihrem Gegensatz und in ihrer Vereinigung im sozialen Leben. *Ethik. Sexual- und Gesellschaftsethik* 6: 341-346, 1929.

However ignored these contributions, their comparative study and analysis deserve attention. Not so much because they exerted direct influence or affected the ways in which science, medicine, and philosophical reflection developed, but because they are useful for an archeology of thought and for uncovering hidden threads of meaning and scholarship that have had covert effects on the ways people consider the humanization of medicine and the humane behavior toward living beings[8].

The medical anthropology of the Heidelberg School, represented by Krehl Siebeck, Weizsäcker and Christian, stressed the fact that a reintroduction of the subject into medicine and biology was necessary[9]. This can be considered in the light of the spirit of what at the time was considered scientific medicine: a truly naturalistic science of objects, governed by the axioms of positivistic science. This attitude, he contended, was responsible for the inhuman use of medical science during the Nazi regime[10]. By recovering an empathic posture to nature and people, the art and science of medicine could be placed in closer relationship to human activity and to the persons involved in the social dialogue. Fritz Jahr, on the other hand, stressed the respect humans owe to other living organisms, and at the same time insisted on the value of action instead of abstractions. The *locus* and origin of ethics were in the acts by which people relate to nature and to other humans.

This primacy of action and the notion of shared responsibility are evident in the contributions of both Jahr and von Weizsäcker. The moral responsibility of humans in the role of partners of other humans and of animals has as a corollary the notion that everyone has to care for him or herself and also show respect for living beings. As a matter of fact, the main ethical concepts that can be evinced in the writings of Viktor von Weizsäcker are *Gegesenseitigkeit* (reciprocity) and *Solidarität* (solidarity). A careful reading of Jahr's contributions uncovers similar principles, albeit expressed in somewhat different forms. His discussion on egoism and altruism needs to be read in this framework.

A further parallel can be established in the use of language. Jahr pleaded for a careful use of terminology, trying to avoid misconceptions and misrepresentations that hamper comprehension. The same can be said for the work of von Weizsäcker, although he was confronted with realities that

---

[8] Jahr, F. *Selected Essays in Bioethics 1927-1934* Fritz Jahr. Medizinethische Materialien Heft 186, Zentrum für Medizinische Ethik, Bochum, 2010. See also in this volume.

[9] Lolas Stepke, F. Paul Christian und die Heidelberger Schule: Persönliche Darstellung eines Werdegangs. *Fundamenta Pychiatrica* 15: 135-138, 2001.

[10] Lolas Stepke, F. *La medicina antropológica y el juicio de Nürnberg. El aporte de Víctor von Weizsäcker*. Monografías de Acta Bioethica Nª5, CIEB, Universidad de Chile, 2010.

needed new forms of expression. Hence, many terms had to be created to account for social phenomena and to describe situations not hitherto encountered. This is the case for the expression *Logophanie*, meaning to imply the transformation of somatic experience into verbal label by means not completely understood. Or the very notion of *Gestaltkreis*, the intimate relationship between action and perception, and the complementariness of the somatic and the psychological aspects of human life.

Ethical reflection permeates the work of von Weizsäcker, although he rarely makes it explicit. The essential point is that medicine and, for that matter, all life sciences cannot derive their morality from the outside, from philosophy as pure speculation. They should be immersed in life, in unreflected life, in the *Lebenswelt* of everyday experience. "*Am Anfang war die Tat*", said Goethe. At the beginning was the act, action, not reflection. In order to investigate life, one should take part in life. There always is something undefined, not amenable to rational reconstruction. This basic relation to nature can be expressed as "*passio*". And, as Jahr would have said, as "*compassio*", that is, the realization of a common fate for animals, humans, and all living things. This primacy of *passio* and *compassio*, the suggestions that the real answer to the quandaries of life does not reside in pure reason but in life itself brings together these two thinkers of the Weimar time, a period in European history so fertile in crisis, intuition, rigor, hope, and ideas.

FROM REVERENCE FOR LIFE TO BIOETHICS: ALBERT SCHWEITZER, A BIOETHICS PRECURSOR

José Roberto Goldim[1] and Marcia Santana Fernandes[2]

An introduction to Albert Schweitzer

Bioethics is a new field of knowledge, it is a meeting ground to share knowledge between disciplines. Bioethics, possibly, is one of the most successful interdisciplinary experiences in our time.

Since the beginning, the founder fathers – Van Rensselaer Potter and Fritz Jahr – proposed Bioethics as this. Even Aldo Leopold, that uses Land Ethics concept instead of Bioethics, used this approach. Interdisciplinarity is a core characteristic of Bioethics.

Many other thinkers made significant contributions to the bioethical theoretical framework, even without using this word. Albert Schweitzer is one of them. He is a precursor of Bioethics reflections, as we know them in present. Many of his writings resemble the ideas of Potter, Leopold and Jahr.

It is important to remember that Albert Schweitzer, as a person and as a thinker, made important contributions to Theology, Philosophy, Music and Medicine in his long life (1875-1965). He integrates Humanities and Medicine. His healthcare practice in Africa was assembled in a humanistic perspective (1).

The aim of this paper is to compare citations of Albert Schweitzer, Fritz Jahr, Aldo Leopold and Van Reseller Potter in order to demonstrate some similar thoughts.

From Reverence for life to bioethics

Reverence for Life (Ehrfurcht vor dem Leben) is the main theoretical contribution of Albert Schweitzer to Philosophy and Bioethics. This

[1] Prof. Dr. Jose Roberto Goldim, Clinical Hospital Porto Alegre, Pontifical Catholic University of Rio Grande do Sul, Faculty of medicine, 90035-903 Porto Alegre, Brazil; jgoldim@hcpa.ufrgs.br.

[2] Prof. Dr. Marcia Sanata Fernandes, Centro Universitário Ritter dos Reis, Porto Alegre, Brazil; marciasantanafernandes@gmail.com

concept emerges as an insight in 1915, when he was in Africa. In one of his biographical notes he describes:

> Late on the third day (September, 1915), at the very moment when, at sunset, we were making our way through a herd of hippopotamuses, there flashed upon my mind, unforeseen and unsought the phrase, "Reverence for Life".

Now I knew that the ethical acceptance of the world and of life, together with ideals of civilization contained in this concept, has a foundation in thought. (1)

This insight was the beginning of a very brilliant thought. Reverence for Life evolved along his long and creative life. He explained and used Reverence for Life in terms of theoretical and practical approach.

The first time he publicly shared this concept was on February 19, 1919 at St. Nicolai´s Church, in Strasbourg, during a Sunday morning service. He said that:

> Reverence for the infinity of life means removal of the alienation, restoration of empathy, compassion, sympathy. (2)

Schweitzer not only proposed, but he really lived with compassion. In the same 1919 sermon, Schweitzer enlarges his concept, saying that:

> Life is feeling, experience, suffering. If you study life deeply, looking with perceptive eyes into the vast animated chaos of this creation, its profundity will seize you suddenly with dizziness. In everything you recognize yourself. (2)

Fritz Jahr, in his foundation article, published in 1927, used the same comprehensive approach to compassion, when he says:

> Philosopher Schopenhauer, openly invoking the Indian ideas, considered, as a special quality of his Ethics, the fact of having claimed also to animals the feeling of compassion. (4)

Aldo Leopold, in his Land Ethics proposal, in 1930s, made a very similar assumption: Land Ethics simply enlarges the boundaries of the community to include soils, waters, plants, and animals, or collectively: the land. (3)

Schweitzer re-affirms, in his book Civilization and Ethics, written in 1923, the wide view, when he said:

> A man is ethical only when life, as such, is sacred to him, and that of plants and animals as that of his fellow men... Only the universal ethic of the feeling of responsibility in an ever-widening sphere for all that lives - only that ethic can be founded in thought. (5)

Jahr made the same assumption:

> From Biopsychology to Bioethics just one step is required, the acceptance of moral obligations to all living beings, not only in relation to humans. (4)

Reverence for life is a commitment to preserve life, to avoid harms, not only to humans, but also to all other living beings. Schweitzer and Jahr had the same thought about the assumption that Ethics could be extended to plants. In another part of the same Sermon, Schweitzer said:

> Reverence concerning all life in the greatest commandment in its most elementary form. We take this prohibition ("Thou shalt not kill") so lightly, thoughtlessly plucking a flower. (2)

Coherently, Fritz Jahr made a very similar reflection when he said

> Most people are naturally not so sensitive as Ed. von Hartmann. Everyone knows that plants are also living beings that are injured when the flower is cut, but the idea that it also feels resentful at that is not familiar to us. (4)

Expanding Ethics to all living beings – humans, animals and plants – was a very creative and a vivacious proposal even in contemporary Bioethics. It is important to remember that Francis of Assisi, cited by Fritz Jahr, extends compassion to the whole nature.

Albert Schweitzer understood that Ethics is intrinsically related to life and deals with good and evil. He linked these concepts when he proposed:

> That is what gives me the fundamental principle of morality, namely, that good consists in maintaining, promoting, and enhancing life, and that destroying, injuring, and limiting life are evil. (5)

Potter, in 1971, justifying Bioethics, in the same way, enhanced this point of view by saying:

> We must plead not for a moratorium on new knowledge, but a coupling of biological knowledge and human values. (6)

Potter, when proposed Deep Bioethics, instead of a shallow Ethics approach to health and science issues says:

> Bioethics as a new ethics science that combine humbleness, responsibility and an interdisciplinary competence, intercultural, that potential our sense of humanity. (7)

From Bioethics to Reverence to Life

Nowadays, is important to know, to understand and to have commitment to all of these primeval concepts of Bioethics. Eve Marie Engels, who was the first author that refers to Fritz Jahr's Bioethics approach in a published text, assumes following:

> Bioethics as an ethical reflection about living beings, including human beings, such as those living beings are presented in everyday relationships in the living world and in theoretical and practical contexts in science and research. (8)

This approach to Bioethics, including human beings as moral partners of all living beings, sharing ethical reflections, in a practical and theoretical framework, is a challenge to our society. Bioethics, these days, could be defined as a complex, shared and interdisciplinary reflection about the adequacy of actions related to Life and Living. (9)

Understanding Albert Schweitzer contributions is fundamental to understanding Bioethics as a whole. Schweitzer is a Bioethics precursor because he anticipates Jahr's bioethical thinking (10). Schweitzer made a contemporary contribution when he said:

> Life demands that we see through to the solidarity of all life which we can in any degree recognize as having some similarity to the life that is in us. (1)

In conclusion, Reverence for Life is a basilar concept to Bioethics. Albert Schweitzer wrote that: Just as the screw which churns its way through the water drives the ship along, so does reverence for life drive the man. (1)

References

1. Schweitzer A. Out of my life and thought. New York: Mentor; 1949.
2. Schweitzer A. Reverence for life: Sermons 1900-1919. New York: Irvington; 1993.
3. Leopold A. Sand County Almanac and sketches here and there. New York: Oxford; 1989.
4. Jahr F. Bio-Ethik: eine Umschau über die ethischen Beziehungen des Menschen zu Tier und Pflanze. Kosmos. 1927; 2(4) 2-4.
5. Schweitzer A. Civilization and Ethics: A&C Black; 1946.
6. Potter VR. Bioethics. Bridge to the future. Englewood Cliffs: Prentice Hall; 1971.
7. Potter VR. Bioética Profunda. O Mundo da Saúde. 1998; 22(6) 370-4.
8. Engels E-M. O desafio das biotécnicas para a Ética e a Antropologia. Veritas 2004; 50(2) 205-28.
9. Goldim JR. Bioética: origens e complexidade. Revista HCPA. 2006; (26) 86-92.
10. Goldim JR. Revisiting the beginning of bioethics: the contribution of Fritz Jahr (1927). Perspect Biol Med. 2009 Summer; 52(3) 377-80.

PHILOSOPHY OF THE WORLD AND PHILOSOPHY OF KARL LÖWITH AS A PRECURSOR AND INCENTIVE TO THE IDEA OF INTEGRATIVE BIOETHICS*

Marija Selak[1]

Introduction. The notion of the world and the philosophy of the world

"We speak about the world in philosophy, as in the common discourse, on many different ways, implying under that notion variety of things. It is accustomed to use expressions such as objective world and subjective world, interior world and exterior world, mental world and the world of objects, macro-world and micro-world, living world and the world of artificial nature, realistic world and the world of fantasy, the world of art and the world of science, the world of good and the world of evil, old world and new world, developed world and poor world, existing world and possible world, etc. If we also say that we all live in the same world or that every one of us has his own world, we can see that the boundaries of the notion of the world are elastic. Once it can be a whole *Kosmos* but, on the other hand, looking in the framework of these astronomic dimensions, man-particle has his own inner world. By all that something that should be the solid groundwork of our existence (based on which we should be able to define our place in the world) is being distributed on numerous significations and avoiding unique meaning."[2]

As in the Greek comprehension the question about the *Kosmos* was suppressed by the question about the Being, accordingly, by further ontologisation, the notion of the world became subordinated to the notion of the Being. Therefore, the crucial idea of the philosophy of the world is to reverse this relation by understanding the world as a necessary groundwork on which epistemological relation can be established. Cognition is not done by penetration (penetration of a subject into the object or vice versa), it is done by construction, equally it is not revealing of 'it is' by which only reduced notion of the world can be expressed. With this the impossibility of

---

* The paper was presented at 1st International Conference „Fritz Jahr and the Foundations of European Bioethics", Rijeka/Opatija, March 11-12, 2011, and published in Selak, M (2011) *Jahr* 2(4) 525-532.

[1] Marija Selak, marija.selak@gmail.com, University of Zagreb - Faculty of Humanities and Social Sciences, Department of Philosophy Ivana Lučića 3, HR-10000 Zagreb, Croatia.

[2] Radman Z (1987) Article "Simboli i svjetovi. O nekim aspektima filozofije E. Cassirera i N. Goodmana", in: Filozofska istraživanja 20, 155-166. (translation M. S.)

metaphysics and ontology to comprehend the notion of the world[3] in its comprehensive character is expressed and philosophy of the world is constructed to correct hierarchical inversion in which the notion of the world came with the notion of the Being.[4] Therefore, the Augsburg-Zagreb discussions (1988, 1990, and 1993) took the notion of the world in the centre of philosophical thinking. The comprehension of the notion of the world in Augsburg-Zagreb discussions was reasoned in German idealism, but also within contemporary philosophy. Although, as above mentioned quotation states, it is very hard, or almost impossible, to define the notion of the world and accordingly to establish any kind of division that is covering all spectrum of meanings, in this paper two comprehensions of the notion of the world are of exceptional importance. First is anthropocentric understanding of the notion of the world and second is cosmological understanding of the notion of the world.

Antropocentric and cosmological understanding of the notion of the world

Zdravko Radman in his paper *Symbols and Worlds. About some aspects of philosophy of Ernst Cassirer and Nelson Goodman*[5], which is one of proceedings of Augsburg-Zagreb discussions, explains when the notion of the world started to be understood in philosophy as man's creation. When traditional philosophy asked the question what is the world it perceived the world by itself and its essence independent from conditions and possibilities by which it exists for us. Not before the appearance of Kant this neutral and indifferent world vanishes and basic becomes the question how is the cognition of the world possible. In the framework of this question the well known turn will happen: the appearance of the world becomes dependent on

[3] Milan Kangrga in his philosophy also denies possibility of metaphysics to comprehend the notion of the world. The reasoning of the world as something that is given and therefore can not be changed, understanding of the world as a stiff object that is independent of subject in active (creative) sense, and only depended in theoretical (penetrating) sense Kangrga is equalising with metaphysical (ontological, which is the same for Kangrga) understanding of the world as something that is given and which finds its accomplishment in science:
„Metaphysics is here (as the one that is non historical) being presented as a positivism and a historical confrontation between positivism and metaphysics is only an illusion and epochal confusion. The only thing that is happening here is consequent finalisation of important presumptions of metaphysics done by positivism and particular sciences. Therefore science as science is just a finalisation of metaphysics." (Kangrga M (1975) Čovjek i svijet : povijesni svijet i njegova mogućnost. Zagreb: Razlog, Zagreb, p. 26. (translation M. S.))

[4] Čović A (1987) Article "Uz temu", in: Filozofska istraživanja 20, p. 3.

[5] Radman Z (1987) Article "Simboli i svjetovi. O nekim aspektima filozofije E. Cassirera i N. Goodmana", in: Filozofska istraživanja 20 (translation M. S.)

conscious, or, as Kant stated, objects must adjust to the capacity of our cognition. By that the cognitive subject was given the creative role by which the reality became dependent on subject's creative potential.[6]

"The world that is being discussed here is overcoming the experience of senses, and also the boundaries of environment. World characterised as human is becoming real only when transcendence of what is directly present is done. That is done not by interaction with the fragments of nature with which the world is coming into touch. It is done by coming into touch with symbolic forms which have made this transcendence possible. That kind of world is a product of our creative practice which is being realised in symbolical languages and is representing a result of cognitive effort of these actions."[7]

No matter are we talking about the notion of the world in relation with art, symbol or technique, practice or production as human essence, or, as another example, self-consciousness as a scientific fundament: "According to philosophical believes of The Modern Age the fact that theexistence of self-consciousness is obvious must represent absolute fundamen of knowledge, a critical instance, constitutive substrate on which all sciences must built or measure their results."[8]

We are discoursing the world which, after Kant, is by human.[9] That kind of understanding of the notion of the world will be defined as anthropocentric. On the other hand we have cosmological understanding of the notion of the world.

"The world is eternal possibility and actuality. It is actualising its eternal possibility in determinative possibility of actuality... The world is in its beginning unchanged and unchangeable. The only thing that is being changed is worlds as historical realisations of this original experience of the world in its beginning of its possibility. Which of this world has been or will be close to this actual world depends on how much did one of these worlds overcome the idea of itself, its inner fixation."[10]

---

[6] For further research see ibid. p.155-166.

[7] Radman Z (1990) Article "'Korijenske metafore' i spoznaja svijeta", in: Filozofska istraživanja 38-39, 1382. (translation M. S.)

[8] Polšek D (1987) Article "'Diskonekcija svijeta' – nužni korak fenomenološkog zasnivanja ontologije?", in Filozofska istraživanja 20, 23. (translation M. S.)

[9] Milan Kangrga in his philosophy also comprehends historical understanding of the world and is concluding that the world is a modern notion that has its historical origin in the France revolution and his philosophical origin in Kant's *Critique of pure reason*.

[10] Pavić Ž (1987) Article "Povijesnost i izvanpovijesnost svijeta", in: Filozofska istraživanja 20, 50.

Every philosophical cosmology, doctrine about the world, relies on the basis of one definition of the world in a sense of some coherence of totality of the being, hierarchically articulated according to the order of a range of Being.[11]

"Opposed to individual frames-notions which are being accomplished in the framework of physiological mechanisms, the picture (or the notion) of the world is transcendenting physiological mechanisms and is being represented as a kind of an over-shape which is being formatted on a level of totality of life experience. Having in mind that the totality of life experience has a universal character of transcendental value – it is avoiding time-space quantification. That means that neither the notion of totality of life experience nor the notion of the world are forms, although they are subordinated to the laws of forming. They are over-forms, in a sense that they are making possible and comprehend every concrete forming." [12]

Interesting example of understanding the notion of the world as *Kosmos* is Eugen Fink's comprehension of the world as Cosmo-ontology. For him world is the place of all places, time of all times and if being would originally be the world, it wouldn't be in its substance joint but un-joint. Also, as Fink emphasises, the notion of the world has the primacy over the notion of the Being because it is, as a part of a concrete life, directly present. In this, as he calls it, confrontation of the world, human, which is positioned between the sky and the ground is a co-fighter, co-player, he is the mediator, he is *ens cosmologicum.* In original experience of the world implicit lies new interpretation of the Being where the substance of the Being finds its place and time in the circulation of the world.

Karl Löwith and philosophy of the world

This cosmological comprehension of the notion of the world can also find reliance in philosophy of Karl Löwith[13]. Löwith sees history as a history of men's falling into anthropocentrism: „It is obvious that Löwith was disturbed by the philosophy of history, eschatology, historical conciseness which

[11] For further research see: Barbarić D (1990) Article "Igra svijeta. Uz Finkov pokušaj novog određenja pojma svijeta", in: Filozofska istraživanja 38-39, 1303.

[12] Užarević J (1987): Article "Svijet i oblik", in: Filozofska istraživanja 20, 120. (translation M. S.)

[13] Karl Löwith (January 9, 1897 – May 26, 1973) was a German philosopher and a student of Heidegger. His most well known works include two books: *From Hegel to Nietzsche,* which describes the decline of German classical philosophy, and *Meaning in History,* which discusses the problematic relationship between theology and history.

penetrated into the core of scientific (in terms of natural sciences) reasoning."[14]

What he claims for is the return of ancient Greek understanding of origin. He sees history as oblivion of nature in ancient understanding of *physis*. He understands the world opposed to anthropocentric orientation of philosophy, the one which represents the apology of subject, subjectivness which finishes in western rationalism in a definition of a man as a being with reason. What he opposes to this understanding of the world is the idea that always was and will be one world for a man and that is the world of nature, ever existing catholically world in which man by nature lives and dies.

"Löwith is trying to persuade us that man is above all *homo naturalis*, he is the nature, he has it as man and his nature is therefore human from the beginning. It is logical to say that this idea has certain truth inside itself which is starting to be of a providential importance in the time when we are questioning ourselves about the conditions and possibilities of one new, ecological civilisation, whose principles is expressing in a multitude of individual motives and differences precisely Karl Löwith."[15]

To explain and therefore to reveal or, it is maybe better to say, to critic contemporary scientific (in terms of natural sciences) reasoning Karl Löwith questions philosophy of history and takes the notion of the world as a central issue of his philosophy.[16] According to Löwith world can be comprehended only in relation with God and man, while the world history is just a manifestation of changes in that relation. Karl Löwith divides stages in world history on theism, deism and atheism.[17] In the God-world-man trinity hierarchy is not propound, but the world history appears as result of hierarchical changes in that relation. First stage of world history is ancient world of Greek *cosmotheology*, where God is in the world, meaning he can't be separated form the world. God and world are one, while the man himself is subordinated to the world and therefore God. In the second, *antropotheological stage*, God is personalised and placed above the world,

---

[14] Löwith K (1990) Svjetska povijest i događanje spasa. Zagreb/Sarajevo: August Cesarec/Svjetlost, p. 15.

[15] Ibid. p. 21.

[16] What Löwith sees as doubtful are dramatic divisions of the world of nature and the world of spirit (history). Although it could look as the world of history is the only world, more powerful and more meaningful than the world of nature which is by itself absolutely self-relevant (od quod substat), the one that moves from itself, we are staying a part of nature even when we are changing it into the dangerous artificial world through science and technique. (Ibid., p. 11-12.) (translation M. S.)

[17] For further research see: Löwith K (1967) Gott, Mensch und Welt in der Methaphysik von Decartes bis zu Nietzsche. Göttingen: Vandenhoeck & Ruprecht.

while the world is redundant and seen as an obstacle in men-God relationship, so it could be said that the world is now subordinated to God and man. In the third, *anthropocosmological*[18] *stage*, God is pushed out of the God-world-man trinity, argues Löwith. Although this is a stage of atheism, taking into consideration the rise of man's power of creation (i.e. penetrating and being able to change the nature of all living beings including himself), God is now inside a man. This is not in the harmonically sense as God pervaded all world in the *anthropotheological stage*, but in sense that man deems himself God:

"Ancient times believed in prophecy, the Church believed in predestination while the modern man, unless he is superstitious, does believe neither in destiny nor in providence. He believes he can alone create future. From big conceptions of ancient times and Christianity – cyclical moving and eschatological accomplishment – we are now coming to the turning point in the history of the world in which man appears on the horizon as soon the only creator of himself and therefore the future. "[19]

It can be seen that Karl Löwith's epochs were framed in a sense of a history of oblivion, estrangement form primordial history, from Greek-Roman notion of *Kosmos* and nature. History of nature and physical world are only been metaphorically spoken about. They are included in one philosophy of world history, in a perspective of one eschatology – history of salvation. What happens by that is something that is fateful for all history and also for present. That is ignoring and unexponentiation of the world of nature, its mystery and order, the world which is not mans act and artificial creation.[20]

In the contemporary *anthropotheological* stage, scientific (in terms of natural sciences) understanding and consequentially exploitation of the world is dominating. The reason for that lies in the fact that the world is perceived as man's creation, something that can and should be empirically explained. On the other hand, cosmological understanding of the notion of the world can be seen as encouragement to contemplate the world otherwise:

---

[18] "Anthropological attribute which philosophy is accepting in the third stage should not be related to the philosophical anthropology or seen as her predominance. What is being demonstrated here is foundation of the whole philosophy on human who is establishing his world and is becoming its subject, in a Heidegger's understanding of the Modern Age metaphysics as 'ontology of subject'. But to accomplish symmetry in denoting and linguistic equivalence term anthropocosmology will be used for the third stage. By doing that, the logic of Löwith's thought will be followed." (Čović A (1993) Article "Aporije Löwithova povratka 'prirodnom svijetu'", in: Filozofska istraživanja 51, 955-965.) (translation M. S.)

[19] Löwith K (1990) Svjetska povijest i događanje spasa. Zagreb/Sarajevo: August Cesarec/Svjetlost, p. 36.

[20] Ibid., p. 13-14.

„By placing analogy with Greek or Roman cosmology we are just pointing out one philosophical/historical benchmark which newer got real chance.“[21]

Bioethics and philosophy of the world

This cosomological understanding of the world for which we can find incentive in philosophy of the world on one hand and which is being used in philosophy of Karl Löwith as an alternative to anthropocentric reasoning of the world, also shares some similarities with contemporary 'movement' of integrative bioethics.

To explain this statement furthermore, it is essential to see when bioethics appears. Bioethics rises in the climax of contemporary *anthropocosmological stage* and precisely in men's encounter with himself in the field of medicine[22]. Gradually bioethical horizon has been widened, from medical ethics and principalism to ethical pluralism and interdisciplinary approach. Finally, bioethics extends its framework to the notion of a life as a whole with corresponding pluriperspectivism and integrative approach.[23]

Both, integrative approach and cosmological understanding of the notion of the world can be seen as an alternative to anthropocentric tradition of modern philosophy, meaning that human is here comprehended and positioned only in co-existence. As a result of integrative approach we can see rise of bioethical sensibility[24] in a concern for nonhuman living beings and responsability for maintenance of all conditions of preserving life in general. Also, methodologically speaking in ancient thinking mesh of the man, world and god was implicated, and in contemporary scientific rationality we still can not find an obverse to it. But we can see modern reflection of ancient non-divided rationality in pluriperspective approach of integrative bioethics.

---

[21] Ibid., p. 15.

[22] Here it is referred on Seattle case in 1962 where it was realised that man can not be simply seen as a scientific fact or mathematically as a count of defined parts – meaning that scientists were not able to decide who shall live and who should die, therefore we could say that the catholicity of a man was awaked.

[23] For further research see: Jurić H, Zagorac I (2008) Article: "Bioetika u Hrvatskoj", in: Filozofska istraživanja 111, 601-611.

[24] "In bioethical discourse the notion of sensibility is describing emphasised delicateness towards environment. It is implying morally reflected referring towards human existence and the existence of other living beings. It is containing enlightened affection towards those who are weak and unprotected, but it is also opening new horizons in 'old' notions of respect, love and sympathy. Bioethical sensibility is appearing as an important integral element of humanity with a high level of respect, finding again its place alongside cold rationality." (Ibid., p. 608). (translation M. S.)

Since integrative reasoning[25] was developed as a resistance to hierarchy of truth and knowledge and as a resistance to exclusiveness of scientific understanding of the world, it can be said that, with its pluriperspective approach which is including non scientific views on contemporary problems, it is also assimilating catholically character.

While in the framework of philosophy of the world we start from the notion of the world and by deduction come to the man as his integral part, in the framework of bioethics, which through integrative understanding widens its horizon from man (medical ethics), form *bios* to *Kosmos*, by induction we come to the notion of the world. Remembering that the aim of integrative bioethics approach is to give a specific kind of knowledge which is supposed to provide orientation in the world, it can be said that integrative bioethics also takes the notion of the world as a central notion in its reasoning and with broadening its activity on life as a whole catholicity as its basic quality. Therefore, we can say that bioethics enforced itself as a 'natural' response to the philosophy of the world. If we see history as a history of man's falling into anthropocentrism (K. Löwith) and take into consideration the appearance of bioethics as an attempt of overstepping narrow anthropocentric perspective the question rises is that kind of integrative bioethics approach leading us on a threshold of a new epoch where, in Löwiths words, *Kosmos* is once more taking over the priority over *Anthropos*.

---

[25] For further research see: Čović A (2006) Article: "Pluralizam i pluriperspektivizam", in: Filozofska istraživanja 101, p. 7-12.

HANS JONAS' INTEGRATIVE PHILOSOPHY OF LIFE AS A FOOTHOLD FOR INTEGRATIVE BIOETHICS*

Hrvoje Jurić[1]

European bioethics and integrative bioethics

If we extract some basic features of "European bioethics" and "integrative bioethics", in order to consider what they are and what they ought to be, we could realize that these two concepts could be used almost synonymously. Namely, one of the basic features of European bioethics is that it is in a permanent dialogical relationship with or even rooted in philosophy, i.e. European philosophical tradition. In other words, if we would try to distinguish European bioethics from "mainstream bioethics" as an Anglo-American "product", it would be easiest to do this by following the line of productive interlacement of bioethical reflection with philosophy. There are two tasks that arise from this fact. First of them could be called the *philosophization of bioethics*. It does not mean that philosophy has "appropriated" bioethics, that bioethics is now to be treated as a mere sub-discipline of philosophical ethics, which would benefit neither philosophy nor bioethics. "Philosophization of bioethics" refers to the activation of the potentials of the Euro-continental ethical or philosophical thought within the bioethical framework, for bioethics to be able to fulfill its original ("Potterian", according to Van Rensselaer Potter, "the father of bioethics") mission that cannot be achieved exclusively with the help of an approach that rests on a reduced understanding of bioethics either in terms of its problems and topics or in terms of its methodology. The second task could be called the *bioethicization of philosophy*. This means bioethically reading the leading authors and works of the Euro-continental philosophical tradition for the purpose of identifying both the footholds of establishing and developing dialogue between bioethics and philosophy, and the incentives to reflect on bioethical problems in partnership. Both tasks imply the demand for *Europeanization of bioethics*, as well as the *wider concept of bioethics* which we call *integrative bioethics*. Integrative bioethics could be defined as an interdisciplinary and pluriperspective consideration of moral dilemmas arising from the new techno-scientific possibilities of manipulation with the life: human and non-human life, living beings and life conditions; in other

---

* This is the revised version of the paper presented at 1st International Conference „Fritz Jahr and the Foundations of European Bioethics“, Rijeka/Opatija, March 11-12, 2011, and published in Juric, H (2011) *Jahr* 2(4) 511-520.

[1] Hrvoje Jurić, hjuric@ffzg.hr, University of Zagreb, Faculty of Humanities and Social Sciences, Department of Philosophy, I. Lučića 3, 10 000 Zagreb, Croatia.

words, life as a whole and each of its parts, life in all its forms, shapes, degrees, stages and manifestations (see Jurić, 2005).

For this purpose, integrative bioethics emphasizes a need for the widened and deepened consideration of the very notion of *life*, which should be taken in account when speaking about ethical dimensions of manipulation with the life at different levels. The same demand and approach – which is developed before or parallel to the "official emergence" of bioethics in 1960s and 1970s – could be traced in the works of "founding fathers" of European bioethics such as Fritz Jahr and Albert Schweitzer, as well as in the works of very important European authors such as Georg Picht, Klaus Michael Meyer-Abich, and especially German-American philosopher Hans Jonas (1903-1993).

Before focusing on the main topic of this article – Hans Jonas' philosophy of life – I will say few words on the very notion of life, i.e. different approaches to it.

Notion of life and different approaches to it

What do we think when we say "life"? Do we all think the same? Hardly. Therefore, one of the main tasks of bioethics is to re-think itself and its fundamental notion. The notion of life should not be used undifferentiatedly, especially in ethical discussions.

Hans Werner Ingensiep, in his article "Was ist Leben?" (What Is Life?), states the following: "It is a fact that ambiguous and diverse notions of life are in use, be it on the purely descriptive or the evaluative level (…). There are divided discourses on life within these discussion areas. Meanings are often ripped out from context and transferred from one area to another, almost drowned between different metaphors, so that the philosophers (…) hold, with good reason, that the notion of life is 'unclear'. But in general we can live with that. Only within the bioethics we are dealing with a particular problem situation, if, for example, the terms 'life' and 'person' should be distinguished. However, the 'persons' also are living beings. Not only because of that, must the biological notion of life be connectable to bioethics. As the examples make clear, the question 'What is life?', which was posed more theoretically up to now, is highly controversial. Although we are dealing with different issues of life, we need the notion of life as a bridge between the separate discussion fields, as well as its integration and communication power, especially if there should be a continuous and constructive dialogue about the 'life' between scientists from natural sciences and humanities." (Ingensiep, 2002: 92-93)

The complexity of life, as well as complexity of dilemmas connected to it, requires therefore an equally complex approach. Discourse on life demands both inclusion of different approaches and dialogical mediation between them, because neither the multidimensionality nor the entirety of life can be embraced from the reductionist point of view.

There are three main traditional approaches to the phenomenon of life and the concept of life: *natural-scientific* approach, *philosophical* approach, and *theological* approach. Each of those approaches could be considered as reductionist if it is exclusive or practiced without taking the other two into consideration.

Natural-scientific approach relies primarily on a biochemical understanding of life, understanding of life as a chemical and physiological structure or process, so that the natural-scientific definition of life offers only the list of separate characteristics that differs living beings and systems from the non-living ones (e.g. organization, metabolism, growth, irritability, adaptation, reproduction, etc.). Due to these features, this approach could also be called a *materialistic* and *mechanistic* and it could be regarded as a reductionist, because it captures only what can be explored by the methods and apparatus of natural sciences, and expressed in terms of natural sciences. In other words, it captures only what fits in predetermined scientific and methodological framework of natural sciences, and thus can be objectified, explored and expressed exactly.

The second approach is the *philosophical approach*. I am not referring to various forms of "natural philosophy" or the philosophical trends such as vitalism or Diltheyian philosophy of life, but the approach which is dominant in modern philosophy and has subjectivist and existentialist features. It is also, in a sense, reductionist. Namely, such a philosophy deals with anthropo-logic of life, plunges into the depths of the human and insists on the specificities resulting from the rational structure of the human subject, and thus in a greater or lesser extent ignores human rootedness in the world of the living.

The third is the *theological approach*. The main features of this conception of life can be read out from the concept of "sanctity of life", whose primary horizon is the religious worldview and different theological conceptions. Life is creation of divine instance, the godlike creation and reflection of god's existence, so that it implies the dignity of life, which, of course, primarily concerns human life. Therefore we can say that the biocentrism specific to the doctrine of the "sanctity of life" – witnessed by the religious myths and legends, as well as theological speculations – is on the one hand based on postulates of theocentrism, while it is on the other hand anthropocentrically founded. In this case we can also speak about a kind of reductionism, since the theological approach also abstracts from non-human life and the natural facts related to human existence. (Of course, the above made division and definitions are, like every other division and definition, imperfect. Namely, there are also many examples of "reflexive natural science", "holistic philosophy" or "biophilic theology".)

Anyway, a kind of reconciliation of these three approaches, or transcending their mutual confrontations and imperfections, we could find in the philosophy of Hans Jonas, which tried to embrace wide spectrum of topics relevant for the discussion on the notion of life, as well as bio-ethical problems.

Hans Jonas' philosophy of life

When talking about Jonas' contribution to bioethical discussions and the very foundation of bioethics, it is expected to take into consideration primarily his *ethics of responsibility*, developed in the late phase of his life and formulated in the philosophical bestseller *Das Prinzip Verantwortung* (or, in English translation, *The Imperative of Responsibility*),[2] which intention was to develop an "ethics for the technological age". Nevertheless, Jonas' contribution to bioethics should be explored in other phases of his work, as well. In his long-term *research of late-ancient and early-Christian Gnostic religion and thought* we can find the very roots of world-views that have marked our Western tradition, including Western concepts of life and relationship to it.[3] But it seems that, in the context of bioethical discussion, the most important part of his philosophy is his attempt to establish the *philosophical biology* as a new philosophy of the human, nature and life. In his work *The Phenomenon of Life* (or, in German translation, *Organismus*

---

[2] Hans Jonas, *Das Prinzip Verantwortung. Versuch einer Ethik für die technologische Zivilisation*, Suhrkamp, Frankfurt a/M 1979; Hans Jonas, *The Imperative of Responsibility. In Search of Ethics for the Technological Age*, The University of Chicago Press, Chicago 1984. Here I am referring to the latter (Jonas, 1984).

[3] See for example: Hans Jonas, *Gnosis und spätantiker Geist*, Part 1: *Die mythologische Gnosis*, Vandenhoeck & Ruprecht, Göttingen 1934, Part 2: *Von der Mythologie zur mystischen Philosophie*, Vandenhoeck & Ruprecht, Göttingen 1954; Hans Jonas, "Gnosis, Existenzialismus und Nihilismus", in: Hans Jonas, *Zwischen Nichts und Ewigkeit. Drei Aufsätze zur Lehre vom Menschen*, Vandenhoeck & Ruprecht, Göttingen 1987, pp. 5-25; Hans Jonas, *The Gnostic Religion. The Message of the Alien God and the Beginnings of Christianity*, Beacon Press, Boston 1958.

und Freiheit),[4] Jonas tried to establish and develop "philosophical biology" as an *integrative philosophy of life* in which centre stands ethically connotated philosophy of nature, based both on the results of contemporary natural sciences and theological speculations.

Jonas' intention was to establish the *philosophical biology* as an "'existential' reading of biological facts", or as "a new reading of biological record" (Jonas, 2001: xxiii), i.e. a new *philosophical* reading of *biological* record. Its main scope is to abolish the "artificial split between spheres of the external and the internal, body and mind, the nature and the human" (Jonas, 1991: 105). This kind of dualism seems to be unsustainable, first and foremost, in the case of *organism*. Organism "is the whole not only in the sense of functioning (…), but also in the sense of body-mind unity", which means that "the inner aspect or the subjectivity of organism is inevitable for the biological understanding as equally as the objectivity of organism" (Jonas, 1991: 105-106). His anti-dualistic enterprise has been made in order "to break through the anthropocentric confines of idealist and existentialist philosophy as well as through materialist confines of natural science" (Jonas, 2001: xxiii). Of course, Jonas promotes, in the same time, the dialogue between natural sciences and humanities, which are today complementary to each other, because none of them can comprise the phenomenon of life in its entirety.

Therefore, a new philosophy of life should embrace in its subject-field both "philosophy of the organism" and "philosophy of mind". According to Jonas, philosophy of the organism starts with the thesis that "the organic even in its lowest forms prefigures mind", while on the other hand, philosophy of mind starts with the thesis that "mind even on its highest reaches remains part of the organic" (Jonas, 2001: 1).

Jonas thinks that everything we find in human has its "rudimentary traces in even the most primitive forms of life" (Jonas, 2001: xxiii). Key concept of Jonas' teleological philosophy of nature is the concept of *freedom*. Since the mind is, from the beginning, prefigured in the organic, it is also the case with the freedom. Freedom exists already on the basic level of organic existence, i.e. in the primal metabolism. Moreover, as Jonas says, metabolism itself is the "first form of freedom", which means that the principle of freedom can be found already in the "dark stirrings of primeval organic substance" (Jonas, 2001: 3).

---

[4] Hans Jonas, *The Phenomenon of Life. Toward a Philosophical Biology*, Harper & Row, New York 1966; Hans Jonas, *Organismus und Freiheit. Ansätze zu einer philosophischen Biologie*, Vandenhoeck & Ruprecht, Göttingen 1973. Here I am referring to a later English edition (Jonas, 2001).

The other key concept of Jonas' philosophy of life is *mediacy* in the relation of organism to environment, i.e. the distance between living being and its environment. It describes the progressive scale of the organic on whose peak stands the human. In addition to the principle of metabolism and the more advanced aspects of mediacy in the relation of living being to environment, in humans there is a wide range of specific human characteristics, in short: the mind. However, this does not imply that the human should be observed in the "metaphysical isolation" from the rest of the living world.

Jonas clearly indicates the ethical implications of his doctrine of life. In that sense, the aim of his philosophy of life is the *affirmation of inherent self-purpose and value of being, life and all living beings*. It is the task which asks for foundation of an explicitly non-anthropocentric ethics, biocentric ethics, or bio-ethics. Fritz Jahr's "bio-ethical demand" or "bio-ethical imperative" strongly affirms such a standpoint: "Respect every living being in principle as an end in itself and treat it, if possible, as such!" (Jahr, 2010a: 4; 2010c: 23), and "Respect every living being, including animals, as an end in itself and treat it, if possible, as such!" (Jahr, 2010b: 8).

Regarding Jonas, we can say that such an ethical conception has been presented in Jonas' *ethics of responsibility*, but it has been announced already in his *philosophical biology*. It implies the following: all living beings and the nature, too, are not only the objects of our moral duties, which means that they deserve moral consideration not only on the basis of "human mercifulness". Our ethical "Ought" arises from the ontological "Is" of the living beings. The nature and the life establish the purposes and the values by themselves. Our moral duty is to recognize and to respect those purposes and values. The ethical formulation of it is – *the imperative of responsibility*. Whatever (or whoever) is exposed to our *power* should be embraced by our *responsibility*, because it is *entrusted* to us. It is the "ontological responsibility". The life itself delivers the purposes and values, which should be only recognized and respected by the human, because the human is "the executor of a trust which only he can see, but did not create" (Jonas, 2001: 283).

Gertrude Hirsch Hadorn briefly presents what the "ontological foundation of ethics" and "ontological ethics" actually means. There is the threefold connection between the *being* and *responsibility*: ability to responsibility as a characteristics of human way of being; being of the human as an object of responsibility; the whole of the being as an instance of responsibility, as well as an instance which makes the responsibility inevitable (see Hirsch Hadorn, 2003: 109-110).

The newly gained insight into the vulnerability of nature results with the new ethical demand. Jonas speaks about the "right of the nature", which is far more than anthropocentric and utilitarian interest of the human and humankind (see Jonas, 1984: 8). Unlike in earlier epochs of humankind, we discover today that the nature is totally exposed to our (techno-scientific) power. Therefore, we should act according to this insight, i.e. we should respect nature's right to be sustained in this state and promote the duty to ensure a future, which both also imply the future existence of the human and humankind. Jonas' categorical imperative – "Act so that the effects of your action are compatible with the permanence of genuine human life" (Jonas, 1984: 11) – is not an anthropocentric one. However, "no previous ethics", says Jonas, "has prepared us for such a role of stewardship" (Jonas, 1984: 8). That is the reason why we need a new ethics, which would be based on the principle of responsibility. Neither traditional scientific worldview nor traditional ethics can tell us what we should do in order to protect the nature as a basis of human existence, as well as future existence of life in general. Moreover, both traditional and current scientific worldview, which made the ethics extremely anthropocentric and short-sighted, "emphatically denies us all conceptual means to think of Nature as something to be honored" (Jonas, 1984: 8). Jonas says, consequently, that "we should keep ourselves open to the thought that natural science may not tell the whole story about Nature" (Jonas, 1984: 8). Similar statement we can find in *The Notion of Nature and Its History* (Der Begriff der Natur und seine Geschichte) by Georg Picht: "The modern natural science destroys the nature" (Picht, 1993: 12), so that "the science which destroys the nature cannot be the true knowledge on nature" (Picht, 1993: 15).

Besides the above sketched biological-scientific and ethical-philosophical way of thinking about the life, there is the third perspective which should be taken into account – the metaphysical and theological one.

The question of life is not only the question about the *development* of life, about its evolution. It is also the question about its *beginning*, which comprises both its biological evolution and successive ontological revolutions. However, there is also a question about the "first cause", which could be helpful while answering questions on purpose and meaning of the great cosmic adventure of life.

Trying to come to the "first cause", as well as to the "final answers" in this field, Jonas included into his philosophy of life the "metaphysical speculations", too, which are certainly inspired by his earlier studies of Gnosticism. Sometimes he did it by using strictly philosophical categories, sometimes he used theological concepts, but sometimes he dared to propose his own cosmogony or even theogony. Jonas explains his "metaphysical speculations" as "an ontological revisioning, fulfilling the concept of 'matter'", i.e. "a meta-physics of the world-substance" (Jonas, 1988: 20).

On the other hand, we cannot say that Jonas was the "creationist", although his theory is close to Judeo-Christian creationism, neither can we say that he was "evolutionist", although his theory is close to the Darwinist evolutionism. His philosophy of life presupposes the spontaneity of life and nature, but he does not deny the concept of "mind in nature" or the "prime mover". However, it should be noted that this weakens both the metaphysical-dogmatic images of "divine plan" and "god-mathematician", and the concept of "pointless and meaningless evolution". Anyhow, there are strong ethical demands arising from this viewpoint, and Jonas develops them following the logic of responsibility.

Conclusion

In brief, Jonas' "philosophical biology" could be explained as an attempt of both "biologization of philosophy" and "philosophization of biology", including "ethicization of the question of life". In its centre stands the ethically connotated philosophy of human, nature and life, which respects the results of natural sciences, but also opens the door to the metaphysical speculations, that means theological or even mythological speculations.

If we follow Jonas, we could show how the synergy of evolutionist-biological, teleological-philosophical and religious-theological approach contributes to answering the question of life. Only such an integratively structured notion of life can be the starting point of *integrative ethics of life*, or *integrative bioethics*, which implies respect and responsibility for the nature and life in general. Only such an integrative philosophy and ethics of life can show us what pluriperspectivity and integrativity in bioethics actually mean, because it is far from any reductionism and tries to include different scientific and non-scientific perspectives, offering at the same time a platform for their dialogical mediation. Fritz Jahr's approach (see, e.g., Jahr, 2010a) which 'naturally' connected the knowledge of natural sciences and psychology with the knowledge of European philosophy, as well as religion and theology (both Christianity and Buddhism), seems to be a perfect example of integrative drive in (re)thinking the question of life.

By emphasizing philosophical approach to the phenomenon of life, I did not want to devaluate other approaches. The point is that the philosophical approach to the life – which we can find in Jonas' works, too – is characterized by exceptional integrativity. On one hand, it activates all the potentials of philosophy (from metaphysics, ontology and anthropology, over ethics, to cultural and political philosophy), while on the other hand, it transcends the boundaries of philosophy as a discipline, because it enables (or even demands) broadening of the perspective on the natural sciences and theology. Philosophy – as a model of integrative thought – presupposes an attempt to embrace and dialogically intermediate empirical-scientific, rational-speculative and psychological-emotional dimensions of the approach to the world and life. This means that philosophy and, especially, philosophical ethics – first and foremost owing to their inherent integrative features – do play a major role in integrative bioethics, which is not in disagreement with the methodological principle of interdisciplinarity and pluriperspectivism. Anyway, neither philosophy nor any other science, neither religious nor non-religious approaches – if they are isolated from other approaches – can have the monopoly on the truth of life. The only ban which is implied by the pluriperspective approach is the ban of monoperspectivism.

"The phenomenon of life itself negates the boundaries that customarily divide our disciplines and fields", says Hans Jonas (Jonas, 2001: xxiv). It is up to us to respect or disrespect this fact. But we should keep in mind that monoperspectivism makes us short-sighted or even blind, while pluriperspectivism enables us to look at the phenomenon of life both with microscope and telescope, as well as with our inner theoretical eye, in order to approach an integrative understanding of and knowledge on life, as well as an integrative ethics of life.

Although Jonas' concept of integrative ethics of life can be viewed as one of the most important footholds for integrative bioethics, we should note that Fritz Jahr's *bio-ethics* (see Jahr 2010a, 2010b, 2010c) represents one of the earliest formulations of this idea. Therefore, comparative reading of Jahr's texts and the texts of other integratively disposed authors (such as Jonas) seems to be an important task of integrative bioethics, because it could show us that a different approach to the question of life is not only needed, but also possible.

References

1. Hirsch Hadorn G (2003) Verantwortungsbegriff und kategorischer Imperativ der Zukunftsethik von Hans Jonas, in: Hans Jonas – von der Gnosisforschung zur Verantwortungsethik, ed. by W. E. Müller. Stuttgart: Kohlhammer, 101-118.
2. Ingensiep HW (2002) Was ist Leben? – Grundfragen der Biophilosophie, in: Jahrbuch Ökologie 2002. Berlin: Wissenschaftszentrum Berlin, 92-103.
3. Jahr F (2010a) Bio-Ethics, in: Selected Essays in Bioethics 1927-1934, ed. by H. M. Sass. Bochum: Zentrum für Medizinische Ethik, 1-4. See also in this volume.

4. Jahr F (2010b) The Relationship of Animal Protection and Ethics, in: Selected Essays in Bioethics 1927-1934, ed. by H.-M. Sass.-: Zentrum für Medizinische Ethik, 5-8. See also in this volume.
5. Jahr F (2010c) Three Studies on the Fifth Commandment, in: Selected Essays in Bioethics 1927-1934, ed. by H.-M. Sass. Bochum: Zentrum für Medizinische Ethik, 19-23. See also in this volume.
6. Jonas H (1984) The Imperative of Responsibility: In Search of Ethics for the Technological Age. Chicago: The University of Chicago Press.
7. Jonas H (1988) Materie, Geist und Schöpfung: Kosmologischer Befund und kosmogonische Vermutung. Frankfurt a/M: Suhrkamp.
8. Jonas H (1991) Erkenntnis und Verantwortung: Gespräch mit Ingo Hermann in der Reihe "Zeugen des Jahrhunderts". Göttingen: Lamuv.
9. Jonas H (2001) The Phenomenon of Life: Toward a Philosophical Biology. Evanston: Northwestern University Press.
10. Jurić H (2005) Stützpunkte für eine integrative Bioethik im Werk Van Rensselaer Potters, in: Integrative Bioethik / Integrative Bioethics, ed. by Ante Čović and Thomas Sören Hoffmann. Sankt Augustin: Academia Verlag, 68-92.
11. Picht G (1993) Der Begriff der Natur und seine Geschichte. Stuttgart: Klett-Cotta.

THE INTELLECTUAL "LAST WILL" OF THE FIRST GLOBAL BIOETHICIST*

Van Rensselaer Potter †

The words of greetings to the „Bioethics and Science in the New Epoch" Symposium[1]

Hello! My name is Van Rensselaer Potter II. For the past 60 years, that is, from February 1st, 1940 to the present, I have been a member of the Department of Oncology in the widely respected McArdle Laboratory for Cancer Research at the University of Wisconsin in Madison. Over hundred graduate students and post-doctoral fellows have been members of my group, engaged in cancer reserach.

And for the past 40 years I have been an amateur biological philosopher on the site with occasional publications beginning in 1962, leading to the invention of the word *bioethics* in 1970, and continuing to the present. I feel greatly honored to be invited to this Symposium on Philosophy and Bioethics here on this charming island of Cres in the Adriatic, near Rijeka.[2] I am greatful to Professor Ivan Šegota, President of the Croatian Bioethical Society, working in co-operation with Professor Ante Čović, President of the Croatian Philosophical society, for this invitation.

It was Professor Šegota's dilligence in pursuing the word *bioethics* that lead him to me and my role in coining the word and to my publications in 1970, 1971, and at intervals up to and including the year 2000. Of course, he contacted me as soon as he learned the connection and we have had a worth-while exchange since that time. I have been pleased to read some of his own

---

[1] This text was originally published in the 30th volume of *Bioetički svesci* (*Bioethical Volumes*) collection (the publisher of which is the Department of Social Sciences of the Faculty of Medicine, University of Rijeka, Croatia), both in Croatian and English, under the title *Hrvatska u grupi bioetičkih zemalja svijeta: Potterova poruka međunarodnom bioetičkom simpoziju na Cresu/Lošinju* (*Croatia in the group of bioethical countries of the world: Potter's message to the International Bioethical Symposium on Cres/Lošinj*), Rijeka, 2001. Van Rensselaer Potter had sent a video tape of his message to the „Bioethics and Science in the New Epoch" symposium to Professor Ivan Šegota, Ph.D., the co-president of the Symposium programme Committee in the Spring of 2001.
The text of the paper has been reprinted in: *Dani Frane Petrića: bioetika i znanost u novoj epohi / Petrić – platonizam – aristotelizam* [The Days of Frano Petrić: bioethics and science in the new epoch / Petrić – Platonism - Aristotlism] (Zagreb: Hrvatsko filozofsko društvo, 2001), 29-38, and in: *10. lošinjski dani bioetike / 10th Lošinj Days of Bioethics* (Zagreb: Hrvatsko filozofsko društvo, 2011), 53-62.

[2] At the time that V. R. Potter taped his message, the Symposium was meant to be held on the island of Cres (instead of Lošinj).

thoughts on bioethics and I followed with interest his great efforts aimed at the development of bioethics in Croatia, while at the same time maintaining a cordial relation with philosophy. I think that his achievements in the field of bioethics education of students at the Faculty of Medicine, University of Rijeka deserves a special praise. I am glad that Croatia also joined the community of bioethics countries of the world whose number is incresing constantly. I hope that this Symposium will give a new impetus to bioethics in Croatia and all of what we think of as the Balkans and neighbouring countries.

Now I have to emphasize that we must constantly examine what we mean by the word *bioethics*. The simplest but all-inclusive image has been captured by my friend [Michael Fox] in his book just published *Bringing Life to Ethics* with subtitle *Global Bioethics for a Humane Society*.

In a few moments you will see on the screen the tiel of my first publication introducing the word *bioethics* and defining it as *The Science of Survival* in 1970.

Listen! Today I declare that it was incorrect and completely inappropriate to call bioethics the Science of anything. I suppose I made this mistake because I am a scientist. But what would I have done with the insight of today?

Here I am influenced by Professor Ursula Goodenough, a biologist at Washington University in St. Louis, Missouri, USA. Her new book *The Sacred Depths of Nature* is deeply involved in the idea of what she refers to as Religious Naturalism. It seems clear that she is calling for a new religion based on the knowledge exemplified in the 'Sacred Depths of Nature' as presented in her book.

I noted that one program at this Symposium on the island of Cres is 'Bioethics and Science in the New Era'. Today I will call on the other pole of the two cultures by suggesting 'Bioethics and Nature-based Religion in the New Era' as a topic.

The tiel of your program does not imply that Bioethics *is* a Science but, rather, that Bioethics is needed as a much-needed monitor of 'Science in the New Era'. Perhaps Bioethics as a system of nature-based morality will serve a useful function in this role as a monitor of science, technology, and politics.
Now hear this! It came to me that bioethics *is* a model of the new religion sought by Professor Goodenough. Bioethics calls for a morality based on beliefs that we have about the relation between humans and biological systems within us and around us. Today, in 2001, I would introduce the word

bioethics in the title of the 1970 publication as *Bioethics, The Morality of Survival* – not as *The Science of Survival.*

Where in the world is there a more needed new religion or morality for survival than here in the shadow of the Balkans, a cauldron of interethnic hatred, with three major religions set off by national boundaries? And what would be more unifying and more needed than a nature-based morality called Bioethics. I say Bioethics is a system of belief because it cannot be proved by rational argument or scientific research that it is right, good, appropriate, and necessary to protect, maintain, and restore the natural environment and to do all we can to assure that future generations have the rights that we have fought for and believe in.

Global Bioethics is a world-view, a *Gestalt*, but most of all, a call for action based on Faith, Faith in the new morality called Bioethics. The question arises, can Bioethics as a new system of morality be the rallying call for a unifying political movement? Of course it can! Every political movement and every political protest is based on beliefs that can't be proved and are often based on the beliefs of particular branches of rhe parent religion. Bioethics as a political organization based on a nature-based morality with 'shall' and 'shall nots' is an idea whose time has come.

I believe that my recent remarks on the 'Bioethical Issues of the 21st Century' are quite relevant to what I have said up to this point. I look forward to any responses that may come to me on my email. I am sure I will learn from these.

Bioethical issues for the 21st century[3]

What are the bioethical problems that we face in the Year 2000 and beyond? Perhaps a look back can help our vision for the future. Please allow me to remind you of my outlook in 1970 when I responded to the disorder and environmentalism of that time.

In that year of 1970 I published two articles that introduced the word ' bioethics' and were interwoven with a dozen reports on cancer metabolism prepared with my students. The titles of those two articles and that of a related third article on bioethics are:

1. 'Bioethics, The Science of Survival' (1970);
2. 'Biocybernetics and Survival' (1970);

[3] V. R. Potter took over this part of the message from the video-tape that he had sent to professor Marcelo Palacios in Gijon (Spain) in mid June 2000.

3. 'Disorder as a Built in Component of Biological Systems: The Survival Imperative' (1971).

These three articles all refered to my book *Bioethics, Bridge to the Future*, which was already in press, appearing in January 1971.

I believe that the appearance of the word *bioethics* in those four publications was a statement of two dire conclusions:

1. Survival in the long-term future boils down to a matter of bio-ethics, not traditional ethics.
2. For the long-term future we will have to invent and develop a political bioethics.

Traditional bioethics is about interaction *between people and people*; *bioethics* is about interaction *between people and biological systems*. Political bioethics is essential for helathy political decision-making and for the creation of sound policies. In other words, action will have to be constrained and guided by biological knowledge as bioethics emerges and is finally turned into voluntary agreements or into regulatory law in specific instances.

Soon after the publications of my 3 articles in 1970 and my book in January 1971, a well-financed Georgetown University Institute for the Study of Human Reproduction added the words 'and Bioethics' to its title and sfifted the focus of bioethics inquiry. It was launched later in 1971 apparently unaware of my four publications.

In the following decades, bioethics was taken over my medical 'bioethics committees' operating at Bioethical Centers in the clinical area and dealing with problems of life and death that are still very contraversial. Today our focus in finally shifted back form clinical ethics to a brosder bioethics.

At this Congress and elsewhere bioethics is beginning to be recognized not only as a medical issue but as environmental and social issues as well.

Although bioethicists have focused largely on clinical ethics from 1970's to the present, work on environmental and other forms of bioethics has continued in some circles. In fact, within a span of two years bioethics, medical bioethics, and a congruent 'new ethic' appeared. It was in 1972 that Lester Brown began a career of fact-finding and publication that led him in 1974 to found The Worldwatch Institute, where a dedicated corps of interdisciplinary fact-finders and writers began to publish an annual book in 1984 and every year since then including the recent *State of the World 1999.*

I urge every member of every Bioethics Association to read this book. This series of reports and the additional publications that supplement them form the bedrock of the facts and intereferences that buttress the ideas that I expresses in 1970, again in 1988 in *Global Bioethics*, and in 1990 in the article ' Getting to the Year 3000'. We are all indebted to Laster Borown for providing the leadeship that makes an annual *State of The World* report available. In addition, Lester Brown has always had a concern for the ethical dimension. We need another 'Brownian? movement – one at the human level.

In 1972, before founding Worldwatch, and acting as a concerned individual, we publish *World Without Borders*. I immediately bought the book. opening the Preface, Lester Borown said he knew no one with all the credentials to write a book with as wide a scope as this one and made no such claim about himself. But then he declared that such an overview was sorely needed, and his purpose was *to provide a global* perspective for considering the *future*. It was a great beginning.

On page 361 he wrote: »Our circumstances require that we become politically active, not in partisan Republican, Democrat, Christian Democrat or Communist terms, but in terms of our self-interest and that of our children. (...) We begin with the formulation of a new ethic. At a minimum, this ethic must seek harmony between man and nature and encompass an ideology of global unification. It must seek to raise everyone above the subhuman conditions of life now prevailing for so much of mankind. (...) You and I must adopt and propagate this new ethic, translating it into political action. Our future well-being depends on how quickly the new ethic now emerging can be translated into a new, more humane set of global priorities, new global institutions, and new levels of global cooperation.«

Lester Brown's 'new ethic' was actually an environmental bioethics combined with a concern for social justice. Over the years I have learned a great deal by re-reading his message. His call for a new ethic in 1972 and subsequently has complemeted my message of bioethics from 1970 to the present.

And it complements my message here today, as I declare that *Global Bioethics must evolve into a politically energized and socially concerned Bioethics Worldwide: Global Bioethics for the 21st Centuraly Calls for Peoplecare with Healthcare and Earthcare with Animalcare.*

To understand the problems in these interwoven domains we need the facts. In putting together the world's best source of information in chapter after chapter on the *State of the World* year after year, it has now been necessary for Lester Brown to choose between political action and credibility. In preserving credibility for its assembled facts the Worldwatch Institute, in effect, places the task of mounting political action in the hands of Associations of Gloabl Bioethics, wherever they may arise.

It is significant that in 1999 the facts again led Lester Borwn to call us to forge a new ethic, actually a social bioethics, in *The State of The World 1999 (Millennial Edition*).

On page 21 of Chapter One by Brown and Christopher Flavin we find »The trends of recent years suggest that we need a new moral compass to guide us into the twenty-first century – a compass that is grounded in the principles of meeting human needs *sustainably*. Such an ethic of sustainability would be based on a concept of respect for future generations.«

Again, I need to emphazise that new moral compass has to be bioethical, and sustainability has to be reached bioethically or it will collapse.

Here I declare that as a bioethicists we see need for a *bioethical sustainability*: sustainability for whom, for *what*, and for *how long*? My answer is bioethical sustainability for diverse world-wide *populations* and *for* the *biosphere* and a *decent society for* the *long-term*. For the *next hundred years* we need a *political* bioethics with a sense of *urgency*. I recall that on the same page in Chapter One that I mentioned earlier, Brown and Flavincall attention to that need, remarking „It is difficult to overstate the urgency of reversing the trends of environmental deterioration", and they point out that this is really a survival issue commenting on the fall of earlier civilizations that were oblivious of what was happening. Political action for long-term social survival is a bioethical mandate. We can hope for a bioethical capitalism while we wonder whether, if it is possible, it will evolve soon enough.

Back in 1932 when I cast my first vote for Franklin D. Roosevelt for the presidency of the U.S., we called for political leadership to end the Great Depression. And when World War II was forced upon us we again needed leadership. In a Fireside Chat on February 23, 1942, Rossevelt declared „Never before have we had so little time in which to do so much", and yet, with his leadership, we got it done. I wish that he could be here today *because* „Never before have we had so little time in which to do so much".

Today we have a different kind of urgency. We need political action. We need to demand that our leadership achieve a humane Global Bioethica aimed at global sustainability for the long-term.

So finally I want to declare that as we face the future we have two possibilities. *The Third Millennium will be either The Age of Global Bioethics or The Age of Anarchy.*

The choice is ours to make. Thank you for having me here.

A global bioethics final message[4]

Dear Global Bioethics Network,

This is a final message and Thank You to all of you. I also want to say that for a long time it has not been appropriate to have a Core Group that excluded so many wonderful people that were in the other group. I think especially of Ivan Segota and Marcelo Palacios but also of Manuel Velasco-Suarez and L. Daniel Otero. The Core Group was really Original Group that grew. Everyone – please excuse all this.

Now a bit of history. I am quite weak and write with difficulty. I hope my memory serves me. For a long time 1970-1990 there was no one who recognized my name and wanted to be part of a mission. In the USA there was an immediate explosion of the use of the word Bioethics by Medical People who failed to mention my name or name any of my four publications 1970-1971. Unfortunately, their image of Bioethics delayed the emergence of what now exists.

It was in 1988 that I published Global Bioethics. Building on the Leopold Legacy. In all the world there was just one person who saw the book, saw the opportunity and contacted me. It was Brunetto Chiarelli, Professor of Anthropology, University of Florence, Italy. In 1990 he invited me to give a lecture "Global Bioethics" in Northern Italy, in Trentino, a progressive community. I accepted the invitation and went there under the care of my son, Carl. It was my last trip outside the USA.

After my lecture before the Italian Society for Bioethics, Prof. Chiarelli asked for permission to use the term Global Bioethics as the title of the transformed local journal to be published in English. I said yes and he got the unanimous approval from the membership.

---

[4] This email is probably the last one sent by Potter to the persons he considered to be in his global bioethics network. V. R. Potter died at the age of 90, September 6, 2001. A.M., H.M.S.

The next recruit was Peter Whitehouse, Professor of Neurology and other specialties at Case Western Reserve Medical School in Cleveland, Ohio. He was Founder/Director of the Alzheimer Research Center for its first ten years. In February 1996 he drove to Madison and wanted to join my mission, with the proposal to change the name to Deep Bioethics. This didn't happen but the word was mentioned later. Peter returned to Madison in 1997 and we spent an entire day together, on a Ferry Boat, at my cabin, in the trees with many photos and a video by Peter. A wonderful day. Peter traveled all over the world on behalf of Alzheimer therapy, and I would get email messages from "30,000 feet about the Atlantic" or from "Helsinki, leaving for Beijing". His subsequent contributions need telling. At about that time Hans Schweinsberg and Ms. Erin Williams, J.D. became No. 3 and No. 4 in less spectacular ways.

The next big event was the 1998 invitation from No. 5, Hyakudai Sakamoto, Prof. Philosophy and President of the 4th World Congress of the International Association of Bioethics which was of course Medical. But as President he chose the theme: "Global Bioethics, South and North, East and West." I declined to travel but offered to make my first videotape, which was accepted. I also sent a copy of the Audio-Script, which he reproduced and placed in every Registration Kit. He offered copies of the tape at cost, and this had world-wide effects.

The following year, 1999, I received an invitation to speak at a Bioethics Congress in Mexico City and I received impressive printed records of events to date from Prof. Manuel Velasco-Suarez. Again, I declined to travel but offered to make a special tape for the occasion. My offer was accepted and the tape has impressive quotes from my host, mentioning "agua, atmosphera" and "opportanidad hoy".

So the next big event was the next year, in 2000, an invitation to speak at an International World Congress in Gijon, Spain, coming from Dr. Marcelo Palacios, President and Founder of the Society of International Bioethics (SIBI). Again, not traveling, I offered a tape and the offer was accepted. This was my third tape and it shows the face of President Franklin D. Roosevelt of USA (1932-1946) with quote. Call it the FDR tape. The Gijon meeting was one of the biggest events in my life, all captured on film with copies sent to me, along with the medal and diploma from the SIBI Prize 2000.

Thus it was year by year from 1998, 1999, 2000, and now in 2001 an invitation from Prof. Ivan Segota to come to a meeting on the Island of Ceres, Croatia in September, and again, a tape instead. But this time three members of my so-called Core Group will speak.

In conclusion I just want to say that I hope you will all send a message of support to Prof. Calvin DeWitt. Also please cooperate with Prof. Thomas Potthast, Max Planck Institute, Berlin who is writing a History of Bioethics. He has spent time with me in Madison a year ago and is coming again this year.

Farewell to all, with love,

Van Rensselaer Potter, Ph.D.
Hilldale Professor Emeritus of Oncology
McArdle Laboratory for Cancer Research
University of Wisconsin-Madison
1400 University Avenue
Madison, WI 53706-1599

Global Bioethics Network
Hans Schweinsberg, Toronto, Canada; Brunetto Chiarelli, Florence, Italy; Hyakudai Sakamoto, Tokyo, Japan; Peter Whitehouse, Cleveland, Ohio; Michael Fox, Washington, D.C.; Erin D. Williams, Esq., Manassas, Virginia; Heinz Herrmann, Storrs, Connecticut; Kelly H. Clifton, Madison, Wisconsin; Calvin B. DeWitt, Madison, Wisconsin; James E. Trosko, East Lansing, MI; Juan Pablo Beca, Santiago, Chile; Arturo Filipowicz, Warsaw, Poland; Alfonso Llano, Bogota, Columbia; Richard Grantham, Lyon, France; Sophie Jakowska, Santo Domingo, Dominican Republic; Eduardo E. Sousa-Lennox Mendoza, Panama City; L. Daniel Otero, Merida, Venezuela; Marcelo Palacios, Gijon, Spain; Giovanni Russo, Messina, Italy; Ivan Segota, Rijeka, Croatia; Manuel Velasco-Suarez, Mexico City; Thomas Potthast, Berlin, Germany; William Atchley, San Francisco, CA; Marcus Duwell, Tübingen, Germany; Peter Price-Thomas, San Francisco, CA; Marcin Jakubowski, Madison, WI; Mark Fisher, Havelock North, New Zealand; Ludwig Schmidt, Caracas, Venezuela; Andrew Jameton, Omaha, NE; Marianna Gensabella Furnari, Messina, Italy; John Hofert, China; David B. Wake, Berkeley, CA; Andres Peralta-Cornielle, Dominican Republic; Svitlana Vekovshynina, Kiev, Ukraine; José Ramón Acosta Sariego, La Habana, Cuba; Garet Lahvis, Madison, WI; Tim Anderson, Madison, WI; Thomas Yuill, Madison, WI.

AHEAD OF HIS TIME
REFLECTING ON FRITZ JAHR'S LATE RECOGNITION

Irene M. Miller[1]

We know very little of the personal side of Fritz Jahr's life. Thanks to the meticulous recordkeeping by the Germans in the archives of the Franckesche Anstalten, the city and the University of Halle an der Saale, Amir Muzur and Iva Rincic provide a skeleton of Fritz Jahr's life on which to hang a few observations and reflections[2]. Halle an der Saale, a university city since 1694, proved to be fertile ground for Fritz Jahr's thoughts. The early exposure to Pietism in his formative primary school years probably sensitized his feelings of compassion towards all forms of life. But what do we make of his frequent job changes and search for employment? Was he unwilling to compromise on political grounds, was it poor health or a combination of these? A student CV and two recently found letters in the Magdeburg Church Archives[3] allow an additional brief glimpse behind a veil we might never be able to penetrate.

What can we glean from the short CV submitted to the Kirchenprovinz Sachsen by Jahr as a student of theology and philosophy? We see a young man, who already as a boy dreamed of becoming a pastor and who was not deterred by his designated course of schooling, which would not permit entry into Theology. Ambitious and highly motivated he embarked on the difficult study of the required languages, but early on suffered setbacks. After his first semester the war broke out. He volunteered for military service and within months was excused from it. The continuation of alternate service (Hilfsdienst) and his full-time studies in theology taxed his health, eventually causing him to miss classes in his favorite subject of ethical issues, due to pneumonia. It is interesting that he mentioned Ferdinand Kattenbusch, Professor of Systematic Theology and Rektor of the University of Halle in 1913/14. Kattenbusch published in 1916, in the middle of WW I, a book 'About Love of the Enemy' (Ueber Feindesliebe im christlichen Sinne'), and in 1934 bravely criticized the Nazi hate ideology in public.

In the two letters we see a person who is unhappy, weak, even desperate, and with whom we would not want to trade places. Yet, he is the 'father of

---

1 Irene M. Miller, celloimm@aol.com, a cellist and a medical professional has worked, after completion of her surgical residency at Yale University, throughout the U.S., also in China and for the U.S. State Department in Uganda.

2 See the article 'Fritz Jahr: on how he had discovered bioethics and how bioethicists have discovered him' by Amir Muzur and Iva Rincic, in this volume.

3 For all three documents see, Evangelische Kirche der Kirchenprovinz Sachsen. Archiv', Rep. A. Spec. P, J 126.

bioethics', a prophet, recognized by none in his time. The letters are neither nice, nor are they pleasant to read. It is almost embarrassingly painful to get such a close look at his feet of clay.

The letter of 1928 is addressed to the Superintendent General [Bishop] of the Kirchenprovinz Sachsen of the Evangelische Kirche der Union. Jahr outlines the many positions he would like to fill and topics to cover in the service of the church. He had addressed the importance of the modern mass media[4] for ethics education, the role of sexual education in changing times[5], also general ethics and the interaction of egoisms and altruism[6] and, being a teacher by training himself, the promotion of character formation versus indoctrination[7].

The letter of 1946 to the Rektor [President] of the Halle University, also in the Magdeburg Church Archive, is signed; this might be an original and perhaps was never sent; a search in the University Archives to find an original is pending. Here, Jahr complains about the low pension and the burden it places on him and his ill wife in February 1946, less than a year after the WW II bombings, when most of the populace of East Germany was hungry, freezing and still digging out from the rubble. If this letter would have been sent and answered, we would not expect to find the response in the Church Archives rather in Jahr's private materials at home which are lost. Anyway, no factual result came out of this letter and Jahr continued to earn additional small money as a music teacher.

This man set the stage for 21st century Bioethics. The letters allow us a glimpse into his privacy, his dreams, his potentials, and his interests. In 1933, when the Nazis took over German politics and life, he wrote: 'My happiness does not depend on money or other outward things, at least not in its deciding factors. Ultimately it always depends on something from within, something spiritual. … Nevertheless, in the most difficult areas and fateful moments of my aspirations for happiness, the presumed recognition of truth, the so-called scholarly truth, most often abandons me. At times it even harms me, forcing

[4] Fritz Jahr: Social and Sexual Ethics in the Daily Press [Soziale und sexuelle Ethik in der Tagespresse], *Ethik, Sexual- und Gesellschaftsethik. Organ des ‚Ethikbundes'*, 1928, 4(10/11):149-150, engl. translation in this volume.
[5] Fritz Jahr: Ways to Sexual Ethics [Wege zum sexuellen Ethos], *Ethik, Sexual- und Gesellschaftsethik. Organ des ‚Ethikbundes'*, 1928, 4(10/11):161-163, engl. translation in this volume.
[6] Fritz Jahr: Egoism and Altruism [Zwei ethische Grundprobleme in ihrem Gegensatz und in ihrer Vereinigung im sozialen leben], *Ethik, Sexual- und Gesellschaftsethik. Organ des ‚Ethikbundes'*, 1929, 6:341-346, engl. translation in this volume.
[7] Fritz Jahr: Character Dictate or Freedom of Thought? [Gesinnungsdiktatur oder Gedankenfreiheit?], *Die neue Erziehung. Monatsschrift für entschiedene Schulreform und freiheitliche Schulpolitik,* 1933, 10:115-116, engl. translation in this volume.

me by civilization and technology into a realm of superficiality, away from the inner and spiritual life, on which my happiness and the happiness of the world depend.'[8] Jahr remains palpably human and honest.

The two very personal letters and, to a lesser degree the student CV, give us only a limited perception of a weak, sometimes even desperate person. He did witness and share the horrors of two world wars. Who is the man behind his well read and forward looking articles? There are many others whose talents go unrecognized, who are prophets without honor in their own countries, - in the words of Dostoyevsky: 'Men reject their prophets and slay them, but they love their martyrs and honor those whom they have slain.'[9]

Curriculum Vitae, stud. theol. et phil. Fritz Jahr of Halle a. S[aale][10]

On January18th 1895, I, Paul *Fritz* Max Jahr, was born the son of office employe Gustav Max Jahr of Halle and received holy baptism on January 21st[11] that same year. My first schooling took place in the Middleschool of the Frankesche Establishment, which I attended until Michaelis 1905.

---

[8]Fritz Jahr: Our Doubts about God [Unsere Zweifel an Gott], *Ethik, Sexual- und Gesellschaftsethik. Organ des ‚Ethikbundes*', 1933, 10:115-116, engl. translation in this volume.

[9] Dostoyevsky 'The *Brothers Karamazov'.*

[10] Lebenslauf des stud. theol. et phil. Fritz Jahr aus Halle a. S[aale] in: Evangelische Kirche der Kirchenprovinz Sachsen. Archiv, Rep. A. Spec. P, J 126.

[11] In the margin a handwritten remark by a church official 'baptized in Catholic rite'. It was not a rare occurrence, that newborns in health emergencies received baptism speedily, prior to any communication with the family. Nevertheless, the Personalakte had to demonstrate that Jahr had a Protestant family background and qualified for church service as a student of theology and as a future pastor. In the Magdeburg Archives, J 126 I, we find the following affidavit: '*Copy: As the insurance agent Max Jahr, also his wife, are of protestant belief [evangelisches Bekenntnis], nothing stands in the way that the son Fritz, who in a totally illegal manner was baptized in the Catholic church and inscribed into the Church Book of the Catholic congregation, was educated and raised in the evangelical confession and at those days also was accepted in the church administered religious education in preparation for the Konfirmation. The child definitely has to be recognized as protestant. - Halle a.d. S., December 5, 1900 J Wuestler, Oberpfarrer zu St. Georgen [seal]. - It herewith is officially confirmed that this Copy is identical with the original Text. Halle, January 30, 1919, [signature] Police Commissioner [seal]'.* - Another document in the Archive attests to Jahr's Confirmation: '*Copy: In Remembrance of the Day of Confirmation.- Fritz Jahr, born January 18, 1895, baptized January 21 1895, was confirmed in the St Georgen church in Halle an der Saale, March 21 (Sunday Laetate) 1909 and hereafter had received Holy Supper.- Confirmation Text: Psalm 23, verse 1 and 2. - Halle a. S., March 22, 1909. - Ev. Pfarramt zu St. Georgen, Leithmann, Provincial-Vikar. - Copy officially confirmed, Halle, January 30 1919. - Witte, Pastor at St Georgen'.- Psalm 23: 1-2 in the* King James version reads 'The Lord is my shepherd, I shall not want; he makes me lie down in green pastures. He leads me besides still waters'.

Thereafter I attended Oberrealschule, which I left Easter 1908 with the Certificate of Maturity.

I was confirmed Easter 1909. Especially the religion classes in school awakened in me the wish to study theology. As the University here offered Latin, Greek and Hebrew courses I could pursue this wish, and did not need to learn banking as initially planned, as a student from Oberrealschule is ineligible for the study of theology.

I studied for a semester when the war began. But in May 1915 I was accepted as a volunteer in the armed forces and was already discharged in August. Therefore I could take the final exams only at Easter 1916 instead of Michaelis 1915.

Since Easter 1916 I am fully enrolled in theology. I am most interested in questions of ethics, even since my former schooldays. Sadly I could not commit as much time to this as I would have liked to, since I was occupied with the alternate service (Hilfsdienst) since 1917 at the Middle and Preschool of the Frankesche Establishment. I had to miss parts of the ethical exercises sponsored by Geheimrat Kattenbusch in which I was enrolled in the 1918 summer semester, as a cold had developed into a pneumonia secondary to the demands of the alternate service. Since New Year's of 1918 I am exempt from the school-service.

Fritz Jahr, stud.theol.et phil.

Fritz Jahr, Letter to the Generalsuperintendent, January 8, 1928

Halle/S., Albert Schmidtstr. 8, January 28. 1928
Not official!

[In margin a note by the Superintendent-General's office: 'Received and filed Feb. 8]

Most esteemed, right reverend Superintendent-General!

Would you most graciously permit that I may approach you with a request, as I already had done a year ago in January. Needless to say, that I would not have wanted to infringe on your very valuable time, safe for the encouragement of the church authorities, that in all modesty I might ask you, most honorable Superintendent-General, to take an interest and grant me a possible consultation and advancement [based on Verf. Urkunde, Art.101.(2);3].

In support of my request I allow myself to direct you most humbly to the enclosed copies of two certificates (Zeugnisse):

In regards to the certificate of my church activities, it is quite evident, that the request for a position in or near Halle at that time was no empty pretext to seek a position in or near Halle, but the deciding factor of medical treatment and homecare (cf. my submission [Gesuch] of February 27. of this year). Furthermore, the mentioned certificate offers proof, that it was not a lack of willingness on my part to serve the church, more to the contrary, I waived any remuneration for my services. An additional indicator of this, most worthy Superintendent-General, is my relatively extensive substitute work for the churches, especially in my home parish of St. Georgen, as well as among the Brothers in Service [Amtsbrüder], without any prejudice against my work, my conduct and my person.

Similar circumstances apply to my certificate from Seydlitz Lyceum, which is also enclosed. Modesty forbids me to comment on its content. Only permit me to humbly request that you may kindly attach greater significance to this second certificate, as directors and staff at this institution consisted entirely of ladies (with a sole exception), who are far more astute than men in the observation of conduct, work-performance, inner and outer attitudes toward work and staff. – I also allow myself, most esteemed Superintendent-General, to draw attention to my scientific and literary work. I already informed you that presently I am working on my dissertation ('About the ethical relationship of men to animals and plants') for my graduation at University Jena. I want to mention in this connection that, on the basis of my activities for the modern animal protection in the pulpit and press (also in the press of the church, comp. enclosed clipping from the January issue of this year "Mut und Kraft" [Courage and Strength], I was elected an honorary member of the Society for the Protection of Animals (see enclosed Animal Protection Calendar, 2nd cover-page). Still of higher value should be the fact, that Geheimrat Prof. Dr. Abderhalden, founder and presiding chair of the 'Ethikbund' [Ethics Alliance], of which I am member – having followed the suggestion of an honor-worthy Consistory in Magdeburg in the 'Kirchliches Amtblatt' [Internal Church Bulletin] – has asked me to participate as an author in the journal he publishes.

I would also like to humbly direct you to my neediness relating to my household. However I already have so often and explicitly reported this, that I believe I can omit going into detail here.

1. After all of this I allow myself to bring to you, most honorable Superintendent General, the following requests, equally as urgent as submissive, for your very kind consideration. I would like very much to be active as *a staff member [Mitarbeiter] at the Institut für Evangelische Weltanschauung* [Institute of Protestant Worldview] in Wittenberg. Even though I do not have higher than normal education in the sciences and theology, nor further training in the natural sciences, I do have the theoretic (Middle School Teacher's Certificate) and practical experience in pedagogy (10 year experience in almost all school settings), which enabled me to be employed by the school system. Besides, I am a member of the Religionspädagogische Gesellschaft [Religious Pedagogical Society]. I would not mind, if the positions I suggested could not be fulltime, resp. not fully paid.

2. I would also like to work in the *service of the church press.* In support of this request I inform you, that I have participated in a press-course of instruction here, given by Ev. Soz. Pressverband [Protestant Social Press League] and also have on many occasions practiced what I have learned.

3. In view of having successfully completed with two theology examinations, ordination, practicum, pedagogical examinations, employment authorization, teaching activities, I would likewise be suited for such *teaching opportunities in the local Diakonissenhaus* [Deaconess House], in the Mädchenschuldienst [Girls' School Service] such as in the Kindergarten Teachers Seminar.

4. On February 18˙ 1926 I requested kind admission to a Taubstummenseelsorgerkursus [Course for Spiritual Care for the Deaf] and I was granted same on February 25. 1926 (VIII Nr. 553). All be it, the course did not take place. However, even without an official course I could, no doubt, acquire the required knowledge with the kind willingness of the director of the Taubstummenanstalt [Institute for the Deaf] to oblige. Naturally I would only do this after assurance of a contract.

5. I would like to make myself available to the *Abteilung für Volksbildung* [Community Education] of the Ev. Konsistorium as contributor in science. May this wish not appear immodest to you, most esteemed Superintendent-General, given my interests and activities (see above) in the field of science and ethics.

6. The above mentioned occupations are by far not all the possibilities that could shore up a person's or even a family's wellbeing, I take the liberty in all humility and modesty to convey the proposition to confer upon me a *pastoral*

[kirchliche] activity in Halle. It was in December 1925 that Reverend Superintendent Meinhof, upper Ev. Konsistorium of the province Sachsen, held this prospect out to me.

7. Lastly I would like to remind you of my application from February 1927 for the position of the *Pastorat of Spickendorf.* The December 28, 1927 writing from the right reverend Konsistorium leads me to believe that my request for a suitable pastoral position, if at all possible, should be heard and I ask you most respectfully to refer to this.

Permit me, most worthy and honorable Superintendent-General, to express my greatest regard and admiration and thank you that I have the great honor of your kind and gracious consideration of my request, which I await in all submissiveness and modesty.

Most respectfully and obediently,
[signature is missing on this page]

Fritz Jahr, Letter to the Rektor of the Halle University, February 23, 1946

Halle, February 23rd 1946
His Magnificence
Rektor of the University
Konsistorialrat Prof. D. Dr. Eissfeldt
Halle (Saale) Seydlitzstr. 14a

Most honored Professor!

Please help me! Please don't leave a former student of the University and pastor in misery, don't let him drown in despair!

I am 51 years old. My pension at present is 84 Mark. This supports me and my wife, who suffers from paralysis of both legs since 7 years. It is obvious that under these circumstances I live in most needy, impoverished and piteous conditions. Together with my wife I am headed for ruin [Untergang] if I don't get help soon.

This help means work for me, a suitable additional work, that together with my pension, I and my wife could live tolerable well. Could you help me find such a work? I entreat you most reasonably and cordially!

I am thinking of an occupation as an assistant pastor or provisional substitute in the church services of Halle[12]. – Formerly I passed good exams, was pastor in Canena, have stood in most pulpits of Halle, repeatedly also in the Markt-, Moritz-, and Ulrichs-Church as well as the Cathedral.

Or is there perhaps a possible occupation at the University, maybe in the theological seminary or the library, or the like? I was always interested and active in the sciences. At present, as a member of the Kulturbund [Cultural Association], I have registered for the Working Group in Theology [Arbeitsgemeinschaft für Theologie], as well as the Working Group in Musicology [Arbeitsgemeinschaft für Musikwissenschaft].

And I also think of possible occupation at the Kirchenmusikschule [Church Music School]. – Until 1945 I had worked successfully for a good two years at the Music School of the former Volksbildungsstätte [Community Educational Center]. Most recently I have been admitted as music teacher at the Kammer der Kulturschaffenden [Chamber of Artists].

I also would like to teach in churches. Formerly I have passed the Middle School Teacher's Exam (Religion, History) and have adequate experience in the service of high, middle and grade schools.

Perhaps a halftime occupation could be found for me in Administration at the University?

I further inform you, that I did not join NSDAP[13] and that there are no scruples about health concerning my employment. So I entreat you again most cordially and urgently: Please do help me, who indeed am unhappy, but certainly not unworthy nor unfit. I am in great need. The health of my sickly wife, respectively her restoration, is also seriously threatened.

Perhaps you are inclined to grant me a personal interview.

In the hopes that it was not in vain to have turned to you with the request for help or at least a consultation, I thank you in advance and cordially salute you.

Yours respectfully,

[12] Professor Eissfeld was a theologian himself, Chair of Old Testament Studies; so he might have had good personal connections to church decision makers in Halle or he might have had forwarded Jahr's original letter to the office of the Superintendent General. sity are pending.

[13] NSDAP, the Nazi Party 'National Socialist German Workers Party'

Your most humble, Fritz M. Jahr, Pastor ret,

Halle (Saale), Albert Schmidtstr. 8

FRITZ JAHR: ON HOW HE HAD DISCOVERED BIOETHICS AND HOW BIOETHICISTS HAVE DISCOVERED HIM*

Amir Muzur[1] and Iva Rinčić[2]

There is little doubt that the discovery of the work of Fritz Jahr was the most intriguing and promising moment in the last fifteen years of bioethics history. The importance of Van Rensselaer Potter's contribution has not thereby been diminished: on the contrary, some Jahr's ideas offer the way out from the peril bioethics was brought into by some our-time authors narrowing down originally broad Potterian concept to mere medical ethics. Expectedly, however, it has to be said that Jahr and Potter do also differ in their teachings: let us not forget that Jahr deduced the word *Bio-Ethik* out of "bios" and "ethics," therefore associating ethics primarily with life (not just of humans, but also of animals and plants), and not with science (as Potter will do almost half a century later, in the early 1970s).

Fritz Jahr: his life and work

A preliminary search of the archives in Jahr's home city of Halle (Germany) has turned up a lot of interesting data, explaining some of the circumstances influencing the "first creation" of bioethics.

Paul Max Fritz Jahr was born on January 18, 1895, in Halle on the river Saale in the Sachsen-Anhalt region of central Germany. Fritz Jahr's father was Gustav Maximilian (1865-1930), insurance official,[3] while his mother was Auguste Marie Langrock (1862-1921): the couple married in 1892.[4] Although both father and mother had been Protestant, Fritz was baptized according to

---

* The paper was presented at 1st International Conference „Fritz Jahr and the Foundations of European Bioethics", Rijeka/Opatija, March 11-12, 2011, and published in Muzur A, Rincic I (2011) *Jahr* 2(4) 385-394. The section on life and work of Fritz Jahr has partly been exploited also in: Muzur A, Rinčić I (2011) Fritz Jahr (1895-1953): the man who invented bioethics, in: *Synthesis Philosophica* 26(1)133-139.

[1] Amir Muzur, amir.muzur@medri.hr, University of Rijeka – Faculty of Medicine, Department of Social Sciences and Medical Humanities, B. Branchetta 20, 51000 Rijeka, Croatia.

[2] Iva Rincic, irincic@medri.hr, University of Rijeka – Faculty of Medicine, Department of Social Sciences and Medical Humanities, B. Branchetta 20, 51000 Rijeka, Croatia.

[3] Actually, his official title was changing from *Schreiber* (scribe), *Magistrats Diätar/Diätar* (municipal clerk), and *Stenograph* (stenographer), to *Assekuranz-Beamter/ Versicherungsbeamter* (insurance officer).

[4] Archives of the Sachsen-Anhalt University and Province Library (*Archiv der Universitäts- und Landesbibliothek Sachsen-Anhalts*), Album No. 40, 1918.

the Catholic ritual.[5] It seems that the Jahr family was often changing home address (always remaining within the city): in 1913, they finally moved to the "famous" Albert-Schmidt-Straße 8, occupying both the basement and the first floor.[6]

Fritz Jahr started elementary school in 1901 (*Mittelschule*), and secondary school in 1905 (*Oberrealschule*). Both schools were owned and run by the Francke Foundation, named after August Hermann Francke (1663-1727), who, together with Philipp Jakob Spener, had founded Pietism, a movement within the ranks of Protestantism, and brought its seat to Halle.[7] The young Jahr's marks were not particularly high: he obtained a "2" (out of 5) for diligence (*Fleißigkeit*), a "2" for behaviour (*Betragen*), and a "2+" for overall results (*Leistungen*).[8] In Easter 1914, Jahr took his A-levels (*Abitur*), and in 1915, an additional exam of Latin and Greek languages. All together, he studied eight semesters: during 1914, mostly philosophy, music, history, national economy, and, from 1915 until 1919, theology (first exam in 1919, the second in 1921). The 1915 summer term Jahr spent as a war volunteer. On March 19, 1921, he took holy orders.

Jahr started to teach as early as May 1917. First he taught for a brief spell at a private school, later at the pre-school (*Vorschule*) and the elementary school (*Mittelschule*) of the same Francke Foundation where he himself had been a pupil. In 1920, he took the elementary-school teacher's exam (*Mittelschullehrerprüfung* in religion and history), and in 1921, he entered the city's Wittekind-School. From 1923-1925, he had been appointed teacher by the Halle Public Schools Administration, but he left that position by May 2, 1925, due to "discord with the then Education Committee." In the summer of 1926, Jahr started teaching again at the Francke Foundation elementary school, and then at the Seydlitz high school for girls from late 1926 until 1928 when he left as "he did not believe the private schooling system would give him long-term satisfaction" (during the period 1917-1925 only, Jahr worked at 11 different schools). After a few years' break, Jahr went to teach briefly at the Hutten School in 1938, while, during the last years of World War II (1943-1945), he taught cello at the National Education Centre Music School.

---

5 Archives of the Francke Foundation (*Franckesche Stiftungen – Archiv*), *Schüleralbum von Ostern 1885 bis Ostern 1914: Realgymnasium der Franckeschen Stiftungen zu Halle a.d. Saale* III, I S B I 7, No. 2072.

6 Halle City Archive (*Stadtarchiv Halle*), *Adressbuch für die Stadt Halle a.d. Saale* (Halle: Hendel, 1895-1900); *Hallesches Adressbuch für 1903* (Halle: Kutschbach, 1903); *Neues Adressbuch: Halle a. S. und Umgebung* (Halle: Scherl, 1905-1918).

7 Lauchert F (1911) Article „Pietism,", in: The Catholic Encyclopedia. New York: Robert Appleton Company. 2 August 2010 http://www.newadvent.org/cathen/12080c.htm

8 Archives of the Francke Foundation, *Schüleralbum*.

Jahr was active within the Church from 1925 onwards.[9] For the first four years, he was a curate at St. John's church in Dieskau (near Halle), later (1929-1930) in Braunsdorf, and finally (1930-1933), a pastor in Kanena.[10] It must have been quite painful, since, due to feeling giddiness, he had to take bromide before ascending the pulpit.[11]

On April 26, 1932, Jahr married Berta Elise Neuholz (born December 31, 1899), "a worker from Bludau," and a daughter of the teacher Franz Hermann Neuholz (1867-1903). Elise and Fritz Jahr had no children and lived at Albert Schmidtstrasse 8, Halle.[12]

At a very turbulent time for Germany, Fritz Jahr applied in 1932 to the Church authorities for retirement on the grounds of "nervous exhaustion." At the age of 38, on March 1, 1933, he indeed did withdraw from service, only a month after Hitler had seized power. During the war, the Jahr family obviously experienced financial difficulties, further aggravated by Jahr's wife suffering from "backbone sclerosis" (she died wheelchaired, on February 1, 1947). Then, a month after the surrender of Germany, on June 18, 1945, Jahr applied for a job in a new elementary school: two weeks before, he had become member of the Workers Association. Unanswered remained Jahr's supplication for job, addressed to the President of Halle University in 1946, mentioning that Jahr and his wife's joint incomes ammounted 84 Marks only.[13] Together with several other pastors, in October 1946, Jahr signed an invitation (published in the *Freiheit* newspaper) to vote for the Socialist Unity Party (*Sozialistische Einheitspartei Deutschlands*).[14] Obviously, he spent his last years working as music teacher (*Musikerzieher*): in a confidential "characteristic" from 1952, signed by certain Luderer-Lüttig and sent to Department of Popular Education (*Dezernat für Volksbildung, Abteilung für Kunstgelegenheiten*) of the Halle City Council (*Rat der Landeshauptstadt Halle*), Jahr's pedagogic qualities are highly appreciated and he is strongly recommended for the instructor of mandolin orchestra.[15]

---

[9] Pfarrerbuch der Kirchenprovinz Sachsen, vol. 4 (Leipzig: Evangelische Verlagsanstalt, 2006), 395.
[10] Halle City Archives. 105/4, 1938/1945.
[11] Sass HM (2011), Postscript, in: Essays in Bioethics and Ethics 1927-1947 by Fritz Jahr. Bochum: Zentrum für Medizinische Ethik (Medizinethische Materialien 188), 46.
[12] Archives of the Sachsen-Anhalt University and Province Library, Album No. 40.
[13] Sass, Postscript, 46.
[14] Ibid.
[15] Halle City Archives (*Stadtarchiv Halle*), A3.21 Kultur-41-4661, January 19, 1952.

Fritz Jahr died at 10 a.m. on October 1, 1953, in his home in Halle, on cerebral apoplexia due to high blood pressure.[16]

The discovery of Fritz Jahr[17]

In 1997, at a conference held in Tübingen, Professor Rolf Löther of Berlin Humboldt University mentioned for the first time the name of Fritz Jahr,[18] whom he credited for having had coined the word *Bio-Ethik* as early as 1927. According to his own account, Löther heard the word "bioethics" in the 1990s for the first time: since the word, nevertheless, seemed to him somehow familiar, Löther started to search through the bundle of old issues of the *Kosmos* journal, left to him by his grandfather and now lying somewhere in the attic. And so he found the editorial of the 1927 volume and the article of Fritz Jahr.

News about the discovery of Jahr eventually was spread mainly thanks to work of Eve-Marie Engels of the University of Tübingen (who had organised the conference Löther attended and eventually edited the proceedings). Engels first mentioned Löther's discovery in an article *sub voce* "Bioethik" in the *Metzler Lexicon* in 1999[19] and in the publication of a series of lectures on *Biology and Ethics* from the same year,[20] then in a paper from 2001.[21] This paper was translated into Portuguese and republished in 2004 in Brazilian journal *Veritas* from Porto Alegre.[22] There, the paper attracted attention of the Porto Alegre clinician and university professor José Roberto Goldim,[23] and of

---

[16] Halle Registry Office, Register of Deaths (*Standesamt Halle, Sterbebuch*) Nr. 433.

[17] For a tentative scheme of the discovery, see: Muzur A, Rinčić I (2011) Fritz Jahr (1895-1953: a life story of the "inventor" of bioethics and a tentative reconstruction of the chronology of the discovery of his work, in: Jahr –Annual of the Department of Social Sciences and Medical Humanities at University of Rijeka School of Medicine 2(4)385-394.

[18] Löther R (1998), Evolution der Biosphäre und Ethik, in: Ethik der Biowissenschaften: Geschichte und Theorie – Beiträge zur 6. Jahrestagung der Deutschen Gesellschaft für Geschichte und Theorie der Biologie (DGGTB) in Tübingen 1997. Berlin: Verlag für Wissenschaft und Bildung, 61-68.

[19] Engels EM (1999), Article "Bioethik," in Metzler Lexikon Religion. Stuttgart: Metzler, 159-164.

[20] Engels EM (1999), Natur- und Menschenbilder in der Bioethik des 20. Jahrhunderts: zur Einführung, in: Biologie und Ethik. Stuttgart: Reclam, 7-42.

[21] Engels EM (2001), Die Herausforderung der Biotechniken für Ethik und Anthropologie, in: Die biologische Machbarkeit des Menschen. Berlin: Wichern, 100-124.

[22] Engels EM (2004), O desafio das biotécnicas para a ética e a antropologia, in: Veritas 50(2)205-228.

[23] Goldim JR (2006), Bioética: origens e complexidade, in: Revista do Hospital de Clínicas de Porto Alegre 26(2)86-92; (2009), Revisiting the beginning of bioethics: the contribution of Fritz Jahr (1927), in: Perspectives in Biology and Medicine 52,377-380.

Leo Pessini, Vicerector and Bioethics Professor at St. Camillus University Centre, who both devoted several papers to the Jahr topic.[24] Under the influence of Goldim and Pessini, "the Jahr news" spread all over Brazil: Ferreira Carvalho da Cruz and Contri Pitton of Saõ Paulo State University (UNESP) in Rio Claro deliver a talk on Jahr in 2009,[25] Mascarenhas and de Oliveira Santa Rosa publish a paper,[26] etc.

As yet, the most thorough analysis of Jahr's 1927 paper and the "bioethical imperative" has been offered by Hans-Martin Sass of the Kennedy Institute of Ethics at Georgetown University in Washington. In 2007, Sass published a first series of papers devoted to Jahr.[27] In the later years, he has continued to promote Jahr's ideas and to present them in various international journals in Europe, Asia, and South America.[28] By the end of 2010, Sass edited a

---

[24] Ferreira Carvalho FM, Pessini L, Campos O (2006), Reflexões sobre bioética ambiental, in: O mundo da saude 30(4)614-618; Leo Pessini and Saad Hossne W (2008), Fritz Jahar (sic!): "O Imperativo Bioético" – nas origens da palavra Bioética, in: Bioethicos 2(1)7-9 (http://www.saocamilo-sp.br/novo/publicacoes/publicacaoEditorial.php?ID=60&rev=b); Pessini L (2008), Bioética na América Latina: algumas questões desafiantes para o presente e futuro, in: Bioethicos 2(1)42-49; Saad Hossne W, Pessini L, de Siqueira JE, de Paul de Barchifontaine C (2010), Bioética aos 40 anos: reflexões a partir de um tempo de incertezas, in: Bioethicos 4(2)130-143.

[25] Ferreira Carvalho da Cruz FM, Contri Pitton SE (2009), A inclusão da disciplina bioética na matriz curricular dos cursos de licenciatura em geografia; paper presented at 9th Seminar of Post-Graduate in Geography of the UNESP at Rio Claro, Brasil (IX Seminário de Pós-Graduaçao em Geografia, UNESP – Rio Claro, November 3-5, 2009 (available at: http://sites.google.com/site/seminarioposgeo/local).

[26] Mascarenhas NB, de Oliveira Santa Rosa D (2010), The teaching of bioethics in the education of nurses: interface with the adopted literature, in: Acta Paulista de Enfermagem 23(3)392-398.

[27] Sass HM (2007), Fritz Jahr's bioethischer Imperativ: 80 Jahre Bioethik in Deutschland von 1927 bis 2007, in: Medizinethische Materialen 175, 1-21; (2007), Fritz Jahr's 1927 concept of bioethics, in: Kennedy Institute of Ethics Journal 17, 279-295.

[28] Cf. Sass HM (2009), European roots of bioethics: Fritz Jahr's 1927 definition and vision of bioethics, in: Od nove medicinske etike do integrativne bioetike: posvećeno Ivanu Šegoti povodom 70. rođendana [From medical ethics to integrative bioethics: dedicated to Ivan Šegota in occasion of his 70th birthday]. Zagreb: Pergamena/Hrvatsko bioetičko društvo, 19-31; (2009), Vom Ursprung der Bioethik aus evangelischer Tradition, in: Zeitschrift für Evangelische Ethik 53(3)177-190; (2009), Asian and European roots of bioethics: Fritz Jahr's 1927 definition and vision of bioethics, in: Asian Bioethics Review 1(3)185-197; (2009), Early 1927 European roots of bioethics, in: Peligros y riesgos en las investigaciones: (bio)ética en la investigación con seres vivos (homenaje a José Alberto Mainetti). Buenos Aires: Editorial Antropofagia, 45-57; (2010), Bioethics as a European innovation: Fritz Jahr's 1927 concept of bioethics, in: Vernunft und Innovation: über das alte Vorurteil für das Neue (Festschrift für Walther Ch. Zimmerli zum 65. Geburtstag). Paderborn: Fink, 369-377; (2011), The Earth is a living being: we have to treat her as such!, in: Eubios - Journal of Asian and International Bioethics 21(3)73-77 (http://eubios.info/EJAIB52011.pdf).

collection of 15 papers by Fritz Jahr (published in German original), as well as a selection of 6 papers of Jahr's in English translation (in a second edition, in May 2011, 15 articles were translated into English).[29] Except for the collection of Jahr's papers in German and English from November/December 2010, Sass has never quoted mentions of Jahr before 2007 (by Löther, Engels, or Goldim). Those authors who got to know about Jahr from Sass, have also been mentioning Fritz Jahr, but not Rolf Löther (or Eve-Marie Engels) in their publications. Such has been the case with Fernando Lolas Stepke, Director of the *Programa Regional de Bioética de la Organización Panamericana de la Salud* (Bioethics Regional Program of the Pan-American Health Organization; PAHO),[30] Juan Jorge Michel Fariña[31] and Natacha Salomé Lima from Faculty of Psychology at University of Buenos Aires,[32] the priest Fernando Pascual from Catalonia,[33] Ricardo Andrés Roa-Castellanos and Cornelia Bauer from Colombia,[34] Hrvoje Jurić from University of Zagreb (Croatia),[35] and others. Unlike them, the „Goldim line" of discovery of Jahr has justfully credited Engels and, sometimes, Löther.

---

[29] Jahr F (2010), Aufsätze zur Bioethik 1927-1938. Bochum: Zentrum für Medizinische Ethik (Medizinethische Materialien 187); Jahr F (2010), Selected Essays in Bioethics 1927-1934. Bochum: Zentrum für Medizinische Ethik (Medizinethische Materialien 186); Jahr F (2011), Essays in Bioethics and Ethics 1927-1947. Bochum: Zentrum für Medizinische Ethik. See also in this volume.

[30] Lolas Stepke F (2008), Fritz Jahr, el 'imperativo bioético y el origen de la palabra 'bioética', in: Boletín Bioética informa 13(45)3; (2008), Bioethics and animal research: a personal perspective and a note on the contribution of Fritz Jahr, in: Biological Research 41,119-123; (2008), El 'imperativo bioético' de Fritz Jahr y la neobioética estadounidense, in: *JANO* 1710,10-16; (2009), Quo vadis bioética?, in: Acta Bioethica 15(1)7-9; (2009), Salud, salud mental, bioética: interfaces dignas de análisis, in: Acta Bioethica 15(2)137-138. It was probably Lolas Stepke to write the article on Fritz Jahr in the Spanish version of Wikipedia (http://es.wikipedia.org/wiki/Fritz_Jahr) and later also in the English version (http://en.wikipedia.org/wiki/Bioethics).

[31] Fariña JJM (2011), Declaración de Rijeka: hacia el futuro de la bioética, in: Ibis News (http://www.ibisnewsletter.org/spip.php?article21) April 2011

[32] Lima NS (2009), Fritz Jahr y el *Zeitgeist* de la bioética, in: Aesthetika 5(1)4-11; (2011), Las raices europeas de la bioética: Fritz Jahr y el Parsifal, de Wagner, in: Etica y Cine (http://www.eticaycine.org/Parsifal) July 2011.

[33] Pascual F (2009), ¿Quién inventó la palabra 'bioética'?, in: ForumLibertas.com (http://www.forumlibertas.com/frontend/forumlibertas/noticia.php?id_noticia=15337&id_seccion=5)

[34] Roa-Castellanos RA, Bauer C (2009), Presentación de la palabra bioética, del imperativo bioético y de la moción de biopsicología por Fritz Jahr en 1929 (sic!), in: Bioethicos 3(2)158-170.

[35] Jurić H (2010), Počeci i pra-počeci bioetike / The beginnings and the primeval beginnings of bioethics, in: 9. lošinjski dani bioetike / 9th Lošinj Days of Bioethics. Zagreb: Hrvatsko filozofsko društvo, 78-80. Also available at: http://www.hrfd.hr/u/dokumenti/Losinj%20knjizica%20-%202010.pdf

Hans-Martin Sass has won many bioethicists (sometimes one by one) to take Jahr's ideas into consideration. And while North-American scientists (no wonder, one might say) have ignored the discovery of Fritz Jahr (one exemption being H. Tristram Engelhardt in a series of his 2009 lectures and papers[36]), his ideas seem to have a better perspective in Asia. In Europe, Jahr has entered the bioethics courses (at University of Rijeka Faculty of Medicine, a lecture on Jahr and European bioethics has been incorporated into each bioethics course; in Spring 2010, Jahr's definition of bioethics was included into the programme of the Basel University "Grundlagen der Ethik" module,[37] etc.).

An interesting issue is how Hans-Martin Sass was attracted to the 1927 Jahr's article. Ivan Šegota, who introduced bioethics into Croatia from the US in the early 1990s, tells in an anecdotal account that he mentioned to Sass "some German" who, allegedly, had coined the term "bioethics" much before Potter, during their dinner in Sydney in 2004.[38] Hans-Martin Sass himself, on the other hand, recalls that he learned about Jahr around 2004 or 2005, from the 1999 paper by Eve-Marie Engels.

The authors of the present paper got to know about Fritz Jahr first from the Croatian translation of Sass' paper from 2007.[39] During an Internet search, while working on her PhD thesis,[40] however, Iva Rinčić discovered the mentioning of Rolf Löther and Eve-Marie Engels in an article by Goldim.[41] The first results of the research on Fritz Jahr's life, gathered only through Internet correspondence with several institutions in Germany, were presented

---

[36] Engelhardt HT, Jr. (2009), Moral pluralism and the crisis of secular bioethics: why orthodox Christian bioethics has the solution, in: Institute of Orthodox Christian Studies lectures, June 5, 2009 (http://www.antiochian-orthodox.co.uk/Publications/engelhardt-bio-ethics.pdf); (2009), Moral Pluralism, the Crisis of Secular Bioethics, and the Divisive Character of Christian Bioethics: Taking the Culture Wars Seriously, in: Christian Bioethics 15(3)234-253; (2009), Global bioethics: taking moral differences seriously, in: Bioethicos 3(1)26-32.

[37] http://www.aeb.unibas.ch/data/pdf/seminars/10_biol/Programm-Ethik-2010.pdf

[38] Šegota I (2008), Predgovor [Preface] to Bioetički imperativ Fritza Jahra: 80 godina bioetike u Njemačkoj od 1927. do 2007. godine by Hans-Martin Sass, in: Bioetički svesci 61,2 (1-4).

[39] Sass HM (2008), Bioetički imperativ Fritza Jahra: 80 godina bioetike u Njemačkoj od 1927. do 2007. godine, in: Bioetički svesci 61,1-44.

[40] Teorijska uporišta, postignuća i perspektive bioetičke institucionalizacije u Europskoj Uniji [Theoretical strongholds, achievements, and perspectives of bioethics institutionalisation in European Union], defended in October 2010 at University of Zagreb Faculty of Humanities and Social Sciences (under the mentorship of Ante Čović and Nada Gosić). An adjusted version of the thesis was recently published as a book (*European bioethics: ideas and institutions*; Zagreb, Pergamena) in autumn 2011.

[41] Goldim, Bioética: origens e complexidade.

at *9th Lošinj Days of Bioethics* conference (Mali Lošinj, 17-19, 2010)[42], as well as in a presentation at Ethics Teacher Training Course held in Dubrovnik (Croatia) in June-July, 2010.[43] In May 2010, the first issue of *Jahr – Annual of Department of Social Sciences and Medical Humanities at University of Rijeka Faculty of Medicine*, was launched, with Amir Muzur as the editor-in-chief and Hans-Martin Sass among the members of the journal Advisory Council. In January 2011, a project was granted to the authors of this paper by Croatian Science Foundation, entitled "Fritz Jahr and the foundations of European bioethics: establishing an international scholars' network". Among other activities, the project included a conference (held in March 2011 in Rijeka/Opatija)[44] and a study trip to Halle, where new materials on Jahr's life and work were collected. The preliminary results were presented in three papers at *13th Rijeka Days of Bioethics* (Rijeka, May 13-14, 2011)[45] and *10th Lošinj Days of Bioethics* (Mali Lošinj, May 15-18, 2011;[46] including invited plenary lecture[47]), as well as in invited lectures delivered at European Academy in Banja Luka (Bosnia and Hercegovina),[48] at Institute for Philosophy and Social Theory in Belgrade (Serbia),[49] and at the symposium on *Bioethics – a Sign of a New Era: Bioethics, Media, Law, and Medicine* (Ohrid, Macedonia, October 21-22, 2011).[50] Papers stressing the importance

---

[42] Rinčić I, Muzur A (2010), Fritz Jahr: prilozi za biografiju osnivača (europske) bioetike / Fritz Jahr: Contributions to the Biography of the Founder of (European) Bioethics, in: 9. lošinjski dani bioetike / 9th Lošinj Days of Bioethics. Zagreb: Hrvatsko filozofsko društvo, 115-116. Also available at: http://www.hrfd.hr/u/dokumenti/Losinj%20knjizica%20-%202010.pdf

[43] Rinčić I (2010), The founder(s) of bioethics: one or more?, presented at UNESCO Ethics Teacher Training Course, Inter-University Center Dubrovnik, Croatia, 28 June – 2 July, 2010.

[44] With participation of some of the first mentioners of Fritz Jahr: Eve-Marie Engels, José Roberto Goldim, and Hans-Martin Sass.

[45] Rinčić I, Muzur A (2011), Od bioetičara-učenika do bioetičara-učitelja: pijetizam i edukacija u životu i djelu Fritza Jahra [From bioethicist-student to bioethicist-teacher: Pietism and education in the life and work of Fritz Jahr].

[46] Muzur A, Rinčić I (2011), Ignaz Bregenzer (1844-1906): preliminarni prilozi životopisu autora *Životinjske etike* (*Thierethik*, 1894.) [Ignaz Bregenzer (1844-1906): preliminary contribution to the biography of the author of *Animal Ethics* (*Thierethik*, 1894)].

[47] Rinčić I, Muzur A (2011), The concept and project of European bioethics.

[48] Rinčić I (2011), Europska bioetika: od otkrića do realizacije" [European bioethics: from discovery to realisation], May 26, 2011.

[49] Rinčić I, Muzur A (2011), One name, two fathers, (too) many midwives: the destinies of F. Jahr's and V. R. Potter's the concepts of bioethics, presented at International Conference on (New) Perspectives in Bioethics, Belgrade (Serbia), October 13-15, 2011.

[50] Rinčić I (2011), Riječka deklaracija o budućnosti bioetike [Rijeka Declaration on the Future of Bioethics]; Muzur A (2011), Bioetika – problemi definiranja i dosega [Bioethics – the problems of definition and reach].

of re-writing a new history of (European) bioethics have since appeared in several editions.[51]

One possible conclusion

Studying the life of Fritz Jahr – or of anyone else – might seem irrelevant, even to the historians of science. However, we often seem to comit mistake by trying to explain the sources of his/her ideas by studying (or overestimating) only the so-called "social circumstances" of a discovery. As we all certainly would agree, personal factors may and do influence the author of a given discovery much more and therefore should never be neglected in a serious search for truth. The (hi)story of the discovery itself, on the other hand, sometimes so curiously indicates the importance of coincidence or of what seems to be coincidence.

---

[51] Muzur A (2011), Evropska i amerikanska bioetika: ima li mesto za obete? [European and American bioethics: is there place for both?], in: Filozofija 10(31)9-18; Rinčić I (2011), Fric Jar (1895-1953): od zaborav do perspektivi na evropskata bioetika [Fritz Jahr (1895-1953): from oblivion to perspectives of European bioethics], in: Filozofija 10(31)33-43; Muzur A (2011), Rijeka na bioetičkoj karti svijeta" [Rijeka on the bioethics world map], in: Sušačka revija 19(73)15-21; Rinčić I, Muzur A (2011), Fritz Jahr: the invention of bioethics and beyond, in: Perspectives in Biology and Medicine 54(4)550-556.

Section 3

Stems of Integrative Bioethics

BIOETHICS, FRITZ JAHR, AND THE CULTURE WARS: MORAL REFLECTION IN THE FACE OF INTRACTABLE MORAL PLURALISM

H.T. Engelhardt jr.[1]

One Term, Two Different Projects

In 1971, the term bioethics was either coined *de novo* or re-appropriated from the past and given a new intension and extension. The biomedically-directed bioethics engendered in the early 1970s has come to identify an academic field established in universities across the world, as well as a clinical field engaged primarily, but not exclusively, in hospitals (Engelhardt 2011b). Bioethics in the biomedical sense of the term is now institutionally established across the world. The term that identifies this latter and later field turns out to have been coined almost a half century earlier to identify a quite different set of concerns, namely, moral concerns, regarding the proper relationship of man with animals, plants, and nature in general. Although there are some family resemblances between the two fields in that both concern moral issues bearing on the use of living beings, they are distinct in content, character, and history. In particular, the biomedically-directed bioethics quickly became established in the academy and in the clinic, unlike the bioethics directed to the biosphere.

This brief study contrasts the early bioethics of Fritz Jahr (1927) with that which emerged in the 1970s focused on biomedical issues. Special attention is given to the circumstance that both fields: are marked by moral pluralism and by an absence of a substantive moral foundation. The result is that it is impossible for either bioethics uncontroversially to become a moral-political movement. Both the bioethics envisaged by Fritz Jahr in the third decade of the 20th century and that bioethics framed in the 1970s focused on biomedicine, although they have a quite different extension and intension. Both confront the intractable moral pluralism that marks secular morality. It is this pluralism that blocks either bioethics from their initial aspiration to become an uncritically accepted moral and political movement. In both uses, as will be shown, such aspirations are impossible, given the fallen human condition marked by both moral pluralism and the absence of foundations.

[1] H Tristram Engehardt, jr, PhD, MD, Professor of Philosophy, Rice University, Houston TX, Professor Emeritus, Baylor College of Medicine, Houston TX, htengelh@rice.edu

Both Bioethics in the Face of Moral Pluralism

Hans-Martin Sass has made a major contribution to the better understanding of the phenomenon of bioethics through his exploration of the earliest use of the term (Sass 2007). His contribution is not just to a better understanding of the term's origin, which often was attributed to Van Rensselaer Potter's (1911-2001) use of the term in 1970 and 1971 (Potter 1970, 1971). Indeed, such accounts often overlooked Potter's use of the term in 1949, in which Potter connects his concept of bioethics to the reflections of the environmentalist Aldo Leopold (1887-1948). In addition and crucially, Sass appreciates Jahr's quite different notion of bioethics, focused not on biomedical / bioethical issues, but on supporting an ethos for interactions with the biosphere. Jahr, like Potter who follows him, ties bioethics to issues that are concerns of environmentalist movements.

Anticipating Potter, Fritz Jahr's first use of the term in 1927 connects bioethics to establishing proper relationships with other living beings, and thus generally with nature (Jahr 1927, 2011). Both Fritz Jahr and Van Rensselaer Potter do not envisage bioethics as a critical, moral-philosophical reflection regarding what ethos should direct man to his proper place in nature. The hope instead was to engender and sustain an encompassing ethos of human responsibility for the world around us, from humans to animals and nature as such. The accent was more on a moral project than on critically addressing moral claims and arguments. Although the broad coalition of moral concerns may in retrospect appear to have had the resources to compass biomedical moral issues as well, Jahr's and Potter's bioethics have failed to achieve this wide scope. Moreover, such was likely impossible. As with much in our morally fragmented culture, such an attempted broad coalition of moral visions would only have been an occasion for deeper disagreement.

The grounds for this state of affairs lie in the intractable character of secular moral and of metaphysical pluralism, which undermines any possibility of a common secular moral vision and therefore of a common ethos to direct man's relationship with nature or to shape a common moral biomedically-directed bioethics, much less of an encompassing ethos embracing both areas of concerns. In addition, the biosphere directed bioethics of Jahr and Van Rensselaer Potter, who want bioethics to embrace a broad range of moral concerns with regard to man, animals, and nature, was advanced more as a social and moral movement rather than as an area of critical philosophical reflection. Bioethics when it began as a biomedically-directed concern in the early 1970s did so in an academic locus where moral and philosophical disagreements were ubiquitous and manifestos rare.

Bioethics in the biomedically-directed use of the term as re-coined or reapplied in 1971 at the Center of Bioethics of the Kennedy Institute at Georgetown University by André Hellegers (1926-1979) and Sargent Shriver (1915-2011) has itself failed, as already noted, to maintain a field with the coherence and focus for which many of its founders had hoped.[2] On the one hand, the Center for Bioethics was at a Roman Catholic university that sustained a general commitment to natural law, even when cut free from the original roots of these views in Roman Catholic Scholastic reflection, presumed that a common morality and bioethics existed for which a rational philosophical grounding could be given. This general view of morality and its foundations had its ultimate roots in the Scholastic faith in reason, which had developed in the 13th century and was expanded in the 16th-century Second Scholasticism. These intellectual commitments lay as a background against which Immanuel Kant (1724-1804) could develop his claims of human dignity. All of this gave force to various late-Enlightenment hopes of finding a common morality and bioethics drawing on a faith in reason that could be traced to the Western medieval synthesis of faith and reason that fashioned Roman Catholicism. This synthesis of faith and reason provided the grounds, when faith in faith was attenuated, if not abandoned after the Thirty Years' War (1618-1648) and the British Civil War (1642-1651), for the rationalist faith in moral philosophy to become the force that marked European modernity and the Enlightenment. That faith in reason that underlies a faith that there is a common morality and common bioethics has proven ungrounded. On the other hand, biomedically-directed bioethics took shape in a cultural area marked by previous millennia of disputes and by being placed in the academy where intellectual disagreement was legion.

The cultures of the West are fragmented by disagreements as to when it is forbidden, licit, or obligatory to have sex, reproduce, take human life, transfer private property, or act on evidence for global warming. These disagreements have prevented the biomedically-oriented bioethics from succeeding as a progressive and united socio-democratic environment on analogy with the moral and social biosphere-directed movement proposed by Fritz Jahr and Van Rensselaer Potter. Thus, one finds a lament on the part of some that bioethics did not become the unambiguous moral movement for which some of its early supporters such as the Kennedys and Shrivers might have hoped. Biomedically-directed bioethics is a theater for disagreement.

---

[2] Controversy exists as to who actually re-coined the term bioethics at the Kennedy Institute. See Reich 1994. Sargent Shriver, it should be noted, claimed that he was the author of the new usage (Shriver 2001).

The rancorousness has been conspicuous in statements made by bioethicists on both sides of the divide in professional publications, blog sites, and the media about matters such as embryonic stem cell research, therapeutic and reproductive cloning, genetic enhancement, other uses of fetal tissues and embryos, end-of-life issues, the place of religious thought and belief in ethical analysis, and about the potential dangers and harms as well as benefits that advances in medical science and technology can bring in their wake (Fox & Swazey 2005, p. 369).

Instead of moral agreement and unity, there are now indeed foundational concerns about the coherence of the field (Engelhardt 2011a). This state of affairs should not be unanticipated. As with all areas of serious moral concern in our contemporary culture, the bioethics of the biomedically-directed genre and the biosphere-directed genre are marked by controversies and disagreement. As Alasdair MacIntyre has noted, "The most striking feature of contemporary moral utterance is that so much of it is used to express disagreements; and the most striking feature of the debates in which these disagreements are expressed is their interminable character" (MacIntyre 1981, p. 6). Bioethics is embedded in a discourse of moral dispute. The possibility of united moral movements and of common bioethical visions is brought into serious question and undermined.

Bioethics after Morality

These concerns should not be considered as merely a function of moral pluralism but also as resulting from a loss of faith in secular morality itself. Even Tom Beauchamp has come implicitly to acknowledge the foundationless character of biomedically-directed bioethics in recognizing the failure of biomedically-directed bioethics to secure a canonical philosophical foundation. The very sense of seeking a sound rational justification to give a foundation for a canonical bioethics is thus brought into question. However, the attempt to ignore this state of affairs is at least materially equivalent to return to a pre-critical innocence before the Fall of moral disagreement and the clear hopelessness of the search for moral-philosophical foundations. Beauchamp is thus apparently at peace with dispensing with the search for a moral-theoretical basis for bioethics. Consider Beauchamp's equanimity in the face of the marginalization of moral theory in the field of bioethics. "This [moral] theory part of the landscape of bioethics I expect to vanish soon, because it is serving no useful purpose" (Beauchamp, p. 210). Theory can neither provide a basis for sound rational argument establishing a single, canonical, content-full moral vision or bioethics nor establish a particular set of content-full moral principles as canonical. Beauchamp is willing under the circumstances to recognize cardinal unclarity regarding central bioethical concerns and principles.

What it is about autonomy that we are to respect remains unclear, and it remains obscure what "respect" means. Most obscure of all is how practice is affected by a theory of autonomy. The contemporary literature in bioethics contains no theory of autonomy that spells out its nature, its moral implications, its limits, how respect for autonomy differs from respect for persons (if it does), and the like (Beauchamp, p. 214).

The problem is general. One cannot expect that philosophy can justify a common secular morality or bioethics, whether for the bioethics directed to the biosphere or for the bioethics directed to biomedicine.[3] An unambiguous account of even such cardinal concepts as autonomy cannot be given. In the face of intractable moral pluralism, disagreement and controversy are the norm.

Against the background of intractable moral pluralism,[4] and given the increased realization that secular morality compasses a plurality of incompatible moralities corresponding to diverse clusters of moral intuitions

[3] Richard Rorty saw the end of the naive view of morality that had come to the West from its Middle Ages. In particular, he recognized the contingent, socio-historically-conditioned character of all secular moral understandings. We can keep the notion of 'morality' just insofar as we can cease to think of morality as the voice of the divine part of ourselves and instead think of it as the voice of ourselves as members of a community, speakers of a common language. We can keep the morality-prudence distinction if we think of it not as the difference between an appeal to the unconditioned and an appeal to the conditioned but as the difference between an appeal to the interests of our community and the appeal to our own, possibly conflicting, private interests. The importance of this shift is that it makes it impossible to ask the question "Is ours a moral society?" (Rorty 1989, p. 64). - Rorty developed this position because, as he recognized, and as philosophers almost two millennia previously had recognized, "there is no way to step outside the various vocabularies we have employed and find a metavocabulary which somehow takes account of *all possible* vocabularies, all possible ways of judging and feeling. A historicist and nominalist culture of the sort I envisage would settle instead for narratives which connect the present with the past, on the one hand, and with utopian futures, on the other" (Rorty 1989, p. xvi).

[4] The recognition that philosophical argument cannot establish a canonical philosophical position, or for that matter resolve foundational moral and metaphysical disagreements has been known since ancient times. Clement of Alexandria (A.D. 155-220) noted this failure in the absence of the ability to reach conclusive philosophical agreement regarding first principles (*The Stromata*, Book 2, chapter IV). On this general issue, the philosopher Agrippa (fl. 3rd century A.D.) showed that philosophical arguments invariably begged the question, argued in a circle, or engaged in an infinite regress. See Diogenes Laertius, *Lives of eminent philosophers*, Pyrrho 9, 88-89, and Sextus Empiricus, Outlines of pyrrhonism I.15.164-169.

sustained by a diversity of moral narratives floating without foundations within the horizon of the finite and the immanent,[5] one can expect only divisive disagreements confronting any attempt to revive the bioethics of Fritz Jahr and Potter. In addition, an attempt to amalgamate Fritz Jahr's bioethics with the bioethics of the Kennedy Institute would only compound disagreement. The project of establishing a particular ethos, whether for the conduct of medicine and the biomedical sciences or to direct man's relationship with nature, runs aground on intractable moral pluralism. The expectation that one can proclaim a morally compelling bioethical vision of whatever genre, and that it would then be generally embraced, is crucially to misunderstand the contemporary moral geography.

One has to rethink the force of secular morality and bioethics. It cannot have the force, as that had been expected within the divine command views of morality, as with Orthodox Judaism or traditional Christianity. The very force of morality and bioethics has been deflated. Secular morality and the various bioethics become just a diversity of moral visions articulated and directed by an agnostic methodological postulate so that in the end all is approached as if all came from nowhere, were going nowhere, and for no ultimate purpose. Morality and bioethics in both senses are as a consequence radically deflated in their moral force, in that it cannot be established that any particular morality or bioethics is canonical, as well as why one ought always to act morally or bioethically, whatever that might mean. If no morality or bioethics is canonical, and if all is ultimately meaningless, why would it always be rational to act morally or bioethically? One should recall that it is for this reason, namely, to maintain the unity of morality and its priority, that Kant, who was likely an atheist,[6] affirmed God and immortality in *Critique of Practical Reason*. Given this background state of affairs, one must critically reconsider the biosphere-directed bioethics of Fritz Jahr, as well as the biomedically-directed bioethics born of the Kennedy Institute (Engelhardt 2010b).

---

[5] It is interesting to note that some fourteen years after his publication of *A Theory of Justice* John Rawls clearly recognized that his much-vaunted account could not be provided with moral or metaphysical foundations (Rawls 1971). His position could at best be defended as the ingredient framework for a social-democratic constitutional regime (Rawls 1985).

[6] Manfred Kuehn states, "Kant did not really believe in God" (Kuehn 2001, p. 391f).

Fritz- Jahr-Style Bioethics in the Absence of God and in the Presence of Moral Pluralism

To sustain a concrete morality or bioethics as canonical, one needs to possess a canonical ranking of basic values and right-making conditions. Absent such a canonical ranking, one is faced with radically different views of the proper relation of humans with humans, and humans with animals, plants, and nature generally. One can better appreciate this state of affairs if one again recalls how in bioethics, biomedically-construed, there are deep conflicts regarding abortion and physician-assisted suicide. While some proponents of access to abortion and physician-assisted suicide see the latter as integral to the reproductive freedom of women and the former as integral to autonomy in end-of-life decision-making, some opponents of abortion recognize that abortion involves the murder of an unborn child and physician-assisted suicide as tantamount to physician-assisted self-murder. Those at disagreement cannot resolve their differences because they do not share basic premises or rules of evidence. Those at controversy are separated by incompatible basic moral and metaphysical premises. The same obtains in the bioethics directed to the biosphere.

As with bioethics biomedically construed, so, too, with bioethics directed to the biosphere there are irresolvable moral disagreements regarding the proper use of animals and of man's relationship with nature. Consider those who affirm the moral good of bullfighting and of the hunt. They will hold that the pain caused to animals is greatly outweighed by the refined pleasure given to humans, which pleasure is enhanced in its quality and refined within complex cultural settings that can count as goods in themselves. This cultural context is not available to non-human animals who have no culture of story-telling and reflection, much less a literature that could bear on bullfighting and the hunt. The proponents of bull fighting and the hunt could argue that the recalling of the fight and the hunt in stories shapes a unique and rich cultural treasure. Moreover, as the proponents of bullfighting and the hunt could observe, humans are persons, while other animals are not. Thus, one is able coherently to hold that humans as persons should be protected from being used as means merely. However, other animals in not being self-conscious, rational agents are not ends in themselves. It is because humans are ends in themselves, the proponents of bullfighting and the hunt will contend, that one can understand the moral appropriateness of protecting humans from being brutally killed by other humans or by lions, but not as easily understand a duty to protect non-rational animals such as zebras from being brutally killed and eaten by lions. If animals had the strong right of moral agents, they should have rights not to be killed by other animals (e.g., zebras by lions). The proponents of bullfighting and the hunt can only consider as incoherent

Fritz Jahr's injunction: "Respect every living being on principle as an end in itself and treat it, if possible, as such!" (Jahr 1927, p. 4).

Of course, the defenders of strong animal rights will invoke one or more among a plurality of moral visions and narratives that are congenial to supporting some form of claims on behalf of the rights of animals (e.g., appeals to the interests of the animals) not to be used to give humans pleasure, as through bullfighting and the hunt. Those in disagreement will surely argue past each other, each holding the other to be a moral reprobate with perverse moral sentiments and a limited moral vision. And this in fact is the state of affairs we encounter: those in disagreement are embedded in incompatible and opposing lifeworlds so that the accounts of what is morally at stake are incompatible. Jahr and his allies will affirm his bioethical imperative, only to be met with a humanistic imperative that recognizes bullfighting and the hunt as integral to living *humaniter*, as living a truly human life.

Similarly, one can have different views regarding the environment. On the one hand, anticipating many contemporary environmentalists, Fritz Jahr can best be interpreted as wishing to preserve some ideal vision of the environment, which he finds compelling and which he wishes to establish as the canonical environment. However, the environment is always in flux, so that to support global warming is not to fail to support or have reverence for the environment. It is rather to affirm a warmer environment, one that existed in the past and that may exist again in the future. Thus, protagonists of global warming may hope to restore to Alaska, northern Canada, and northern Russia the possibility of having two or three harvests a year. To those not of like mind who object that they are losing their beloved pre-global-warming environment, the advocates of global warming might retort: it is now our turn to have our environment of choice (de Sherbinin et al. 2011). The cardinal problem is that in order to identify the "bioethically" appropriate or canonical environment, one must already know which value perspective is canonical. The difficulty is that without already knowing the right answer ahead of time, that is, without possessing a canonical standard one cannot know what the right answer should be about which environment is appropriate or normative. As a consequence, the bioethics of Fritz Jahr fails to identify a canonical morality and bioethics of the environment, just as the biomedically-oriented bioethics born of the Kennedy Institute failed to discover *the* canonical morality or bioethics to guide health care and the biomedical sciences.

The Higher Truth of Fritz Jahr's Bioethics

Kant recognized that, without an effective God's-eye perspective, morality would fracture into an intractable plurality and would no longer necessarily trump the claims of self-interest (Engelhardt 2010b). He hoped to block the collapse of Western morality and moral philosophy that the West because of its Western Christian synthesis of faith and reason had taken for granted. Hegel was willing to face this state of affairs: a culture after the death of God and of Western Christianity.[7] He recognized that the canonicity to which secular morality could aspire could only be that of being established at law and in enforceable public policy (Engelhardt 2010a). After Kant and in the spirit of Hegel,[8] the force of secular morality and bioethics is then radically deflated. The force of morality, as well as of both senses of bioethics, becomes simply that the morality and bioethics are politically successful and legally established, but not norms the breach of which require one to consider oneself no longer unworthy of happiness, but only in some danger of legally enforced punishment. Morality and bioethics both as focused on the biosphere and on biomedical issues are thus socio-historically-conditioned, regional, and ultimately contingent sets of norms that depend for their "canonical force" on being established at law.

Bioethics in both senses are then best understood as a cluster of competing socio-political movements with diverse agendas, none of which has a foundation in a conclusive sound rational argument. Such moral proposals become at best biopolitical proposals. This is in the end the most charitable interpretation of the bioethics of Fritz Jahr, Van Rensselaer Potter, and the Kennedy Institute. As already observed, absent a canonical foundation, each of the particular moral and bioethical visions represents a particular cluster of moral intuitions, supported by a particular moral narrative, floating without a canonical foundation, all within the horizon of the finite and the immanent. To achieve anything more than the status of one among a great plurality of moral, biomedical, and environmental proposals and movements, a moral or

[7] Already in 1802, Hegel spoke of the death of God. God for Hegel was dead because He was no longer acknowledged as central in the vanguard culture of Hegel's time. That culture, characterized by a failure to acknowledge God as a transcendent noumenal entity, was after God. As a result, this culture became characterized by "the feeling that 'God Himself is dead'" (Hegel 1977, p. 190; Hegel 1968, p. 414).

[8] There are numerous accounts of Hegel's project. The author of this essay builds on the general philosophical position of Klaus Hartmann's (1925-1991) non-metaphysical account of Hegel's project. Hartmann recognized that Hegel's mature project could only coherently be understood as post-metaphysical. As Hartmann's best-known article in English puts it, Hegel's position was "non-metaphysical" (Hartmann 1972). Klaus Hartmann's view of Hegel's project is available in Hartmann 1999, 1988, 1981, 1976, 1970, 1966. For an exploration of Hartmann's account of Hegel, see Engelhardt & Pinkard 1994.

bioethical proposal or movement must achieve the legal establishment. Again, the "moral" status is a legal status, not the status of a moral vision embedded in reason as such.

None of this is a direct criticism of the bioethics of Fritz Jahr. It is rather an attempt to locate its significance within the geography of secular moral reflection "after God" and in the face of secular moral pluralism. Also, none of these reflections addresses what a bioethics, including an environmental bioethics, should look like, carried out in the face of the God Who commands.[9] Traditional Christians know that God gives man rulership over all animals, declaring, "Rule over the fish of the sea and the birds of the air and over every living creature that moves on the ground" (Gen 1:28). God then expands this authority in giving to Noah and his sons an even stronger rulership. "The fear and dread of you will fall upon all the beasts of the earth and all the birds of the air, upon every creature that moves along the ground, and upon all the fish of the sea; they are given into your hands. Everything that lives and moves will be food for you. Just as I gave you the green plants, I now give you everything. ... I now establish my covenant with you and with your descendants after you and with every living creature that was with you - the birds, the livestock and all the wild animals, all those that came out of the ark with you – every living creature on earth" (Gen 9:2-3, 9-10). Against this background, this vision of Fritz Jahr is substantively post-Christian in a traditional sense of Christian. Absent commonly acknowledging Who the true God is, the theological turn itself generates further controversy.

A Skeptical Postscript

These reflections on the status of morality and bioethics lead to the conclusion that the future of bioethics both in Fritz Jahr's sense and in that sense born of the Kennedy Institute is intractable disagreement. To judge the contemporary force of Fritz Jahr's contributions, one will therefore need to re-interpret the force of bioethics in a sense incompatible with the universalist aspirations of classical Kantian and utilitarian theories. One will indeed need in part to acknowledge how Fritz Jahr in an innocent, indeed naive fashion held his intuitions to be the canonical intuitions regarding how man should be related to animals, plants, and nature. In these circumstances, it will be best to regard Fritz Jahr as one of the many who contributed to the cluster of environmental movements and their various ethics. He in addition coined a term that came to have a significant influence, although for a quite different field with a quite different intension and extension. One can better appreciate his contribution to the history of the moral-political controversies in the 20th

---

[9] For an account of a bioethics grounded in a recognition of morality derived from the commands of a transcendent God, see Engelhardt 2000.

century and afterwards. The fact that his term "bioethics" through historical accident became the term that now identifies the controversy-laden field of biomedical bioethics should also remind those reflecting on his work that its contributions have been transformed by the moral pluralism and lack of foundations that mark all secular morality. Jahr's bioethics, if it is revived, will be received within a public space defined by moral disagreement in general and with regard to man's relationship to animals, plants, and nature in particular.

References

Beauchamp, Tom L. 2004. Does ethical theory have a future in bioethics? *Journal of Law, Medicine, and Ethics* 32: 209-217.

de Sherbinin, A., M. Castro, F. Gemenne, et al. 2011. Preparing for resettlement associated with climate change, *Science* 334 (October 28): 456-457.

Engelhardt, H. T., Jr. (ed.). 2011a. *Bioethics critically reconsidered.* Dordrecht: Springer.

Engelhardt, H. T., Jr. 2011b. Core competencies for health care ethics consultants: In search of professional status in a Post-modern world, *HealthCare Ethics Committee Forum* 23.3 (September): 129-145.

Engelhardt, H. T., Jr. 2010a. Kant, Hegel, and Habermas: Reflections on "Glauben und Wissen", *The Review of Metaphysics* 63 (June): 871-903.

Engelhardt, H. T., Jr. 2010b. Moral obligation after the death of God: Critical reflections on concerns from Immanuel Kant, G.W.F. Hegel, and Elizabeth Anscombe, *Social Philosophy & Policy* 27.2 (Summer): 317-340.

Engelhardt, H. T., Jr. 2000. *The foundations of Christian bioethics*. Salem, MA: Scrivener Publishing.

Engelhardt, H. T., Jr. & Terry Pinkard (eds.). 1994. *Hegel reconsidered.* Dordrecht: Kluwer.

Fox, Renée C. & Swazey, Judith P. (2005). 'Examining American bioethics: Its problems and prospects,' *Cambridge Quarterly of Healthcare Ethics* 14: 361-373.

Hartmann, Klaus. 1999. *Hegels Logik*, ed. Olaf Müller. Berlin: Walter de Gruyter.

Hartmann, Klaus. 1988. *Studies in foundational philosophy*. Amsterdam: Rodopi.

Hartmann, Klaus. 1981. *Politische Philosophi*e. Munich: Alber.

Hartmann, Klaus (ed.). 1976. *Die ontologische Option*. Berlin: Walter de Gruyter.

Hartmann, Klaus. 1972. Hegel: A non-metaphysical view. In *Hegel*, ed. Alasdair MacIntyre (pp. 101-124). Garden City, NY: Anchor Books.

Hartmann, Klaus. 1970. *Die Marxsche Theorie*. Berlin: Walter de Gruyter.

Hartmann, Klaus. 1966. On taking the transcendental turn, *Review of Metaphysics* 20.2: 223–249.

Hegel, G.W.F. 1977. *Faith & knowledge,* trans. Walter Cerf & H.S. Harris. Albany: State University of New York Press.

Hegel, G.W.F. 1968. *Jenaer kritische Schriften*, in *Gesammelte Werke*, volume 4. Hamburg: Meiner.

Jahr, Fritz. 2011. *Essays in bioethics and ethics 1927-1947*. Bochum: Zentrum für medizinische Ethik. See in this volume.

Jahr, Fritz. 1927. Bio-Ethik. Eine Umschau über die ethischen Beziehungen des Menschen zu Tier und Pflanze, *Kosmos. Handweiser für Naturfreunde*, 24.1: 2-4.

Kuehn, Manfred. 2001. *Kant: A biography*. Cambridge: Cambridge University Press.

MacIntyre, Alasdair. 1981. *After Virtue*. Notre Dame, IN: University of Notre Dame Press.
Potter, V.R. 1949. *Global bioethics - building on Leopold's legacy.* East Lansing, MI: Michigan State University Press.
Potter, V. R. 1970. Bioethics: the science of survival, *Perspectives in Biology and Medicine*, 14: 127-173.
Potter, V.R. 1971. *Bioethics: bridge to the future.* Englewood Cliffs, NJ: Prentice Hall.
Rawls, John. 1971. *A theory of justice*. Cambridge, MA: Belknap Press.
Rawls, John. 1985. Justice as fairness: Political not metaphysical, *Philosophy and Public Affairs* 14.5 (Summer): 223-251.
Reich, Warren. 1994. The word 'bioethics': Its birth and the legacies of those who shaped it, *Kennedy Institute of Ethics Journal* 4.4: 319-336.
Rorty, Richard. 1989. *Contingency, irony, and solidarity.* New York: Cambridge University Press.
Sass, Hans-Martin. 2007. Fritz Jahr's 1927 concept of bioethics, *Kennedy Institute of Ethics Journal* 17.4 (December): 279-295.
Shriver, Sargent. 2001. Letter to author, January 26.

THE EUROPEANIZATION OF BIOETHICS*

Ante Čović[1]

It seems that the ultimate goal at which our bioethical discussions, research, and projects should aim is most clearly represented by the idea of the Europeanization of bioethics. However, in order to properly understand that idea, it is necessary to explain three key preconditions upon which it rests: the first reveals a global-historical reason of the emergence of bioethics; the second reveals its character, while the third precondition uncovers the immanent logic of bioethics development.

The main and most important reason of the birth of bioethics should be looked for in the fact that the contemporary science has lost its orientation authority and existential competence for a contemporary human. Therefore, bioethics should be primarily understood as intellectual and existential field which creates orientation within this, in principle, new situation which is highly determined by factors of scientific and technological development. The orientation can still be reached in differently set situations, different means (methods) and different levels of elaboration. So far, bioethics has provided orientation in a „new medical situation" by means of direct moral reflections (called *principalism*), and then in the more broadly set situation of globalized subject field by the means of ethical reflection within the interdisciplinary frame.

But since moral and ethical reflections, although necessary moments in establishing all forms of orientation, have proved to be insufficient methods for achieving orientation in the circumstances of threatening completion of scientific and technical civilization and the situation at the turn of the world historical epochs, the bioethics – following it orientation role, has become more aimed at the *philosophical historical reflection*. With this, it enters its new phase of development which will enable it to show all its orientation potentials. However, it implies the activation of not only history of philosophy but also the total Euro-continental philosophical tradition in the methodology pattern of bioethics. This clearly shows that the Europeanization

---

*This paper resulted from a "fusion" of two programmatic texts. The first one was written at the beginning of the project of developing the bioethical cooperation and founding the bioethical institutions in South-East Europe (2004), while the second one summed up the achieved results at the end of the project (2010). For the complete versions, see: Čović A (2011) Bioethics in South and Southeast Europe, in: Jahr 2(3)13-14 and Čović A (2010) Vorwort des Herausgebers, in: Integrative Bioethik und Pluriperspektivismus. Sankt Augustin: Academia Verlag, 1-3.

[1] Ante Čović, ante.covic1@zg.t-com.hr, University of Zagreb, Faculty of Humanities and Social Sciences, Department of Philosophy, I. Lučića 3, 10 000 Zagreb, Croatia.

of bioethics is not an internal nor institutionalized problem of European bioethics, but a constitutional and developmental imperative of bioethics as such. This raises the question: why should the extensive task of Europeanization of bioethics and establishing the European bioethical horizon be carried out by those who live or those who gather at bioethical conferences on the geographic periphery of Europe. The answer is simple: because the horizon is the most open on the periphery.

The historical development of bioethics towards the stage of integrativity proceeded on the methodological plan as a gradual expansion of the circle of perspectives on addressing bioethical problems, from accepting ethical plurality to scientific interdisciplinarity and pluriperspectivity. On the other hand, the processes of both expanding and deepening of problem areas in bioethics were taking place in parallel, ranging from healthcare problems and biomedical research to issues of non-human living beings and general conditions of life sustainment (ecology), and exploring the profound philosophico-historical dimension of these problems.

The history of bioethics can be divided into three development stages in which major breakthroughs were achieved in the methodological constituting of bioethics and in determining its scope of research. Development changes were competently confirmed in the *phase definitions of bioethics,* which can be found in the first and the second edition of the *Encyclopedia of Bioethics* and in newer texts that emerged within the project of "integrative bioethics". The precise period delimitations of these development stages are rendered more difficult by the fact that the *phase forms of bioethics* continued to exist in parallel. Let us briefly summarise the development stages of bioethics with their corresponding phase definitions and key publications that articulate the features of its methodology and subject-matter in the respective periods:

1. *The development stage of "new medical ethics".* In its initial stage, bioethics was methodogically defined by immediate moral reflection (principlism), while its subject was restricted to healthcare and biomedical research. In the Introduction to the first edition of the *Encyclopedia of Bioethics* (1978), W. T. Reich defined bioethics in the first stage of its development as "… the systematic study of human conduct in the area of the life sciences and healthcare, insofar as this conduct is examined in the light of moral values and principles."[2] A year later, T. L. Beauchamp and J. F.

[2] Reich WT (1995) Introduction, in: Encyclopedia of Bioethics, vol. I. New York: Simon & Schuster Macmillan, XXVIII.

Childress published the influential book *Principles of Biomedical Ethics*,[3] which became the fundamental work of principlism;

2. *The development stage of "global bioethics".* In the second stage of development, bioethics was methodologically defined by means of *ethical pluralism and scientific interdisciplinarity,* while its scope of research expanded to the problems of life in general and to the socio-political and global-ecological context. In the second edition of the *Encyclopedia of Bioethics* (1995), W. T. Reich re-defined bioethics, noting the methodological transformation it underwent in the meantime and designating it as "… the systematic study of the moral dimensions – including moral visions, decisions, conduct and policies – of the life sciences and health care, employing a variety of ethical methodologies in an interdisciplinary setting."[4] A decisive role in the expansion of the scope of research in bioethics on global scale was played by V. R. Potter, "the father of bioethics", who published his second book in this period, which was significantly entitled *Global Bioethics: Building on the Leopold Legacy* (1988);[5]

3. *The development stage of "integrative bioethics".* From the methodological point of view the current development phase of bioethics is marked by the idea of *pluriperspectivity,* while its scope of research is deepened by analysing its philosophico-historical dimension which sheds light on the scientific-technical character of the phase and the role of modern science. It considers the changes in the fundamental relationships of man towards historical givens and detects the processes of turns of the world-historical epochs. In the dialogue framework of bioethical co-operation which started developing in 2004 from a centre in Croatia in the region of South-east, South and Central Europe, bioethics is defined as "… a pluriperspective area in which the interplay of heterogenous perspectives establishes the footholds for orientation in the issues related to life or conditions of its sustainment".[6] A large number of papers have been published within the project of "integrative bioethics", the most notable of which for our topic is

---

[3] Beauchamp TL, Childress JF (1979) Principles of Biomedical Ethics. New York: Oxford University Press.

[4] Reich WT, op.cit., XXI.

[5] Potter VR (1988) Global Bioethics: Building on the Leopold Legacy. East Lansing: Michigan State University Press.

[6] Čović A (2005) Bioethik unter den Bedingungen des Postkommunismus – Fallbeispiel Kroatien, in: Bioethik und kulturelle Pluralität: Die südosteuropäische Perspektive. Sankt Augustin: Academia Verlag, 150–151.

the paper by Hrvoje Jurić entitled “The Footholds of an Integrative Bioethics in the Work of Van Rensselaer Potter”.[7]

While bioethics in its first two development stages formed mainly in the United States, its most recent breakthrough in development moved to Europe, to the area of South-east Europe where it became the centre of an intense and productive bioethical discussion.

It was in this period of the rounding off of the concept of integrative bioethics that Hans-Martin Sass published a series of papers dedicated to the work of the German theologist and pastor Fritz Jahr who, in 1927, not nly introduced the term *bio-ethics* (“Bio-Ethik”) in the popular science journal *Kosmos* but also elaborated the original concept of that discipline,[8] which can be considered as the predecessor of the idea of integrative bioethics because of its methodological and subject-matter aspects. This coincidence opened the horizon of European bioethics, paving the way for an appearance in full splendour of bioethical contributions in Eurocontinental philosophical thought and spiritual tradition in general, ranging from Plato and Francis of Assisi to Kant and Albert Schweitzer.

7 Cf. Čović A, Hoffmann TS (2007) Integrative Bioethik: Beiträge des 1. Südosteuropäischen Bioethik-Forums, Mali Lošinj 2005. Sankt Augustin: Academia Verlag, 68-92.

8 Cf. Sass HM (2007) Fritz Jahr's bioethischer Imperativ. 80 Jahre Bioethik in Deutschland von 1927 bis 2007, in: Medizinethische Materialien des Zentrums für Medizinische Ethik 175.

FRITZ JAHR AND EUROPEAN VALUES IN BIOETHICS

Matti Häyry[1] and Tuija Takala[2]

Introduction

Fritz Jahr's notion of the bio-ethical imperative, first presented in his 1927 essay "Bio-Ethik: Eine Umschau über die ethischen Beziehungen des Menschen zu Tier und Pflanze",[3] can be seen as the earliest European expression of some of the ideas governing the field that has come to be called bioethics. The notion will be explained right after this introductory section. In this paper, we approach Jahr's imperative as a mid-level principle in the sense defined by Tom Beauchamp and James Childress in their book *Principles of Biomedical Ethics*.[4] In other words, we proceed from two assumptions. First, the bio-ethical imperative is either intuitively acceptable to all as a commonsense truth, or it is open to theoretical justification in all major moral frameworks.[5] A few words will be said below in defence of this assumption. Secondly, the imperative is a useful tool in defining more practical rules for everyday use in administrative and professional activities.[6]

In exploring the validity of the second assumption, we will take up an idea presented by Raanan Gillon in his analysis of Beauchamp and Childress's four principles. Gillon argued that while autonomy, non-maleficence, beneficence, and justice are all equally important values in medical and healthcare ethics, autonomy (perhaps in a more profound sense) still remains

---

[1] Matti Häyry, PhD, matti.hayry@manchester.ac.uk, Professor of Bioethics and Philosophy of Law, University of Manchester, UK.

[2] Tuija Takala, PhD, Academy Research Fellow at the Academy of Finland and Adjunct Professor in Social and Moral Philosophy at the University of Helsinki, Finland.

[3] F. Jahr, "Bio-ethics: Reviewing the ethical relations of humans towards animals and plants," [1927], in this volume.

[4] T. L. Beauchamp and J. F. Childress, *Principles of Biomedical Ethics* [1979], fifth edition (New York: Oxford University Press, 2001).

[5] Beauchamp and Childress have kept changing their minds about this – see P. Herissone-Kelly, "The principlist approach to bioethics, and its stormy journey overseas," in M. Häyry and T. Takala, eds, *Scratching the Surface of Bioethics* (Amsterdam and New York: Rodopi, 2003): 65-77.

[6] As an example, we have shown how mid-level principles can be used in generating and justifying what we called mezzanine rules for ethical decision-making in the regulation of human genetic databases. See M. Häyry and T. Takala, "American principles, European values, and the mezzanine rules of ethical genetic data banking," in M. Häyry, R. Chadwick, V. Árnason and G. Árnason, eds, *The Ethics and Governance of Human Genetic Databases: European Perspectives* (Cambridge: Cambridge University Press, 2007): 14-36.

the "first among equals".[7] In the same vein, we will consider some contemporary European values in, or principles of, bioethics, and investigate the potential role of Jahr's rule among them. What interpretations should be given to the principles of dignity, solidarity, and precaution – all suggested as important European ideals – under the assumption that they are all governed or qualified by the over-arching bio-ethical imperative introduced by Jahr?

In what follows, we will first explain Jahr's imperative and its potential justifications. We will then describe the main three European principles of bioethics and their main differences from their American counterparts. These steps should allow us to analyse Jahr's imperative and its possible role as the first among equals in European bioethics.[8]

Jahr's bio-ethical imperative and its justification

The original formulation of the bio-ethical imperative is: "Respect every living being on principle as an end in itself and treat it, if possible, as such!"[9] Jahr observes that the scientific development started by Darwinian biology and then continued in physiology and psychology has made it difficult to uphold a strict boundary between humanity and other living beings.[10] He goes on to point out that this fact, made apparent by modern science, has much older roots in moral thinking. Many prominent philosophers and schools of thought have recognised the bio-ethical principle long before we had systematic knowledge concerning the physiological and psychological similarities between all living beings. Jahr takes as his examples Francis of Assisi and his great love for animals; Michel de Montaigne who granted all living beings an entitlement to be treated according to moral principles; Johann Gottfried von Herder who saw animals as man's older brothers; and Jean-Jacques Rousseau's romantic awe towards the nature as a whole.[11] Among major world religions, Jahr thinks that certain "fanatic" forms of Hinduism and Buddhism have taken the bio-ethical principle too far, striving to respect all life without considering the costs to humanity and other beings. But the more balanced respect for all living beings embedded in these religions is hailed by him as a healthy manifestation of the bio-ethical

---

[7] R. Gillon, "Ethics needs principles – four can encompass the rest – and respect for autonomy should be 'first among equals'," *Journal of Medical Ethics* 29 (2003): 307-312.

[8] Our analysis is doubtlessly influenced by the pioneering work done by Hans-Martin Sass, e.g. in H.-M. Sass, "Fritz Jahr's 1927 concept of bioethics," *Kennedy Institute of Ethics Journal* 17 (2008): 279-295.

[9] Jahr 1927, in this volume.

[10] Jahr 1927, in this volume.

[11] Jahr 1927; "Death and the animals: Contemplating the 5th commandment" 1928; "Three studies on the fifth commandment" 1934; all in this volume.

imperative.[12] Within the Christian faith, Jahr reminds us, already Apostle Paul recognised a commonality between man and animal; and Martin Luther believed that animals will be accepted into the Kingdom to come.[13] As a theologian himself, Jahr held that "if we want to honor the creator, we must regard with awe and respect his creatures" thus making the bio-ethical imperative a cornerstone of Christian belief.[14]

Jahr firmly believes that the bio-ethical imperative is intuitively accepted on the level of common sense and common morality. According to him, we have a long history of taking the welfare of animals into consideration in our dealings with them, and more recently laws have been enacted to make animal protection more practical. As for respecting plants, Jahr believes that "our intuition will lead the way, when it hinders us to kill plants right and left with a cane when doing a stroll or picking flowers only to throw them away after a short while, or when we detest the blind acts of destruction by uncivilized guys cracking the crowns of young trees along the road or in the woods."[15] Both animals and plants are, then, acknowledged as worthy of some moral consideration already.

As for philosophical theories, Jahr offers many partial justifications for his imperative. One line of argument is that expanding the sphere of care to animals and plants will make us more sensitive to human needs, as well. The moral obligation to do so is, according to Jahr, recognised by Kant who thought that compassion and mercy towards animals is a duty that man owes to himself.[16] Shifting from the vocabulary of duties to that of rights, Jahr notes that the nineteenth-century German philosopher Karl Christian Friedrich Krause argued that all living beings, including humans, have similar rights, although "not Equal Right", depending on their specific needs to reach their specific destinies.[17] This seems to give all living creatures moral entitlements that are based, in modern language, on their interests, an idea that is not too far removed from the notions advocated by teleological ethics and natural law theories. And finally, although Jahr would probably have shunned away from the epithet "utilitarian", there are notable consequentialist elements in his account. The avoidance of harm to all living beings can be classified under this rubric as a consideration of negative outcomes; and he

[12] Jahr 1927, in this volume.
[13] Jahr 1928, in this volume.
[14] Jahr 1928, in this volume.
[15] "Animal protection and ethics" 1928, in this volume.
[16] Jahr 1928b, in this volume.
[17] Jahr 1927, also 1928b, in this volume.

also stressed the importance of happiness as something that we all value over money, possessions, or other external things.[18]

Making an absolute and positive commitment to the wellbeing of every living creature would, however, be too much for Jahr.[19] He notes that "we content ourselves with the avoidance of unnecessary suffering",[20] thereby implying two norms. The happiness of all living beings need not guide our decisions if this would mean disregarding our own interests – we all struggle to live and this must be accepted and respected as a fact of life.[21] And, for the same reason, we do not have to be fanatic about avoiding harm to others, either. The bio-ethical imperative requires us to respect all living beings as ends in themselves and to treat them as such, but only "on principle" and "if possible". Living beings, including human beings, are entitled to look after their own interests first, since doing otherwise would mean disrespecting life in oneself. But the important addition that Jahr offers to anthropocentric ethics is that no living being should be harmed without prior consideration.

It is not clear from Jahr's 1927 article whether he meant his principle to be applicable only, or primarily, to human relations towards animals and plants (as the title of his article seems to suggest) or whether he intended it to guide all human interactions, including those among humans. There is some evidence supporting the latter reading in the original text: Jahr speaks, for instance, of "Bio-Ethics as the assumption of moral obligations not only toward humans, but towards all forms of life".[22] He could also be interpreted as saying that what bio-ethics does is simply to expand the moral duties that we have to our fellow humans to all living beings. In this sense his principle of bio-ethics can be seen to cover all our dealings with living beings, human and otherwise. In his essay "Animal protection and ethics" from 1928 Jahr seems to say that the primary function of his principle of bio-ethics is to guide our moral dealings with other human beings: "And if someone does not accept the validity of this principle, as far as it is concerned with animals and plants, then, in repeating what already was said, one nevertheless should follow it in recognition of the moral obligation toward human society in general."[23]

In this paper we will follow Hans Martin Sass in assuming that Jahr's imperative is an all-embracing ideal for bioethics – it is not limited to

[18] Jahr "Our doubts about God," 1933, in this volume.
[19] Jahr 1934a, in this volume.
[20] Jahr 1927, in this volume.
[21] Jahr 1928a, in this volume.
[22] Jahr 1927, in this volume.
[23] Jahr 1928b, in this volume.

relations between humanity and the rest of the natural world, but can and should also be extended to guide our moral dealings among our own species.[24]

European values in bioethics[25]

Jahr's works present, no doubt, the earliest European account of a moral doctrine in which "bio" and "ethics" are combined (although similar ideas were expressed at about the same time by, for instance, Albert Schweitzer[26] in his "reverence for life" view). But the idea was somehow lost, and practical ethics had to be re-invented in the context of human rights, justifiable warfare, medicine, and science.[27]

The last decade has seen the rise of something that can be called European values in bioethics. This has, to a considerable extent, been a reaction to external influences. Many philosophers and ethicists have viewed the four principles presented by Beauchamp and Childress, as well as other American attempts to deal with moral issues, with suspicion.[28] European – and probably also global – policies, they have argued, should be based on European values, which are more reflective, communal, and ethical than the pragmatic codes imported from the United States.

The fear that American values suppress their European counterparts if the four-principle model is accepted is most probably misplaced. Beauchamp and Childress originally argued that the ideals of autonomy, nonmaleficence, beneficence, and justice can be based both on the duty-based moral philosophy of Immanuel Kant, and on the outcome-based ethics of Jeremy Bentham and John Stuart Mill.[29] Since these doctrines originate from

---

[24] Sass 2008.

[25] The contents of this section have been adopted from M. Häyry, "European values in bioethics: why, what, and how to be used?" *Theoretical Medicine and Bioethics* 24 (2003): 199-214.

[26] A. Schweitzer, "The ethics of reverence for life," *Christendom* 1 (1936): 225-239.

[27] M. Häyry, *Liberal Utilitarianism and Applied Ethics* (London and New York: Routledge, 1994), p. 147 ff.

[28] S. Holm, "Not just autonomy – the principles of American biomedical ethics," *Journal of Medical Ethics* 21 (1995): 332-338; J. D. Rendtorff and P. Kemp, eds, *Basic Ethical Principles in European Bioethics and Biolaw* Vols 1 & 2 (Copenhagen and Barcelona: Centre for Ethics and Law & Institut Borja de Bioètica, 2000); T. Takala, "What is wrong with global bioethics? On the limitations of the four principles approach," *Cambridge Quarterly of Healthcare Ethics* 10 (2001): 72-77.

[29] I. Kant, *Metaphysische Anfangsgründe der Rechtslehre*, in *Metaphysical Elements of Justice* (Indianapolis and Cambridge: Hackett Publishing Company, second edition 1999); I. Kant, *Grundlegung zur Metaphysik der Sitten* in *Ethical Philosophy* (Indianapolis and Cambridge: Hackett Publishing Company, second edition 1994); J. Bentham, *An*

Germany and England, they can hardly be seen to impose an external threat to indigenous European moral thinking (although utilitarianism has, for some reason, always been a bitter pill to swallow for many Continental moralists). Beauchamp and Childress have, in later editions of their renowned work, replaced ethical theories with common morality as the proper framework for their principles, but even with this move they cannot be claimed to have abandoned the Western roots of their model.

Having said this, there may be a sense in which the threat cannot be dismissed as entirely imaginary. When the four-principle model was devised in the late 1970s, it was customary to think, at least in English-speaking academic circles, that the "deontological" and "consequentialist" moral views championed by Kant, Bentham, and Mill were the only viable options in normative ethics.[30] This is why the authors of the *Principles of Biomedical Ethics* could claim a fair degree of universality for a set of principles which could be justified by appeals to outcomes as well as to duties. But however well Beauchamp and Childress may have combined the two doctrines they had in mind, there is a third alternative that they initially missed altogether, namely virtue ethics, which started to (re)emerge in the 1980s and became instantly popular in Europe.[31] This "teleological" approach has its roots in the ethical teaching of Aristotle, and the work of Thomas Aquinas introduced it, in the thirteenth century, into Roman Catholic moral philosophy and theology.[32]

This oversight may be the reason for the less-than-universally-enthusiastic reception of the American four-principle model in Europe. By ignoring moral (and religious) virtues, and thereby many deliberations about the ideal nature of a good, virtuous human being, Beauchamp and Childress left their views wide open to accusations of *short-sighted hedonism* (the view that people's happiness here and now is the prime moral concern); *excessive individualism* (the doctrine that people are always more important than the values prevailing in their communities); and *sneaking nihilism* (the view that all our inherited

---

Introduction to the Principles of Morals and Legislation [1789], J. H. Burns and H. L. A. Hart, eds (London and New York: Methuen, 1982); J. S. Mill, *On Liberty*, in *On Liberty and The Subjection of Women* (Ware, Hertfordshire: Wordsworth, 1996); J. S. Mill, *Utilitarianism* (Indianapolis and Cambridge: Hackett Publishing Company, 1979).

[30] E.g., W. K. Frankena, *Ethics* (Englewood Cliffs, New Jersey: Prentice-Hall, 1963); J. Rawls, *A Theory of Justice* (Oxford: Oxford University Press, 1972).

[31] A. MacIntyre, *After Virtue: A Study in Moral Theory* (London: Duckworth, second edition 1985); E. D. Pellegrino and D. C. Thomasma, *For the Patient's Good: The Restoration of Beneficence in Health Care* (New York: Oxford University Press, 1988).

[32] Aristotle, *Nichomachean Ethics* (Cambridge, Massachusetts: Harvard University Press, 1982); T. Aquinas, *On Law, Morality and Politics* (Indianapolis and Cambridge: Hackett Publishing Company, 1988).

values may be wrong). Hence the search for new values among European ethicists.[33]

Dignity

An important value in ethics and bioethics that has considerable support in Europe is dignity. Many think that the notion is unambiguous, and that appeals to it provide, without further argument, sufficient grounds for banning activities like abortion, euthanasia, assisted reproduction, and research on human stem cells. But this is not self-evidently the case. There are at least five mutually conflicting interpretations for the concept of dignity, or *human dignity*, and they can all yield different answers to difficult bioethical questions.[34]

First, the Kantian reading says that dignity is based on *rationality*. Kant's own authoritative words in 1797 on this were: "[Man] as a person, i.e., as the subject of a morally-practical reason, is exalted above all price. For as such a one (*homo noumenon*) he is not to be valued merely as a means to the ends of other people, or even to his own ends, but is to be prized as an end in himself. This is to say, he possesses a dignity (an absolute inner worth) whereby he exacts the respect of all other rational beings in the world, … and can esteem himself on a footing of equality with them."[35]

Kant's original German word, translated here as "dignity," is *Würde*, which

---

[33] Rendtorff and Kemp 2000; M. Kaiser, "'The precautionary principle and its implications for science' – Introduction," *Foundations of Science* 2 (1997): 201-205; P. H. Martin, "'If you don't know how to fix it, then please stop breaking it!' The precautionary principle and climate change," *Foundations of Science* 2 (1997): 263-292; S. O. Hansson, "The limits of precaution," *Foundations of Science* 2 (1997): 293-306; M.-A. Hermitte, "Le Principe de Précaution à la lumière du drame de la transfusion sanguine en France," in O. Godard, ed, *Le Principe de Précaution dans la Conduite des Affaires Humaines* (Coéditions INRA, MSH et Association Natures, Sciences, Société-Dialogues, 1997): 179-189; M. Häyry and T. Takala, "Genetic engineering and the risk of harm," *Medicine, Health Care and Philosophy* 1 (1998): 61-64; M. Häyry, "Precaution and solidarity," "Precaution and solidarity," *Cambridge Quarterly of Healthcare Ethics* 14 (2005): 199-206; K. Bayertz, ed, *Solidarity and the Welfare State*, *Ethical Theory and Moral Practice* 1 (1998): 293-392; R. Houtepen and R. ter Meulen, eds, *Solidarity in Health Care*, *Health Care Analysis* 8 (2000): 329-411; T. Takala, "Justice for all? The Scandinavian approach," in R. Rhodes, M. P. Battin and A. Silvers, eds, *Medical and Social Justice: Essays on the Distribution of Health Care* (New York and Oxford: Oxford University Press, 2002): 183-190.

[34] E.g., M. Häyry, "Another look at dignity," *Cambridge Quarterly of Healthcare Ethics* 13 (2004): 7-14.

[35] I. Kant, *Metaphysische Anfangsgründe der Tugendlehre*, §§ 434-435, in *Ethical Philosophy* (Indianapolis and Cambridge: Hackett Publishing Company, second edition 1994), p. 97.

means "worth" or "value." But he also used, from time to time, the Latin term "dignitas" in brackets to explain what he meant by this, so the conceptual connection with dignity is justified.

The second interpretation links dignity with the *sanctity-of-life* doctrine prevalent in Christian theology, especially Roman Catholic thinking. Pope Leo XIII expressed the idea in an encyclical letter in 1891 like this: "[All] men are equal; there is no difference between rich and poor, master and servant, ruler and ruled, for the same is Lord over all [Rom x, 12]. No man may with impunity outrage that human dignity which God Himself treats with great reverence, nor stand in the way of that higher life which is the preparation of the eternal life of heaven. Nay, more: no man has in this matter power over himself ...; for it is not man's own rights which are here in question, but the rights of God, the most sacred and inviolable of rights."[36] It has been argued that the references made to human dignity in the United Nations Declaration on Human Rights are historically based on the Catholic teaching.[37]

The third interpretation is quite recent, and it makes a connection between *genes* and dignity. This is expressed in UNESCO's 1997 Universal Declaration on the Human Genome and Human Rights, which states: "The human genome underlies the fundamental unity of all members of the human family, as well as the recognition of their inherent dignity and diversity. In a symbolic sense, it is the heritage of humanity. ... Everyone has a right to respect to their dignity and for their rights regardless of their genetic characteristics. ... That dignity makes it imperative not to reduce individuals to their genetic characteristics and to respect their uniqueness and diversity. ... Practices which are contrary to human dignity, such as reproductive cloning of human beings, shall not be permitted."[38] This declaration has received considerable attention since its publication, and its ban on human cloning has been especially heatedly debated.[39]

---

[36] N. Tonti-Filippini, "The concept of human dignity in the international human rights instruments," in J. de Dios Valdo Correa and E. Sgreccia, eds, *Identity and Statute of Human Embryo* (Città del Vaticano: Libreria Editrice Vaticana, 1998): 381-404, p. 388.
[37] Tonti-Filippini 1998, p. 389.
[38] *Universal Declaration on the Human Genome and Human Rights*, adopted by the General Conference of UNESCO at its 29th session on 11 November 1997, §§ 1, 2, 11.
[39] J. Harris, "Is cloning an attack on human dignity?" *Nature* 387 (1997): 754; K. Labib, "Don't leave dignity out of the cloning debate," *Nature* 388 (1997): 15; A. Kahn, "Cloning, dignity and ethical revisionism," *Nature* 388 (1997): 320; D. Shapiro, "Cloning, dignity and ethical reasoning," *Nature* 388 (1997): 511; J. Harris, "Cloning and bioethical thinking," *Nature* 389 (1997): 433; C. Dennis, R. Gallagher and P. Campbell, "Everyone's genome," *Nature* 409 (2001): 813.

When ethicists make appeals to dignity, they often make two assumptions. They believe that the rational, religious, and genetic readings exhaust the uses of the word in ethical debates. And they think that their philosophical, theological, and biological interpretations assign dignity to the same group of living beings.

The first assumption is false, because nothing prevents people from different schools of thought, and from other cultures, from employing the term. Proponents of utilitarianism, for instance, can argue that prohibitions of voluntary euthanasia deny people the right to die with dignity. According to them, unnecessary suffering should always be avoided, but this moral norm is violated by keeping people alive against their considered wishes. As for more traditional cultures, dignity has often been regarded as an extraordinary quality, possessed only by individuals of specific importance. Seen from this fifth viewpoint, the insistence on equal dignity, shared by the first four readings, can seem immoral and absurd. It is, of course, possible that some usages of the term are more legitimate than others, but, pending further analysis, the notion remains contested.

The second assumption is false, because all five readings of the concept assign dignity to slightly different groups of living beings. In the Kantian model, dignity belongs to rational agents. In the Catholic doctrine, all human beings created by God, including the unborn and the irrational, are embraced. The genetic reading should, logically speaking, expand the sphere of dignity to artificially produced human beings, whether or not they are seen as an original part of God's creation. Most utilitarians exclude the unborn and the irreversibly comatose, because these human beings cannot suffer, and some of them include other sentient beings, because they, in turn, do have this ability. And people from other cultures have entertained a variety of ideas concerning the worthiness of human and nonhuman beings.

Jahr does not use the term "dignity" in his works, but he does seem to employ the neighbouring concepts of "holiness" and "sacredness" to support his view that all life is worthy of our respect.[40] We will return to the possible interpretation of dignity in light of his bio-ethical principle in the concluding section of this paper.

---

[40] Jahr 1928a, also Jahr, "Faith in the hereafter and ethics in Christianity" 1934b, in this volume.

Solidarity

Solidarity is often portrayed as the European counterpart of justice.[41] It is related to, but should not be confused with, the liberal and utilitarian accounts of fairness and equality. Liberal ethicists usually emphasize the protection of the rights of individuals, and utilitarians focus on the equal consideration of, and equal respect for, the needs and interests of individuals.[42] In both models, the state is seen as a primary force behind the coercive organization of social life. Solidarity, in contrast, is in European debates linked with communities rather than official state functions, voluntary rather than enforced activities, spontaneous rather than organized events, and reciprocal rather than contractual exchanges.[43]

Solidarity in its non-liberal and non-utilitarian meaning was introduced to ethics by Christian thinkers, who in the early twentieth century invented the doctrines of "solidarism" and "personalism". According to these, individuals and societies cannot, or should not, be distinguished too sharply, since they are fundamentally thrown on to each other. Instead, the interconnectedness of people should be taken as the cornerstone of ethics.[44]

"Solidarism" and "personalism" are both Christian doctrines, but it has been argued that religion is not a necessary part of understanding the true nature of solidarity. All that is needed is the conviction, shared by the followers of Aristotle, Hegel and Marx (but not necessarily Kantians or utilitarians), that individuals are "embedded in social contexts". From "the fundamental social embeddedness of individuals" it can, according to two contemporary writers, be deduced, for instance, that the European welfare state is "the incarnation of solidarity as a pre-rational fundamental value."[45]

An important distinction can be made between the *universality* of justice and the *local scope* of solidarity. This raises a question regarding the distribution and redistribution of goods in and across societies. How far should solidarity extend in matters concerning material welfare? Should we see ourselves as "thrown on to" only our compatriots, or are we also interrelated with people in the developing countries, in which case we should also think about them in our political decision making?

---

[41] Bayertz 1998; Houtepen and ter Meulen 2000.

[42] Rawls 1972; Häyry 1994.

[43] R. Houtepen and R. ter Meulen, "New types of solidarity in the European welfare state," *Health Care Analysis* 8 (2000): 329-340.

[44] O. von Nell-Breuning, *Baugesetze der Gesellschaft* (Freiburg: Herder, 1990); A. McFaden, *The Call to Personhood: A Christian Theory of the Individual in Social Relationships* (Cambridge: Cambridge University Press, 1990).

[45] Houtepen and ter Meulen 2000, p. 335.

Proponents of solidarity have responded to questions like this by saying that ethical concepts do not always need to be universally defined. The liberal notion of justice, they argue, is impersonal, cold and calculating, because it is supposed to be the same everywhere. A better, more caring, way to deal with the variety of the human condition, they say, is to let ethical concepts "breathe", to let them find their own expressions in the interaction between people in their shared "lifeworlds". A lifeworld (from the German word *Lebenswelt*)[46] is not the physical environment in which we coexist with inanimate objects, plants, and animals, but a social realm where we live in shared embeddedness with other human beings who belong to the same culture. Solidarity makes justice, or fairness, dependent on the values shared in *our* lifeworld and *our* society.[47]

A word of qualification is required at this point. Although the foregoing description captures an important sense of the notion of solidarity, it is by no means the only possible interpretation. Solidarity is a rich concept, and its contents cannot be exhausted by any one analysis of its historical background and implications. Many other equally legitimate readings can be found in contemporary ethical and political discussions.[48]

As a term, "solidarity" is not a part of Jahr's vocabulary, but his ethos is, of course, quite solidaristic. We should see all human and other living beings as our brothers,[49] and by protecting ourselves against vices such as alcoholism and loose manners we not only prevent harm to ourselves but also do good to our neighbours, "actually to one's entire country."[50] The extent of solidarity in Jahr's model will be considered in the concluding section.

Precaution

The principle of precaution, or the precautionary principle, is best known in the context of climate change, and the protection of our natural environment. Formulations of it can be found in many related treaties and declarations.[51] According to one commentator, the principle "embodies the idea that public and private interests should act to prevent harm. Furthermore, the precautionary principle suggests that action should be taken to limit, regulate,

---

[46] E. Husserl, *Cartesian Meditations: An Introduction to Phenomenology* (The Hague: Martinus Nijhoff, 1960).
[47] Houtepen and ter Meulen 2000.
[48] Bayertz 1998; Houtepen and ter Meulen 2000; Takala 2002; M. Black and G. Mooney, "Equity in health care from a communitarian standpoint," *Health Care Analysis* 10 (2002): 193-208.
[49] Jahr 1928a and 1934b, in this volume.
[50] Jahr 1934a, in this volume.
[51] Martin 1997, pp. 282-283.

or prevent potentially dangerous undertakings even in the absence of absolute scientific proof."[52]

Or, as another commentator has put it: "[The Precautionary Principle] is designed to address the existence of scientific uncertainty in areas where our failure to anticipate … future harm may lead to disaster. The principle says that in these cases we have to be prepared to act [in a cautious manner] even though our fears of disaster may turn out to be unfounded in the long run. The Precautionary Principle appeals to our sense of controlling risks, and it assigns responsibility to present generations to think about the consequences of their action for future generations."[53] As the principle of precaution has gained popularity in environmental ethics and policy, it has also found its way into European discussions concerning medicine and health care. The first indication of this was an appeal to precaution in two European Council directives on genetically modified organisms in 1990. Since then, issues like blood banking and human genetics have also been addressed in similar terms.[54]

The proponents of the principle argue, in the quoted passages, that its aim is to prevent harm, to limit potentially dangerous undertakings, to anticipate future harm, to control risks, to manage responsibly the world's resources, and to be accountable for the consequences of our actions for future generations. These goals could easily be shared by liberal and utilitarian ethicists. But since the point of precaution, as a specifically European principle at least, is to provide an alternative to the ideals underlying the four-principle model, a distinction should be made between cautiousness in the sense intended here, and the straightforward assessment of risks of harm.

Risk can be defined as the possibility or probability of harm or damage.[55] But ethically relevant harm and damage can be defined in many ways, and attitudes towards the acceptable level of probability can vary considerably. In one corner of the conceptual map only clearly predictable concrete injuries to people or property are seen as serious risks. More cautious views also take into account the indirect and uncertain effects of our choices, psychological and social as well as physical harm, and prevailing attitudes concerning what is good and what is bad. At the other end of the continuum, theorists have argued that when our actions could conceivably have catastrophic consequences, the evaluation of probabilities is superfluous. We should

---

[52] Martin 1997, p. 264.
[53] Kaiser 1997, p. 202.
[54] Martin 1997, p. 290.
[55] Häyry and Takala 1998.

simply refrain from activities which could have extremely bad outcomes.[56] Alternatively, it can be held that damages and violations which seem merely symbolic and immaterial can, in fact, inconspicuously pave the way to unspeakable disasters.[57] We should not "play God", or test the limits of what is "natural", in fear that we could meet with divine or cosmic punishment.[58]

What is the location of precaution on this conceptual map? There are four main options. The first is to identify cautiousness with common prudence, and to note, for instance, that we cannot always rely on risk analyses prepared by parties with vested interests. The second possibility is to resort to the theory of games and decisions, and to argue that the strategy called "maximin" approximates reasonably well the precautionary principle. According to this strategy, we ought to choose the action or policy that guarantees, as far as we can tell, the best outcome in the worst scenario we can think of.[59] The third option is to assert that uncertainty as such can be used as an argument against technological development, and for protective regulations. If the unforeseen consequences of our actions could be disastrous, precaution dictates that we should not undertake them.[60] The fourth alternative is to say that it is not wise to challenge or endanger prevailing moral views, because they provide the cement of social life and human interaction.[61]

The popular appeal of the precautionary principle is based on the ideas of common prudence and maximin decision making. The actual policy recommendations made by the proponents of the principle are, however, sometimes based on the more controversial interpretations. In environmental matters, the argument from unforeseen outcomes is regularly evoked, and in health policy, the protection of moral values seems to be more important than the avoidance of more concrete harm on individuals.[62] This can be problematic, as everything we do *could* have disastrous consequences, and

---

56 H. Jonas, *Das Prinzip Verantwotnung: Versus einer Ethik für die technologische Zivilisation* (Frankfurt am Main: Insel, seventh edition 1987).

57 B. Pascal, *Pensées* (Paris: Le livre de poche, 1972), § 233.

58 R. Chadwick, "Playing God," *Cogito* 3 (1989): 186-193; M. Häyry, "Categorical objections to genetic engineering – A critique,", in A. Dyson and J. Harris, eds, *Ethics and Biotechnology* (London and New York: Routledge, 1994): 202-215; T. Takala, "The (im)morality of (un)naturalness," *Cambridge Quarterly of Healthcare Ethics* 13 (2004): 15-19

59 Rawls 1972; Hansson 1997.

60 Jonas 1987.

61 P. Devlin, "The enforcement of morals," in *The Enforcement of Morals* (Oxford: Oxford University Press, 1965); M. Häyry, "But what if we *feel* that cloning is wrong?" *Cambridge Quarterly of Healthcare Ethics* 10 (2001): 205-208.

62 Kaiser 1997; Martin 1997; Hansson 1997; Hermitte 1997; Häyry and Takala 1998; Häyry 2005.

agreement on the values which *should* be safeguarded by precaution is not easy to come by.

One possibility is to argue that the concept of precaution simply shows the limitations of rational decision making. Since we cannot predict all the outcomes of our actions, nor agree on the most important moral values, we should not pretend that we have scientific grounds for our ethical and political choices. The practical strength of this line of argument depends, however, on the alternatives that can be given to the assessment of risks of harm.[63]

The precautionary principle as such is a product of later twentieth-century developments in science, technology, and control over nature, and it is understandable that these issues are not explicitly on Jahr's agenda. He does, however, advocate caution in preserving the good religious and related practices of the societies in which we live, while also recognising the manifold dangers inherent in a lack of consideration towards non-human living beings.

The bioethical imperative and European values

Including Jahr's bio-ethical imperative, the principles of European bioethics could, then, be:

1. Respect for all life, first introduced by Jahr in 1927.
2. Respect for (human) dignity, formulated in the late nineteenth century and revived in subsequent international declarations.
3. Commitment to solidarity, articulated at the turn of the twentieth century but open to varying interpretations.
4. Observance of precaution, suggested during the latter half of the twentieth century and amenable to different readings.

So assuming, like we do in this paper, that the bio-ethical imperative – "Respect every living being on principle as an end in itself and treat it, if possible, as such"[64] – can be seen as the "first among equals", or as the principle that qualifies and informs the other three, how should dignity, solidarity, and precaution be understood? All three can be construed in many ways, but can Jahr's commandment give them unity and fibre as European values?

---

[63] On precaution and related principles, see also M. Häyry, *Rationality and the Genetic Challenge: Making People Better?* (Cambridge: Cambridge University Press, 2010), pp. 190 ff.

[64] Jahr 1927, in this volume.

Jahr's demand to respect all life could be seen as an expansion to the reverence for (human) dignity as it has developed during the last two centuries. Kant's theory requires us to treat all rational and autonomous moral persons as ends in themselves; Thomism (the late nineteenth-century Catholic doctrine) stresses the sanctity of all naturally conceived human life regardless of rationality and sentience; and the UNESCO declaration on the human genome potentially extends the sphere of protection to artificially created individuals like clones as well. Jahr's model could mean taking the next logical step by including animals and plants in the circle of moral consideration.

The idea of expanding our duties and obligations to non-human forms of life, when suggested by contemporary thinkers, especially of a utilitarian persuasion,[65] has not found unequivocal support. The main criticisms include the following. If the threshold of moral status is set "lower" than we have done with the criterion of humanity – say, at sentience – this enables us to reduce cruelty to animals, but at what cost? People become morally comparable to other living beings on the basis of their capacities and abilities, and not all of them will fare well in the comparison.[66] Unborn human beings, infants, and some individuals with severe mental disabilities will fall below gorillas, chimpanzees, orang-utans, and dolphins in moral considerations, and that is not acceptable to all.[67]

Jahr's formulation of the bio-ethical imperative allows, however, an alternative reading, which concentrates on the restrictions that he makes, instead of the extension. Although Jahr believed that all living beings should be respected as ends in themselves, he did not expect human beings to forgo the pursuit of their own genuine interests to safeguard the interests of animals and plants. The imperative demands our respect "on principle" and "if possible", and this can be the key to understanding its best effect on (human) dignity as a European value. Jahr's message could be that balancing and moderation are as important to morality as the need to make all human – or living – beings the focus of moral consideration. Treating other people as ends in the Kantian sense does not mean ignoring one's own needs in the

---

[65] E.g., P. Singer, *Animal Liberation: A New Ethics for our Treatment of Animals* (New York: New York Review/Random House, 1975); P. Singer, *The Expanding Circle: Ethics and Sociobiology* (New York: Farrar, Straus and Giroux, 1981).

[66] As already noted by Jeremy Bentham: "But a full-grown horse or dog, is beyond comparison a more rational, as well as a more conversible animal, than an infant of a day, or a week, or even a month, old. But suppose the case were otherwise, what would it avail? the question is not, Can they *reason*? nor, Can they *talk*? but, Can they *suffer*?" –Bentham 1982 [1789], p. 282.

[67] E.g., E. Feder Kittay, "At the margins of moral personhood," *Ethics* 116 (2005): 100-131.

process. Initial moral attention can be given to people – or animals or plants – although further analysis may reveal that they have to be overlooked or sacrificed in the eventual practical decisions.

When it comes to solidarity, the "expansion" version of Jahr's model seems to contradict the idea of protecting and preserving specific social arrangements and lifeworlds. Despite our budding tendencies towards inter-species sympathy, the primary object of our solidarity has been humanity or a part of it. Widening the scope to animals and plants would not fulfil the criteria of mutuality, reciprocity, and voluntary participation, and it would mean abandoning the "local" essence of solidarity and moving to the direction of "global", impartial justice.

Jahr's qualifications can, of course, come to rescue yet again. It would be good to extend our solidarity to all people, animals, and plants, but since solidarity in more confined forms protects some of our vital social and personal interests, we cannot always apply the moral principle to practice. The important thing here would be to remember that all living beings are worthy of initial moral consideration as our fellow travellers and occupants of planet Earth.

Precaution is not an explicit topic in Jahr's writings. He talks about avoiding harm to ourselves and to our fellow beings, but does not directly address issues such as technological change and environmental decay. He does, however, forcefully defend the existing religious and social traditions, and he does have his precautionary doubts concerning science and technology. Writing about the useful, if not always fully grounded, acceptance of scientific and scholarly "truths", he notes: "At times it even harms me, forcing me by civilization and technology into a realm of superficiality, away from the inner and spiritual life, on which my happiness and the happiness of the world depend."[68] This can be read as a call for caution in the face of technology draining our lives of cultural and spiritual meaning. We may not know what distinct innovations will draw our attention away from the sources of our happiness, but we should beware of developments unless there is evidence that they are harmless.

In sum, Jahr's qualified defence of all life could work quite well with the principles of dignity, solidarity, and precaution as a basis for a new European ethics and bioethics. According to this, no living being should be morally ignored or destroyed without careful consideration. The need for equality and solidarity between human beings may override the bio-ethical imperative in practical situations, but this leaves the imperative, "on principle", intact.

[68] Jahr 1933, in this volume.

Caution should be exercised lest we lose the deeper contents of our lives and succumb to trivialities. Above all, we should keep in mind that we are not alone in the world, and that others in it deserve the same moral considerations that we would like to be extended to us.

BIOETHICS, LAW AND EUROPEAN INTEGRATION*

Judge Christian Byk[1]

Bioethics is always described as implying a multidisciplinary and pluralistic approach of the issues encompassed. No doubt that the law and lawyers have deeply contributed to its origin and present development. However, conversely to the United States, it seems that bioethics has opened some new perspectives to the law in Europe. First, it forced the law to move out of its own frontiers and to apply its reasoning to life sciences issues with the consequence that some consider that the legal norms are used to legitimate unlawful practices while others believe legal norms have imposed binding conditions to the free development of science and technology. Second, Europe is the only region of the world where biomedical techniques are subjected to legal and sometimes binding harmonisation. In some way, we may conclude that the role played by the law in the elaboration of a European bioethics is just an example of the important role of the law in the European integration.

The political dimension of Bioethics and Law in Europe

There is a specific dimension of the transformation of the patient-physician relationship in Europe. It concerns the fact that most European countries have developed social security systems and that consequently, the new biomedical development raises the issue of access to health care as a global socio-political issue.

Therefore, law more than ethics plays in Europe a role in defining the conditions of accessing to the new medical technology and in elaborating a public policy in this field.

When looking back to the recent history of Bioethics and Law in Europe, we see the emergence of mechanisms to produce new norms outside the medical and scientific community. What is interpreted as the end of a paternalistic ethical system, also common to the bioethics movement in the USA, is also

---

* The paper was presented at 1st International Conference „Fritz Jahr and the Foundations of European Bioethics“, Rijeka/Opatija, March 11-12, 2011, and published in Byk, C (2011) *Jahr* 2(4) 431-444.

[1] Christian, Byk, christian.byk@gmail.com, Court of appeal, Paris, Secretary General, International Association of Law, Ethics and Science**.

** The Association constitutes an international and multidisciplinary network in the field of science, ethics and society (www.iales.org). It publishes since 1990 the International Journal of Bioethics (www.eska.fr), a bilingual quarterly.

the symbol that law is viewed in Europe as an appropriate way to bring bioethics issues at the forefront of societal debate.

The counterpart of this "political" appropriation of Bioethics – in the traditional meaning that it has become a question for citizens' discussion- is that the law might be perceived as imposing a new "paternalistic" approach, not the one of the medical ethics but the one of the State imposing to individuals rules based on a collective idea of good, happiness or what is a normal social behaviour.

A Bioethics, law and democracy

The idea that bioethics will stop the confiscation by the medical profession of the decision making process on the ethical issues raised by the progress of medicine implied that Society would be able to define democratic ways according to which individual and collective choices could be made. Naturally, the process of making law in Western democracies could serve as a practical example: public debate in Parliament and courts are good references and rational way of reasoning is a good methodology. So the idea came that once appropriate institutions would have been set up, they might be incorporated or their work might be used in the process of elaborating the norms necessary to organise and regulate the new techniques.

Bioethics, law and the (new) institutions

A significant outcome of the end of the XX century Bioethics success story is certainly the growing role of new and renewed institutions in producing and applying bio law.

Among the newly created institutions, we may clearly distinguish those in charge of assessing and discussing bioethical issues with those in charge of applying the legislation.

- The first category encompasses both **"the pilgrim mothers"** of the bioethics institutions, those created in the 1980's in the different European countries (the Warnock Commission in the UK, the Benda Commission in Germany, the Braibant Committee in France or the Santesuosso Committee in Italy) to deal with reproductive and genetic technologies and the now expanding category of National Bioethics Committees (France 1983, Denmark 1986…).

Although those commissions have not been instituted on the framework of a law reform commission, they all add in mind the issue: should we legislate?

Even the new National Bioethics Committees which were set up as standing committees with a mandate to organise ethics discussion very early raised the same question and were influential in suggesting to governments and parliaments to move to legislation while their opinions also served as references in court disputes.

Therefore parliamentary committees reorganised themselves to tackle with bioethical issues: special commissions were set up; offices for technology assessment were incited to deal with such issues.

- The second category gathers **bodies which are part of the regulating process** adopted to rule specific biomedical technologies. Some do it on a broad scale at the national level such as the Human Fertilisation and Embryology authority in the U-K or the Biomedicine Agency in France. Others do it on a case by case review: IRBs for research on human beings or transplant or genetic therapy committees… Both produce byelaws or "case law" according to their mandate. They merely interpret or apply the existing legislation more than they create it but they have anyway a very concrete role for those who are either practising the technologies concerned or applying to benefit from them.

Whatever is the ethical issue concerned, a clinical case, a biomedical protocol or a broader societal issue; there is today at least one body to look at it. Public ethical institutions have clearly replaced professional associations at the forefront of the bioethics debate. This change in dealing with bioethical issues raised another important feature: the capacity of the mandated institutions to work democratically. For that, they have to be inspired by the legal practice and develop a due process of law in the specific area of bioethical issues.

Bioethics institutions and the due process of law

Ethical issues have moved everywhere from the closed professional world of doctors and scientists to the global world of public arena. But contrary to the United States, Europe is not essentially relying on the concept of autonomy to institute more democracy in the decision making process in bioethics. Legal procedure plays an important role in bringing some legitimacy to the institutions in charge with bioethical issues.

- **The search for sanitary democracy** is a recurrent issue in reframing the health systems in Europe by the turn of the Millennium and bioethics even poses more difficult problems.

For example, how a national bioethics committee can be inspired by the due process of law?

It certainly means that contradictory arguments have to be identified and discussed, that the process of elaborating an opinion should be transparent and that the opinion should be argued.

Regarding the agencies in charge with the regulation of technologies and practices, how can we be sure that they do not recreate a closed world for experts that will leave outside lay persons and public representation? How can we be sure that their decision will not be biased by conflict of interests?

A more sensitive question would be:

- **How can we prevent** that the review of research protocols or individual clinical cases by ethics bodies is giving birth to an "**ethicocracy**"? Transparency, rational reasoning and possibility to appeal are necessary as a counterpart of the authoritative role of those new bodies in making decisions. Efforts have been made in different European countries to improve this process but this is too often a task which is the consequence of major dysfunctions of the system.

We may say that the XXIst century marks the triumph of bioethics institutions because we have now five categories of such institutions in Europe: national bioethics committees, ethics research committees, high technology ethics committees (on genetics, organ transplant, biotechnology…), clinical (or hospital) ethics committees and academic or professional ethics committees. At least, the first 3 categories are usually regulated by law. But, we do not have yet a global perception of what is the result of this integration of bioethics institutions in law policy making. Does it really lead to more democracy in the field of biomedicine and biotechnology or are we entering into a "Brave new world"?

Bioethics, law and the "Brave new world"

The marriage of law and bioethics is a feature of the techno-scientific society. The law is necessary to facilitate social transformation and to alleviate the fear of the progress. But in doing so, the law might raise some ambiguity: it could either support a positivist approach to consumerism in the field of biomedicine or create a new dogmatism which will limit individual freedom, scientific creativity and entrepreneurship.

Biolaw and the positivist approach to consumerism in the field of biomedicine

A common point between bioethics and biolaw could be in the fact both have put biomedical issues in the public arena where choices are not only made by the physicians following the traditional medical ethics. While admitting that new biomedical technologies raise issues that concern the individual rights of the patients as well as some societal choices, bioethics and biolaw however keep their eyes wide open on the practice. They try to be references for existing practices in suggesting and sometimes imposing limits and conditions to health care providers and patients. Globally, they offer a set of norms to rule the different techniques and to make them socially acceptable rather than to challenge the legitimacy of such techniques.

In doing so, bioethics and biolaw may be viewed as derived products from biology rather than sub branches of ethics and law. In supplying the traditional medical ethics which most patients criticized for its paternalist attitude and which many physicians thought it was no more in capacity to bring appropriate answers to the ethical legal and social issues generated by the reproductive and genetic revolutions, biolaw developed a practical and concrete approach to facilitate the access to the new techniques in the respect of individual rights. But it did not really challenge the philosophy and organisation of the techno-scientific society.

It therefore contributed and reinforced some of the main characteristics of our post-modern society: individualism and subjectivity, on one hand; materialism, consumerism and reification of the human body, on the other hand.

For this reasons, opponents to this positivist approach think that it is not enough to introduce some ethical questioning in the way we produce our law. They argue about the necessity to substantiate our legislation on strong fundamental values rather than on bioethical discussion and procedural norms.

Biolaw and the risk of a new dogmatism

What may be obvious in the convergence of ethics and law in the field of biomedicine is a common will to found the set of rules governing the new technologies on strong values. To guarantee that the development of medicine and science will not serve to build a Brave new world, we need to recall what is perceived as symbols of a humanist philosophy: Judaeo-Christian references and human rights principles. The implementation of existing rights as well as the emergence of new rights is totally constructed for the benefit of the protection of what is human either as an individual person (and parts of it) or as a collective body (the human species).

Globally, we may proclaim that ethics and law contribute together in the designation of these fundamental values. Some times, ethics has a determining role in this consecration although law does not ignore the principles which are commonly designed as fundamental values. This is the case with the informed consent principle but also with the concept of dignity. In other cases, the process is working the other way around as with the principles of privacy and non discrimination which have been broadly applied in the field of genetics and access to health services.

But, the idea and practice of establishing a set of fundamental rights may be confusing.

It might mean that the benefactors of such rights are entitled to claim the application of the rights in the different areas of biomedicine and that those rights are universal, authorise limited exceptions and may serve as references to develop derived principles.

The fundamental rights approach could also be a way to introduce in the legal system the idea that the rights designed as fundamental are at the top level of a new hierarchy of norms, no more grounded on a formal distinction but essentially based on their substance.

Due to the fact they contribute to define what is human, one of the main characteristics of such rights is their transcendence which means that they surpasses all other rights and principles. They offer some objective definition of the "human nature" and find references in the concept of *jus naturalism* and also in religious believe. We may therefore fear that they will reintroduce some form of dogmatism and absolutism in our law. The controversy rose by the utility and applications of the concept of dignity but also the ongoing discussion about embryo research are good illustrations of such risk.

Although the consequence of this ethically-legal approach is well known in US Bioethics, it might be more influential in Europe because in most countries biomedical research and their applications are largely relying on a State and Society support. Law therefore plays in Europe an important role in implementing the different national policies in the field of life sciences.

As some of these policies are restrictive while other are more permissive, there is in Europe a specific geography for Biolaw which differs considerably from the US traditional distinction between federal regulation (only binding for the researchers asking for federal funds) and State law which may be inexistent in many area of bioethics. Consequently harmonising legislation in Europe is the only way to make acceptable the idea that fundamental values could rule biomedical issues without infringing individual human rights.

II Legal harmonisation of biomedical legislations: an original method to resolve contradictions

The legal approach to bioethics issues is also an interesting example to demonstrate how the diversity of European approaches may live together and move to substantiate fundamental principles. Although the margin of discretion of the national legislations is still great in deciding how far the beginning of life should be protected or what is the positive duty of a State in providing death with dignity, European States have accepted in different ways to join a competitive dynamics in which law is a key instrument.

A The law as an instrument of competitive dynamics

Originally conceived as an economic community based on coal, steel and uranium, the European Community progressively embraced broader areas of jurisdiction to facilitate the emergence of a global European market which products and services could easily compete abroad. Legal harmonisation became a way to integrate this global market but life sciences issues demonstrated soon the limits of technical harmonisation without the support of common European values.

The industrial and economic need is a strong incentive to regulate

It seemed obvious that if Europe would like to play a significant role in the international competition its industry had to benefit from rules that would facilitate the circulation of goods & services. This task has been accomplished for drugs and blood products and for medical professionals. However, in 1995, it failed to be realized for biotechnology and patents but after a new proposal was put on the table in 1996, finally a European regulation was adopted in July 1998. Obviously, this demonstrated that it was no more possible to answer the industrial approach without looking also after the ethical issue.

Let us put aside one moment the quarrel of words and ideologies to revisit the European policy on biotechnologies, and more particularly the role of the law by asking a double question: what is resulting from 20 years of European legal rationalization in this field? And does this result bring a satisfactory response to the concerns expressed by the public opinion? Our feeling is that the extraordinary development of the regulation on biotechnologies did not make us escape a deep social crisis of confidence between scientists and political decision makers, on one hand, and the citizens, on the other hand, with the fear of a biotechnological apocalypse. And in this frontal opposition of views, the law had a key role for each party, which was to legitimate the social choices and to transpose them in concrete reality. This is the marriage between social values and reality that constitutes the legitimacy of the European legal system.

The ethical dimension is necessary to make techno-scientific regulation socially acceptable

The new biomedical technologies are often presented as the only real challenge to our social organization, creating a risk for the human person and even the human species by offering more and more sophisticated possibilities to manipulate, for example, procreation, heredity and in a near future the human brain.

Let me explain why I believe it is not a good approach to bioethics issues. The problems which are posed to us are in fact due to the conjunction of different social attitudes, the importance of which will depend on the cultural context. These attitudes can be summed up as follows: the unlimited search of new scientific knowledge, the irresistible pursuit of individual happiness, the importance of money and business, the role of institutional structures and ideology.

- The first two attitudes concern primarily **the physician-patient relationship**. Traditionally, the Hippocratic medical practice was based on the specificity of the relationship between the patient and the physician. The physician benefited from the patient's confidence because his/her duty was to act only for the direct and personal benefit of his/her patient.

Of course, we know for long time ago that physicians had to face in some circumstances opposite interests: either private interest – should a physician tell the family the nature of the illness a patient is affected with? – Or public interest when it is necessary, for example, to prevent the spreading of epidemics?

But the major changes in modern medicine came from the fact that biomedical research has now been integrated as a normal aspect of the medical progress: "ethically necessary but necessarily immoral" as observed

Prof. J. Bernard the honorary chairman of the French National Bioethics Committee. Consequently, as soon as the patient became the subject of biomedical research, healing the patient was no more the unique goal of the medical practice.

Another element of the transformation of this relationship is to be found in the greater role played by the patient himself. Higher education and further development of individual rights have led to a greater consideration of the autonomy of the patient. Paternalism is less and less accepted but moreover patients, as they are now better informed of the new medical advances, have stronger views about their wishes and would like them to be satisfied by the physicians. People want to decide about their own treatment but they also want to decide about their own quality of life and death, about the moment they will procreate and the characteristics of their offspring.

What they require from the health care providers is more and more the satisfaction of their personal desire and happiness than it is a therapeutic treatment.

Therefore, the pursuit of further knowledge by the physician on one hand and the pursuit of happiness by the individual on the other hand could progressively transform the physician patient relationship on a purely legal convention loosing its specific characteristic which implies a mutual respect of each partner.

- Such a consequence has probably already occurred in **circumstances when business is prevailing over medical considerations:**

We could certainly ask why so many physicians are interested in industrial countries to develop new reproductive technologies the financial cost of which is high. But we all know what a couple could do to get a desired child. We are sometimes surprised to hear the existence of very modern genetic centres providing with genetic medicine for a few people in countries where general hospitals cannot usually face with their normal duties.

New technologies are also a good manner for advertising the ability in medicine or simply the ambition to be viewed as a person who is at the forefront of the new medicine.

Sometimes these events are good successes: it was the case with Prof. Chris Barnard's first heart transplant or with the birth of Louise Brown due to B. Edwards & A. Steptoe, but, it is not always the case so it is the reason why it

appeared necessary to include an ethical approach to the European regulatory policy.
It largely concerns the ethical review of biomedical research. Since 1991 the European Union Biomedical Research Program has included specific research incentives for bioethical issues related to medical research.

It covers also the assessment and the regulation of a wider range of bioethical problems at the European level.

After having set up two *ad hoc committees* (on embryo research and human genome research) the European Union instituted in 1991 a standing group (on ethics of biotechnology) to advise the Commission on ethical aspects of draft European regulations.

The other European organization, the Council of Europe, which is an intergovernmental institution for cooperation, has such a standing body since 1983: this is the now called steering committee on bioethics (CDBI) which prepared many recommendations (on genetic issues, reproductive technologies, human experimentation...), the European convention on biomedicine and human rights and is presently pursuing its mission to elaborate protocols to the European convention on bioethics.

The law as a key instrument to substantiate European values

The European methodology

The funding basis of the European regulation appears very different, not to say contradictory.

The pragmatic approach

This is mainly the approach of the European Union which jurisdiction is essentially concerned with economic affairs. In this perspective although the bioethical approach cannot be regarded as subsidiary, its necessity is imposed, not by theoretical considerations, but by the fact that some industrial applications of research have raised great concerns in the public. European authorities are now aware that any specific policy could pose major political issues in term of protecting the consumers, the environment or simply assuring the public that these questions, but also fears and anxieties are taken into account.

This is the meaning of the European Commission statement declaring that "the Commission has expressed a clear wish to build a Europe for science and technology which should both promote the European development and be respectful of the rights of each European citizen".

This is the reason why the Commission has set up the above mentioned Group of advisers on the ethics of biotechnology the mandate of which can therefore be regarded as complex and difficult. The group should have indeed a dynamic approach taking into account the acceleration of scientific knowledge which is a permanent incentive to reform our regulation.

As proved by the first years of activities of the group this task, although it does not ignore the general human rights perspective, is merely accomplished through a case by case approach in which political aspects are sometimes more important than legal aspects.

Till the turn of the millennium, the ethical implications of some technical or economic issues were less apparent in the European Union regulatory process.

Although biomedical research funded by the European Union has by contract to obey some ethical guidelines (good clinical practices, or specific guidelines for embryo research or human genome research), these guidelines are not, strictly speaking, European rules imposing obligations to the Member States. They are only conditions a contractor, the European Union, is imposing to a co- contractor in order for this co- contractor to benefit from a European grant.

Regarding what is properly called the European regulations; only a few texts are specifically referring to ethical issues: we can quote the 1989 directive on blood products which mentions the ethical rules adopted by the Council of Europe and the 2001 directive on clinical trials which mentions the role of ethics review committees. But for many, the 1989 regulation was not deemed as being very efficient because the ethical rules were mentioned as an objective to fulfil with and not as an obligation to compel with.

Another text, a draft one, attempted to include ethical considerations. It concerned the issue of patenting biotechnology but as these considerations did not appear as making ethical issues as prominent, the European Parliament rejected the bill. The new one introduced in 1996 and adopted in 1998 was more explicit about ethical issues specially the prohibition to patent the human body and its components and to ban germ line gene therapy.

Indeed, the recent history of these European regulations proves how illusory it could be trying to incorporate in the same text different preoccupations. It could mean that the enforcement of Human Rights provisions in the field of biomedical sciences should probably follow a different process of harmonization.

The human rights approach

It presupposes that bioethics is not an isolated problem and should be treated as part of a common heritage encapsulated in the European Convention on Human Rights – the common European heritage.

The concept of human person which has such a tremendous importance in bioethics and in law is naturally a theoretical category invented by philosophers but it is also largely the heritage of the three funding elements of the European culture: Greek philosophy, Roman law and Judeo-Christian believes.

Since it was formulated at the end of the XVIIIth century, the principle that each man is a human person with equal rights was used as a key political operative concept which was progressively incorporated into the legal system. The European Convention on Human Rights is since 1950 the living example of this recognition of the central value of the human person in "die Europäische Weltanschauung".

The status of the human body but also our concept of filiations is deeply related to this perspective.

The human body cannot be separated from the human person: man is both a physical and a non-physical entity. Therefore, the legal protection granted to the human person does also apply to the human body as far as the person has a legal existence.

The concept of filiations plays also an important role in contributing to the identification of each man as a single person having his/her own biological and social origins.

Since it is an individual physical integrity and an individual private life which are primarily at stake, the Human Rights are in a prominent position, and particularly articles 2 (right to life), 3 (right to be protecting against treatment contrary to human dignity), 8 (right to privacy), 12 (right to marriage) and 14 (right not to be discriminated) of the European Convention.

However, the way these principles can be implemented and enforced in the biomedical field is not so easy because of the divergences existing in national legislations or practices as we mentioned above. So, there is a need for a process which allows some European harmonisation. It could be done through two ways: the jurisprudence of the European Court of Human Rights and the elaboration of new Human Rights instruments.

Case law derived from the European convention on Human Rights

Far from ignoring questions resulting from the development of the biomedical sciences, the case-law of the Convention contains some "surprises". These involve both the replies to questions submitted for examination and the approach used to analyse the cases considered.

- Thus, as regards the first aspect, the principle of free and informed consent by the person concerned has been recognised explicitly in connection with medical experiments. It is also known that, to the extent that everyone's right to life also applies to the foetus, this protection is not absolute.

Moreover, while the case-law does not yet directly provide a solution to all problems connected with the life sciences, guidelines may be identified. They are based, on the one hand, on the definition of certain concepts – inhuman treatment, private life or interference – to which reference can clearly be made for the purpose of clarifying our discussion.

On the other hand, some "case-law policies", such as the recognition that, in certain circumstances, the State has positive obligations and the importance given to a realistic approach towards family ties, are likely to create a move towards the development of the Convention, the more so because the development of the legal and moral environment is also taken into account as the case-law develops.

Too much should not be expected therefore from strengthening existing rights in the Convention, but some aspects of the Convention favour a development of certain rights. It is up to the Convention organs of control to determine the outline of these on a case-by-case basis, while distinguishing what is simply a new application of a positive right from the start of new rights.

New instruments specific to the biomedical field

Within the Council of Europe, after the Parliamentary Assembly called for it, this task was mandated in 1993 to the steering committee on bioethics. The committee, which is a multidisciplinary committee, was however mandated with a regulatory objective: to propose principles that could be applied to regulate the different fields concerned with bioethics.

- It spent a 7 year period in drafting recommendations, which although not binding, were accepted by Member States to implement and guide their domestic rules.

Genetics, human research, prenatal diagnosis, medical data and privacy, end of life, reproductive technologies and embryo research were the areas for which the Committee produced important works. Except the last two fields, all its recommendations were adopted by the Committee of Ministers and we can hope they have progressively inspired the different national legislations.

- However, as all the scientific areas were not covered and as it became important to set up a link between the above recommendations and the European Convention of Human Rights, the Committee was entrusted in 1990 to prepare a draft European convention on biomedicine.

This instrument is a framework convention which means that it gathers in one text a set of fundamental principles that should be applied to all biomedical technologies: respect of human dignity, free informed consent, protection of the vulnerable, equal access to services, protection of medical privacy, ban on germ line gene therapy, right to be compensated from damages due to medical or scientific activities... This clearly shows the bridge existing between Human Rights and biomedical issues.

But to enforce the above mentioned principles, it was also agreed that annexed protocols to the Convention with detailed provisions for each technology would be elaborated. At present four protocols have been adopted (human cloning, biomedical research, organ transplants and genetics).This methodology is very flexible because it allows to add a new protocol each time it appears necessary. And in the meantime, it is always possible, if there is no specific protocol, to refer to the general principles encapsulated in the text of the Convention itself.

If such a methodology can permit reaching some kind of consensus on substantial issues – although conflicts of views are still existing and important – it raises more difficulties when the question of a following up procedure is coming. In fact, the Members States rejected the idea to have the Court of Human Rights as direct judiciary recourse but they also refused to create any *ad hoc* body to follow the implementation of the Convention. But as the convention is binding for member States, it may be used as a reference in court to support the recognition of new rights and to solve individual cases.

Conclusion

Bioethics is not only the ethics of life; it is the ethics of society.

I mean that as a responsible person, we exercise our own autonomy but we also share with our co- citizens a common duty towards the community.

Consequently we have to find the appropriate rules to regulate conflicts of interests but also to promote a common interest which should rely on the respect of human dignity, personal rights as well as social solidarity.

Therefore, the policy approach to bioethics is less difficult that it could appear. We do not need new values, new rules to govern those issues because there is no reason not to use the existing rules which govern democratic society. But I agree that they are still major difficulties in implementing those principles to the different technological fields. It will take time and will imply respect for the diversity of cultures and opinions. However, it should not lead to forget the political will to promote an open and non authoritarian European legislative framework for biomedical sciences.

EUROPEAN BIOETHICS INSTITUTIONALISATION IN THEORY AND PRACTICE*

Iva Rinčić[1] and Amir Muzur[2]

Introduction

The major part of this paper brings out the results of the research undertaken for the purposes of doctoral thesis entitled *Theorical strongholds, achievements, and perspectives of bioethical institutionalisation in European Union*[3].

Orientation towards the European issues in bioethics stresses the new trends in the history of bioethics (with emphasis on Fritz Jahr's work), but also opens new chapters in research of bioethics itself. Although bioethics today represents one of the most prominent part of scientific community and unique phenomena of different discipline collaboration, research on bioethics usually omits the sociological perspective. Attempts of strenghtening chains between social sciences and bioethics (De Vries 1998, 2004, 2006, 2007) often remain at exploring social sciences methods in bioethical research, lacking the interests and efforts in deeper sociological research of bioethics. Seen in sociological perspective, the institutions are facts and products of social reality, points of social stability and creators of new trends. In this sense, understanding the phenomena of bioethical instutions and process of institutionalisation is an important part of undertaking research on bioethics. Exploring the specific characteristics of bioethical institutionalisation in Europe, calls for well known position in theory of institution (Arnold Gehlen), but also for new (bioethical) perspectives regarding institutions (Fritz Jahr).

Conclusion of the article presents preliminary results of the project 06.05/17 »*Fritz Jahr and European Roots of Bioethics: Establishing the International*

* The paper was presented at 1st International Conference „Fritz Jahr and the Foundations of European Bioethics", Rijeka/Opatija, March 11-12, 2011, and published in Rincic I, Muzur, A (2011) *Jahr* 2(4)415-429.

[1] Iva Rincic, irincic@medri.hr, University of Rijeka – Faculty of Medicine, Department of Social Sciences and Medical Humanities, B. Branchetta 20, 51000 Rijeka, Croatia.

[2] Amir Muzur, amir.muzur@medri.hr, University of Rijeka – Faculty of Medicine, Department of Social Sciences and Medical Humanities, B. Branchetta 20, 51000 Rijeka, Croatia.

[3] Under the mentorship of professors Ante Čovic and Nada Gosic this doctoral thesis was prepared and defended on 12 November 2010 at University of Zagreb - Faculty of Humanities and Social Sciences.

Network of Scholars«, EUROBIO*N*ETHICS) financed by the Croatian Foundation of Science[4].

Bioethical institutionalisation in the United States

Although today we know that the first mention of bioethics is the one by a German protestant theologist Fritz Jahr (1895-1953)[5], several decades later, new history of bioethics in USA was on stage.

Coming out from the ruts of other American 1960s social movements in the 1970/71 V. R. Potter' works (1911-2001), plea for institutionalisation of *bioethical* ideas was even earlier issue of Potter's concern. Potter's activites out of his vocation expanded after reading Margaret Mead work[6], recognizing the need of establishing institutional university units (*Future Departments*). In 1961 Potter took a part in a organisation of Interdisciplinary Seminar for the Future of Men (for university teachers) and participated at the American Academy of Arts and Sciences' conference, and in 1962 at the University of South Dacota held the lecture ***Bridge*** to the ***Future***. The Concept of Human ***Progres.*** By the same year, Committee for Interdisciplinary Research of Future of the Man has become an official unit of his affiliation (University Wisconsin – Madison)[7].

1960s traces of bioethics hide one even more important institutional reference of future bioethical development – establishing the first Ethical Committee in Seattle (Washington State) (1962-1967), »... trying to set up ethical standard for the distribution of a scarce medical technology to dying patients (Renal Dialysis).«[8] Establishing new form of institutionalised decision making in medicine (interdisciplinary ethical committee vs. physican's paternalism) represent real birth of bioethics and the moment when the development of bioethical institutionalisation, sensibility and scientific – professional discussion can be traced later[9].

---

[4] For more informations regarding this project: Rincic I (2011) Introduction, in: Jahr – Annual of the Department of Social Sciences and Medical Humanities, University of Rijeka - Faculty of Medicine 2(4)381-382.

[5] Jahr F (1927) Bio-Ethik. Eine Umschau über die ethischen Beziehungen des Menschen zu Tier und Pflanz, in: Kosmos. Handweiser für Naturfreunde, 24(1)2-4.

[6] Mead M (1985) Toward more Vivid Utopias, in: Science (126)957-961.

[7] Potter VR (2007) Bioetika – most prema budućnosti/Bioethics - Bridge to the Future. Rijeka: Katedra za društvene znanosti Medicinskog fakulteta Sveučilišta u Rijeci, Hrvatsko društvo za kliničku bioetiku, Hrvatsko bioetičko društvo i Međunarodno udruženje za kliničku bioetiku (ISCB), p. 24.

[8] Drane JF (2010) A Liberal Catholic Bioethics. Münster: LIT Verlag, p. 35.

[9] Zagorac I, Jurić H (2008) Bioetika u Hrvatskoj/Bioethics in Croatia, in: Filozofska istraživanja 28(111)601-610, p. 602.

Important turning point in many ways in American bioethics are the phenomena of establishing scientific institutes and centres. First among then, Institute of Society, Ethics and the Life Sciences (today's Hastings Center, New York), originates from March 1969, due to Daniel Callahan i Willard Gaylord. As before Potter's work the term and definition of bioethics were unknown in American society, in explaining the mission of the Center, founders quoted: ».... to address fundamental ethical issues in the areas of health, medicine, and the environment as they affect individuals, communities, and societies.«[10]

Bioethical institution in the capital of the USA is even more important due to the vision of Andrè Hellegers and Sargent and Eunice Kennedy Shriver. During 1967 the Kennedy Foundation and Harvard Divinity School had already cosponsored international conference on abortion »... to which both European and North American scholars were invited. In October 1971, within months of the Kennedy Institute's opening, the Foundation sponsored an international conference with the theme "Choices on Our Conscience." Speakers from Europe joined their North American collegues in exploring some of the major issues of that time.«[11]. Despite controversies regarding the genesis of the term and name of the Institute (according to Engelhardt it is still not clear whether Hellegers and Shrives *de novo* invented the term or re-aplied the one already coined by V. R. Potter), the Kennedy Institute of Ethics »succesfully applied a name to a social phenomenon that by the end of 20th century had transformed the moral context within which medical and science policy were framed. ... The result was a cultural turning point: the establishment of a socially recognized body of moral experts in authority to give moral decision regarding moral decision-making and conduct in health care and biomedical sciences.«[12]

In the year 1987 American Hospital Association published a description of 77 bioethical organisations in USA; since than, such number tripled[13].

In the field of government involment of regulating biomedical practice and scientific research, after Second World War different attempts were constantly reminding American public sphere on misconduct of scientific

[10] http://www.thehastingscenter.org/About/Default.aspx (Accessed by 19 August, 2011).

[11] Walters L (2003) Article "The Birth and Youth of the Kennedy Institute of Ethics", in: The story of bioethics: from seminal work to contemporary explorations. Washigton, D.C.: Georgetown University Press, 215-231, p. 215-216.

[12] Engelhardt T (2008) Article "Introduction: Bioethics as a Global Phenomenon", in: Regional Perspectives in Bioethics. London – New York: Taylor & Francis Group, XIII-XXI, p. XV.

[13] Drane F, A Liberal Catholic Bioethics, p. 41.

practice or even more, unsatisfied ethical standards in health care (Beecher, 1967).

Even before the establishing of first hospital ethics committees in 1960s, American government in the year 1953 proposed Group Consideration for Clinical Reserach Procedures Deviating from Accepted Medical Practice or Involving Unusual Hazards to be used within newly open NIH reserach hospital (Bethesda, Maryland)[14]. During 1960s U.S. Public Health Service also has started to establish permanent ethical standards for research practice, and during 1970s formed the National Commission for the Protection of Human Subjects of Biomedical and Behavioral Research. Beside 125 recommedations for improving the protection of human rights and welfare of human subjects, this Commission published the Belmont Report, identyfing basic ethical principles (respect for person, beneficience, justice)[15].

The first public national body to shape bioethics policy in the U.S. was created by the Congress in 1974 and was under the Dept. of Health, Education and Welfare (now known as the Dept. of Health and Human Services), followed by:

[1974-1978] National Commission for the Protection of Human Subjects of Biomedical and Behavioral Research
[1978-1983] President's Commission for the Study of Ethical Problems in Medicine and Biomedical and Behavioral Research
[1988-1990] Biomedical Ethical Advisory Committee
[1994-1995] Advisory Committee on Human Radiation Experiments
[1996-2001] National Bioethics Advisory Commission (NBAC)
[2001-2009] President's Council on Bioethics (PCBE)
[2009 -] Presidential Commission for the Study of Bioethical Issues (Executive Order 13521, November 24, 2009.).

By the end of XX century, despite the time delay, bioethical institutionalisation has achieved international character[16].

---

14 Levine RJ (1995) Article "Research Ethics Committees", in: Encyclopedia of Bioethics. New York: Simon & Schuster Macmillan, 2266-2270, p. 2266., according to Lipsett MB, Fletcher JC, Secundy M (1979) Research Rewiew at NIH, Hastings Center Report 9(1)18-21.

15 Drane F, A Liberal Catholic Bioethics, p. 39.

16 Engelhardt TH, Smiths Iltis A, Jotterand F, Bioethics: Institutionalization of, eLS - Encyclopedia of Life Sciences (http://www.els.net, Accessed by 22 January, 2009).

Achievements of bioethical institutionalisation in Europe[17]

In the field of bioethical centres and institutes the leadership belongs to South-European countries (Spain – Institut *Borja de Bioetica*[18], 1976; France – Center for medical ethics, 1984; Italy – Center for bioethics, 1985; Germany – Center for medical ethics and Academy for ethics in medicine, 1986; Belgium - Center for biomedical ethics and law, 1986; Croatia – Center for bioethics, 1986 etc), but today bioethical centres are spread all over Europe[19].

France was also the first European country to establish a national ethics committee: National advisory ethics committe for the life sciences and health (Comité Consultatif National d'Ethique pour les sciences de la vie et de la santé) in the year 1983. The French model of permanent and pluridisciplinary national institution served as a reference to many other countries[20].

Following the political trends of strenghtening the union of European countries, other social phenomena occured, like establishing bioethical institution at the level of new European community. In July 1988 the European Commission had set up a working group Predictive Medicine Working Party, since 1989 known as Study Group on Ethical, Social and Legal Aspects of the human genome analysis programme (ELSA)[21]. By the

---

[17] More detailed distinction between american »Union of States« and european »Federation of States« see Rincic I, Muzur A, European bioethics institutionalisation in theory and practice, p. 423-433.

[18] »The Borja Institute of Bioethics ... was founded in 1976 on the initative of its founding president, Dr. Francesc Abel, after his return from a 5-year stay in United States. ... Dr. Abel witnessed the creation and growth of the first and most important two bioethical centres in the world: The Kennedy Institute (KI) and The Hastings Center. He returned to Spain determined to establish a bioethics institute modeled as closely as possible to the ones he had known in the United States, allowing for suitable adaptations and guided by three operative objectives: (1) to create a specialized library in bioethics; (2) to establish forums for bioethical dialogue in hospital settings by creating bioethics committees; (3) to maintain institutional independence from ecclesiastic and civil institutions.« (Abel F, Terribas N (2010) Article "The Dynamics of the Bioethical Dialogue in Spain", in: Ibero-American Bioethics. Dordrecht, Heidelberg, Londin, New York: Springer, 245-260, p. 246.).

[19] More on bioethical politics in Europe see in Lafond F (2005) Article "Towards a European bioethics policy? Institutional structuring and political responses", in: Health governance in Europe: issues, challenges and theories. New York: Routledge, 152-173, p. 155-156.

[20] Ibid., p. 157.

[21] This Group was officaly recognized at the informal meeting of the Ministers of Research of twelve member states and representatives of Commision in Kronberg (Germany), March 1990. At the same meeting, another working group was organized - for the research

end of 1991 the European Group of Advisers on the Ethical Implications of Biotechnology was established by Commission to identify ethical questions regarding biotechnology development, to evaluate ethical activities within the Community and to identify impacts to broader society. In 1998, this body was replaced by a European Group on Ethics in Science and New Technologies, EGE), active in different mandates today.

Beside the European Comission, the Parliament also became involved in bioethical issues from 1984 (it was following the rumours concerning the trafficking in fetuses and their possible misuse, as a concequence of law absence). The Parliament has used its new competences in research (contained in Single European Act, 1987), in the occasion of the adoption of the research programme »Anaylsis of the Human Genome« (1989-1991).

Considering the tradition Council of Europe has had in human right protection (European Convention for the Protection of Human Rights and Fundametal Freedoms, 1950), since late 1970s this institution has started to deal with bioethical questions, setting up resolutions and recommedations and being involved in »third-generation human rights«.[22] The need for new bioethical institution within Council of Europe culminated in the year 1983 with Ad Hoc Committee on Genetics Experts (1985 this body was transformed into the Ad Hoc Committee of Experts on Bioethics; 1992 into the Steering Committee on Bioethics, CDBI). The main role of CDBI was identifying priorities in bioethics, with special emphasis on the possibility of preparing and adopting common European *bioethical convention*.

Not only by its name[23], this document was for the several following years point of interest of European structures. After years of negotiating and lobbying, the final version of the document Convention on Human Rights and Biomedicine, CETS No.: 164. has been signed in Oviedo on 4 April 1997 by 22 of 40 members, and came into force on 1 December 1999 after six countries (Denmark, Spain, Greece, San Marino, Slovakia, Slovenia) had initially ratified it[24]. Since then, four more Additional Protocols[25] were

---

on the human embryo (HER) (Lafond F, Towards a European bioethics policy? Institutional structuring and political responses, p. 166.).

[22] Ibid., p. 159.

[23] The name proposal - *bioethical convention* - was an issue of great pressure and critics, as the document can only be an instrument of law, not of bioethics (Gevers JKM (1994) De Bio-Ethiek Conventie: kanttekeningen bij een ontwerpverdrag van de Raad van Europa, in: *Tijdschrift voor Gezondheidsrecht* (8)456-464, p. 456.).

[24] Comprehensive analysis of this document see in Rincic - Lerga I (2009) Article "Deset godina Konvencije o ljudskim pravima u biomedicini Vijeća Europe (1997-2007.) Postignuća i perspektive/Ten years of the Convention on Human Rights and Biomedicine

adopted, with no possiblity to be signed and ratified without the main document.

Institution – from social fact to institutionalisation process[26]

Although mostly present in every day life and conversation, there is still no such thing as universal definition of the term, meaning and purpose of institution. Going beyond narrowed terminological explanations, in terms of *institute*, *establishment*, *association*, *foundations* or *rule*, *law*, *decree*, *regulation* etc.[27], few sociological founders offer much broader explications[28]. According to Emilé Durkheim (1858-1917), the precondition of forming and existence of institution is society's own entity, existence of social life, above pure sum of its individulas[29]. Being new *sui generis* category, society gains new potential: creating its own (social) forms – institutions. In this sense, institutions are facts of society or, even more, social facts, realisation of social collectivity independent of individual wishes, having their own collective existence[30]. Such position still omits other important postulate of Durkheim sociology, the one regarding the role of institutions: »Social fact is any kind of agency having outwarded coercive measure toward

---

(1997-2007). Achievements and perspectives", in: Integrativna bioetika i interkulturalnost. Sarajevo: Bioetičko društvo u BiH, 297-310.

[25] The list of the Additional Protocols to the Convention on Human Rights and Biomedicine includes the following documents: on the Prohibition of Cloning Human Beings (CETS 168); on Transplantation of Organs and Tissues of Human Origin (CETS 186); concerning Biomedical Research (CETS 195); concerning Genetic Testing for Health Purposes (CETS 203).

[26] For most comprehensive guide through philosophical elaboration of institution (and institutionalism) issues see the book of collected papers (2010) Institucije i institucionalizam/Institutions and Institutionalism. Beograd: JP Službeni glasnik, edited by Petar Bojanic and Ivan Mladenic.

[27] Rincic I (2010) Teorija institucija Arnolda Gehlena: prilog istraživanju bioetičkih institucija/Arnold Gehlen's Theory of Institutions: A Contribution to the Analysis of Bioethics Institutions, in: Filozofska istraživanja 117-118(1-2)141-159., according to Klaić B (1986) Rječnik stranih riječi/Foreign words dictionary. Zagreb: Nakladni zavod Matice Hrvatske, p. 596.

[28] According to Stanford Encyclopedia of Philosophy, contemporary sociologists even today accept narrow definition of institutions »... to refer to complex social forms that reproduce themselves such as governments, the family, human language, universities, hospitals, business corporations, legal system.« (Article "Social Institutions", in: Stanford Encyclopedia of Philosophy, http://plato.stanford.edu/entries/social-institutions, Accessed by 13 June, 2011).

[29] Durkeim E (1999) Pravila sociološke metode/The rules of sociological method. Zagreb: Naklada Jesenski i Turk, Hrvatsko sociološko društvo, p. 12.

[30] Ibid, p. 19.

individuals.«[31] In this sense, compulsion and force are not the essence of social facts, but exterior trait to be recognised and generally accepted by individuals.

Modern definitions of institution even more emphasise their enduring character and society stabilization role: for Jonathan Turner institution is »... a complex of position, roles, norms and values lodged in particular types of social structure and organising relatively stable patterns of human activities with respect to fundamental problem in producing life-sustaining resources, in reproducing individuals, and in sustaining viable societal structures within a given environment.«[32], Anthony Giddens claims that »Institutions by definition are the most enduring features of social life.«[33] while Jon Elster adds that »Institutions are mechanisms imposing rules and protecting society from decay (as long as there is something protecting institutions from decay).«[34]

It is obvious that we here stand at the position of holistic – including structuralist-functionalist account, stressing the inter-relationship of institutions themselves (structure), as well as their function in broader society[35]. Seen in this way, institutions have their own structure, organisation and universally known procedures of maintaining, but their mutual connection with other parts of society is also not a question. Without being well incorporated and accepted by the individuals from outside, institution lose their own stability and (in the long run) reasons of existence.

---

[31] Kalanj R (1999) Article "Predgovor hrvatskom izdanju: Emile Durkheim i metodološko utemeljenje sociologije/Preface to the Croatian translation: Emile Durkheim and methodological foundation of sociology", in: Pravila sociološke metode/The rules of sociological method. Zagreb: Naklada Jesenski i Turk, Hrvatsko sociološko društvo, I-XVII, p. VIII.

[32] Turner J (1987) The Institutional Order. London: Longman, p. 6.

[33] Giddens A (1984) The Constitution of Society: Outline of the Theory of Structuration. Cambridge: Polity Press, p. 3.

[34] Elster J (2000) Uvod u društvene znanosti – Matice i vijci za objašnjavanje složenih društvenih pojava/Nuts and bolts for the Social Sciences. Zagreb: Naklada Jesenski i Turk, Hrvatsko sociološko društvo, p. 175.

[35] »Funcionalist theories in the social sciences seek to describe, to understand and in most cases to explain the orderliness and stability of entire social system. In so far as they treat individuals, the treatment comes after and emerges from analysis of the system as a whole. Funcionalist theories move from understanding of the whole to an understanding of a part of that whole, whereas individualism proceeds in the opposite direction.« (Barnes B (1995) The Elements of Social Theory. Princeton, N. J.: Princeton University Press, according to Article "Social Institutions", in: Stanford Encyclopedia of Philosophy, http://plato.stanford.edu/entries/social-institutions, Accessed by 13 June, 2011).

Although the crucial moment of creating institution is the one when earlier agreed and accepted *contract*[36] acquire institutional form, broader view of understanding institution impose not only pure act or signature, but complete process, period of time in which all needed elements become institutionalised. Even the dynamics of institutions itself, would be more understandable if we accepted institutionalisation as a variable, not a constant or nominal category, points Peters[37].

The phenomena of establishing bioethical institution could be, on theorethical level, compared with similar examples. According to Arnold Gehlen's (1904-1976) philosophical-anthropological theory of institution, a man was always burdened with overcoming dualism of his mind and body. Being always jeopardised by natural environment, at one stage of his development, man had become aware of his own weakness and biological character, of his incompleteness and a lack of own possibilities and resources to gain his full integrity. As a fact of social life and a product of collectivity, institutions originate as the mediators in creating man meaning and need, force of his stabilisation and »shelter« for vulnerable one[38].

Gehlen published his most important works in Gemany just several years after the Second World War, fact probably relevant for his theory of institutions. His orientation toward archaic institution still needs to be explored in more detail regarding modern institutions as bioethical ones. This is, of course, even more important when we know that the first notion of bioethics comes from Fritz Jahr (1927), Gehlen's contemporary, who also has several notes regarding institution.

Beside Jahr's view upon professional press to be »... a medium to establish or at least strongly influence public opinion, ... it becomes even an obligation to take part in this type of character formation (Gesinnungbildung) ...«[39], (is a

---

[36] According to Searle, *contacts* are *conventions*. In his »theory of institutional factors«, social fact have their own ontological status, reality in a material world. Money, property or points in premium are not important just for the reason of their material characteristics, but because of the agreement of society upon their other funcion or value. Institutional fact are the issue of conventional formation (Pharo P (2008) Sociologija morala/The Sociology of Moral. Zagreb: Masmedia, p. 69.).

[37] Peters G (2000) Institutional Theory: Problems and Prospects, in: Political Science Series (69)1-18, p. 13.

[38] The main part of Gehlen's work in Croatia was published in a book Gehlen A (1994) Čovjek i institucije/Man and Institutions. Zagreb: Nakladni zavod Matice Hrvatske, Filozofski fakultet, Humanističke i društvene znanosti – Zavod za filozofiju (edited by Hotimir Burger), comprising of Gehlen's book Urmensch und Spätkultur (1956) and his public lecture Man and institutions (University of Freiburg, 1960).

[39] Jahr F (2010) Social and Sexual Ethics in Daily Press, in: Medizinethische Materialien (186)9-10, p. 10. See also in this volume.

press, for Jahr, a trace of an institution?), Jahr offers the most interesting remarks within different forms of social life.

»Our entire life and activities in politics, business, in the office and in the laboratory, in the workshop, in the field farms are ... not based on love in the first place, but many times on competition with other competitors. Quite often we are not cognizant of this fight as long as we proceed without hate and in a fair, legal and accepted manner. Similarly, as we cannot avoid the fight with our fellow humans, we cannot avoid the struggle for life with other living beings. Nevertheless, we will not want to lose the ideal of responsibility as a guiding point, neither for the first nor for the latter.«[40] Jahr's search fot the ways of implementing his *bioethical imperative* (1927, 1928) in all aspects of life is not limited to persons, or natural living environments but also to cultural environments. Being artificially established, such cultural artefacts are product of society, or in other words, institution. Their struggle for life with other forms of institutional life is not out of other demand and need to follow the same principle and virtues of responsibility. Limitation of bioethical imperative only to individual relations, weakens its strength and reach, as well as chances of success. Only its universal application in all forms of life (including social one, like health care institution, corporations, and even more committees) set up the grounds and preconditions of full realisation of bioethical imperative in a broader contex of European bioethics.

Opportunities and perspectives of practical bioethical institutionalisation in Europe

The process of bioethical institutionalisation represents an important step in the process of bioethical development. Before this stage, the 1960s were characterized by spurn (due to terrifying scandals in biomedical reseach); followed by the period of *institutionalisation* (provided by the first documents and committees at different levels). This phase of the 1990s proceeds another one – the level of implementation, when earlier established rules became more concrete and standardised[41].

Beside historical facts on bioethical institutionalisation in Europe, recent trends in research and promoting of European bioethics points out also need

---

[40] Sass HM (2009) Article "European Roots of Bioethics: Fritz Jahr's 1927 Definition and Vision of Bioethics", in: From New Medical Ethics to Integrative Bioethics – Dedicated to Ivan Šegota in Occasion of His 70th Birthday. Zagreb: Pergamena, Hrvatsko bioetičko društvo, 19-31, p. 24., according to Jahr F (1928) Tierschultz und Ethik in ihren Beziehungen zueinander, in: Ethik. Sexual-und Gesellschaftsethik. Organ des ‚Ethikbundes' 4(6-7)100-102, p. 101; see in this volume.

[41] Fagot - Largeault A (2004) L'émergence de la bioéthique, in: Revue philosophique de la France et de l'étranger 129(3) 345-348.

for new approaches in bioethical institutionalisation. One of the first steps in promoting European bioethics was the establishing of the journal *JAHR – Annual of the Department of Social Sciences and Medical Humanities of University of Rijeka Faculty of Medicine*, with the international Editorial and Advisory Board's, Editor-in-Chief Amir Muzur (first issue published in May 2010, the 4th in October 2011, 5th planned for Spring 2012).

After the first conference dedicated to Fritz Jahr and European bioethics, held at the University of Rijeka Faculty of Medicine (Croatia) in March 2011, the EUROBIONETHICS group was established[42] and the next conferences are already planned to be held during 2012[43]. Department of Social Sciences and Medical Humanities of University of Rijeka Faculty of Medicine, the host of the project »*Fritz Jahr and European Roots of Bioethics: Establishing the International Network of Scholars*«, continuously works on collecting documents and literature for *Documentation Center Fritz Jahr* to be a database for students, but also scholars in the field of European bioethics.

The most recent, first phase of the new project (European Bioethics in Action – EUROBIOACT) in October 2011 was prepared and applied for founding to the Croatian Science Foundation.

Finally, the idea on establishing Fritz Jahr Award for Promoting and Research of European Bioethics is still open. For the first time presented at Rijeka conference in March 2011, this initiative presents one of the possibile manners not only in promoting bioethics in European context, but also in institutionalising it on permanent basis[44].

Although it is obvious that European bioethics institutionalisation has a lot to accomplish (making formal institutions is only a first phase, followed by high standards of bioethical awareness)[45], only long term planned and (institutionaly) grounded institutionalisations opens new possibilities, cooperation challenges and prosperity.

---

[42] For more details see http://www.eurobionethics.com.

[43] According to current information, next conference will be hosted by Martin Luther's University Halle-Wittenberg's Institute of Medical Ethics and History of Medicine, under the supervision of Florian Steger. In May 2012, the meeting »The Legacy of Fritz Jahr in Ibero-America and the Future of Bioethics« will be organizing as a satellite round table to 8th International Conference on Clinical Ethics and Consultation (São Paolo, Brazil).

[44] In addition to the official ceremony (every year on occasion of Fritz Jahr birthday, 18 January), it is planned that several months professorship at Martin Luther's University Halle-Wittenberg would constitute a part of this Award. More details regarding current parameters and chronology of this initiative can be found on www.eurobionethics.com.

[45] Covic A (1999) Bioetika je znak nove epohe/Bioethics is a sign of a new epoch, in: Vjesnik (13 April).

Conclusion

Institutionalisation is the important and necessary phase in the bioethical development, and a precondition of further trend of implementing bioethical principles and ideas in every day life. In the USA, bioethics was »saved« by institutionalisation in different aspects of social life, but at the same time narrowed in its application (medical and clinical ethics).

Until recently, European history of bioethics has offered restricted approach to institutionalisation of bioethics. New trends in research (Fritz Jahr's concept of bioethical imperative) both open new chapters in the bioethics history and calls for reconsideration and setting up of new theoretical and practical approaches to European bioethical institutionalisation.

AN ASIAN PERSPECTIVE ON FRITZ JAHR AND INTEGRATED BIOETHICS

Michael Cheng-tek Tai[1]

Introduction

Who coined the word Bioethics? Most people would answer Van Rensselaer Potter of USA. Undoubtedly, Dr. Potter published his book bearing the title "Bioethics, Bridge to the Future" in 1971 in which he said "human ethics cannot be separated from a realistic understanding of ecology in broadest sense" and argued that " the science of survival must be built on the science of biology and enlarged beyond the traditional boundaries to include the most essential elements of the social sciences and the humanities" [1:1f]. He also said that a holistic biology should be concerned with the whole animal and the whole situation" [1:7]. These ideas sound pretty familiar when we read Jahr's works published long before Potter's.

Only recently the emerging concept of a 'Bioethics Imperative' by Fritz Jahr from 1927 has come to light [2]. Fritz Jahr argued in his works that the concerns of bioethics should be extended from human to all forms of life. Van Potter 's appeal to a holistic ethics reflected Fritz Jahr's theory. Hans-Martin Sass thus wrote that Fritz Jahr is the Father of Bioethics. [9]

The bioethical imperative

Jahr proposed that moral obligation of human kind cannot only be toward humans but should be towards all forms of life. He traced to Francis of Assisi (1182-1226) who showed an example of great love towards all forms of life and the entire nature. Jahr also indicated that Indian philosophy, especially the school of Sankya taught that under no circumstance should human kind be allowed to live at the expense of co-creatures. For instance, the followers of Jainism in India wear veil to prevent inhaling even a small living being. Theologian Schleiermacher declared that it is immoral to destroy any life and formation as they are if there is no reasonable cause to do so. Philosopher Krause requested a respect to every being including plants, animals and asserted that all forms of life hold the similar rights as human beings to survive. Schopenhauer claimed a sentiment of compassion toward animals as well. Jahr also quoted the Scripture "wolves shall dwell with lambs and the leopard shall lie with the young goat and the calf and lion and the fattened calf together and a little child shall lead them" (Isaiash 11) to describe a

[1] Michael Cheng-tek Tai, taimichael@hotmail.com is Distinguished Professor of Bioethics & Medical Humanities, Chungshan Medical University, Taiwan.

peaceful bioethical garden of the world in a multiworld model. When God made the covenant with Noah, He included all creatures as the Scriptures say in the book of Genesis. It was a covenant not only with human kind, it was a covenant God made with all nature.

In his article 'The Relationship of Animal Protection and Ethics', 1928, he wrote: "If we have a compassionate heart toward animals then we will not withhold our compassion and help towards suffering humans. If someone's love is great enough to go beyond the borders of human only and sees the sanctity even in the most miserable creature, he or she will find this sanctity as well in the most poor and lowest fellow human, will hold it high and will not reduce it to class of society, interest group, one party or what else may be considered" [3].

Jahr in his seminal article 'Bio-Ethics' therefore proposed his Bioethical Imperative to be: "Respect every living being in principle as end in itself and treat it, if possible, as such" [3]. Surely enough, Jahr was not an extreme thinker proposing a total ban of animal killing but voiced for a total kindness toward all forms of life. He said that all plants and animals have similar rights as human but it is not equal right, thus when there is reason when human rights are to be observed more precious than the animal rights, human kind may be allowed to exert what is reasonably justified but it must be carried out in a manner and sense of compassion toward other beings. He used an example of what Eduard von Hartmann said "whenever I see a rose in a glass of water, I cannot fight the unpleasant thought that a human being has murdered a flower life for the sole purpose to enjoy his/her eye, while dying, heartless enough eyes, not to sense an unnatural death under the appearance of life [3]. What Jahr upheld was that unless there is a rational reason, humankind should not destroy any forms of life. He hinted that there rational reason must be presented when any forms of life should be sacrificed for human's sake. Thus he said in his bioethical imperative, "respect every living being on principle as an end in itself and treat it, if possible as such". Like Schleiermacher, Jahr requested to kill animals only when a reasonable cause demanded it. This bioethical concern is not a single-minded endeavor but extended from human to animal, from individual to society, from society to the whole environment. When applying this to human health, the bioethics proposed by Jahr will therefore be a integrated concern, not only individual's physical wellbeing but also psychological, mental and spiritual wellbeing,

The Bioethical Imperative, as described by Sass, in a broad perspective includes "individual health responsibility and public health obligations but position it into concrete actual challenge of his times and into the wider vision of global, universal and integrated bioethics." [9]

The inspirational thought of Fritz Jahr for bioethics

Three aspects of bioethics

Jahr's concern is that bioethics should not only be a discipline with human individual's wellbeing but asking for an ecological justice covering all forms of life on earth. If we regard bioethics is only a biological or medical concern toward human body, our understanding will be short-sighted. According to Jahr, bioethics has a broader scope, that include biomedical, environmental and social concerns. Therefore bioethics is in the same time, a land ethics, environmental ethics, animal ethics, plant ethics, sexual ethics, corporate ethics, social ethics.

This view echoes my own description of bioethics in my book "Medical Ethics and Humanities", in which I argued that bioethics should at least include three aspects, namely, biomedical, social and environmental. Since ethics, according to Asian understanding, is a science of relationship aiming to enhance the balance and restore the harmony of different forces in cosmos [4:18] and its approach laid on the foundation of mutual respect among all relations including man to man, man to self, man to nature and man to God, bioethics, therefore, has to be holistically integrated. Jahr's insight coincide this Asian belief of ethics that extends from man to man, to nature, to whole universe and to the ultimate reality. An integrated bioethics cares not only about human's health, communitarian well-being, social harmony, but also the condition of the earth including all forms of live therein. We respect life and will try to make sure that all relationships in life are mutually enhancing and cherished. In Jahr's words, respect every living being in principle as an end and treat it, if possible as such.

Biomedical ethics cares about how a medical decision is made regardless if it is within clinical settings or experimental engagements for new drug and renovated medical procedures. We must consider whether or not those undertakings fair to all concerned? Is the motivation for the public good? Are the approaches taken carefully considered for the benefit of all concerned? Will they enhance the relationship and are the endeavor and application based on mutual respect? These concerns echo the biomedical principles of non-maleficence, beneficence, autonomy, justice, compassion, integrity, dignity, responsibility.

Environmental ethics is a major part of bioethics because if our living environment becomes hostile to our survival, no other bioethics concerns can be needed. In other words, how to take care of our living environment is the uttermost important issue in all ethical concerns. Jahr's insightful emphasis of the importance of extending our ethical concerns to all living forms on earth is exactly an expression of this environmental ethics. If bioethics intends to promote the harmonious relationship among all, we have to start to learn how to live peacefully with our nature, - in harmony. Bioethical relationship must be integrated among the relationship among men, and men to nature of all within the universe. Human are but the stewards of the earth rather than the master of nature. We have to humble ourselves and take care of the earth in which we are placed. Human kind is but a creature, same as all other beings and must live in harmony with all existences.

Social relationship is an extension of human relationship beginning from one to self to one to others. None can not live in isolation and must learn to relate peacefully with others. Social ethics has always been the emphasis of Confucian relationship. Each person has been given a certain responsibility and must comply. Confucius based his teaching of benevolence in human relation. Each person is given certain duty and obligation and must exercise accordingly. The so-called Five Relations is the basis of this social relationship in Confucianism, namely, father-son, ruler-subject, husband-wife, brother-brother and friend-friend [5: Analects 12:11]. Everyone plays a different role in different situation and must act accordingly based on compassion, righteousness, propriety and integrity.

B. Bioethics has to be holistically integrated

Bioethics at least is three-dimensional as we have just discussed above therefore it must be integrative. Jahr inspires us to extend ethical concerns from human to animal. In 'Bio-Ethics' 1927, he first looked into the modern psychology that includes all living beings in it research. He called it a Bio-psychik [3] that calls for moral duties not only towards humans but also towards all living beings. In other words, bioethics is progressively inclusive as a person is not only a physical being, he, in the meantime, is also a social being bearing responsibility toward whatever he/she attempts. Thus bioethics, when attempting to humanize human relationships and to bring wellbeing to human kind, physical aspect must not be the only concern. Health cannot be the absence of diseases but rather the wellbeing of whole existence and must be cared for holistically. Thus, the bioethics that promotes human wellbeing needs to be integrative.

Asian bioethics emphasizes the importance of the balance of a person's body, mind and spirit [5: Analects 13:19]. In other words, Human kind is not only a biological beings, he is also a psychological and social being never being isolated or insulated from other influences. The biomedical ethics of an integrated medicine emphasizes not only physician-patient relationship, it extends to include patients' family and his surroundings because a complete health in the sense of the balance of body, mind and spirit within a person should be the focus of medical attention. A person is not only affected by his own physical condition, but also by his living environment, social association and family.

The discovery of the importance of bio-medical ethics in the late 20th century proved that human kind has come of age for recognizing the health of a person has to be holistic and integrative. With the rapidly changing technology, the distance between west and east, north and south have been greatly abridged. The communication among people has also become instant. These progresses however come with a price. The ecological crisis, the deteriorating human relationship, the commercialization of social activities, the privatization or socialization of health care…. have down-graded human closeness and trust. No wonder the term "dehumanized human relation" or "the lost generation" have been used to describe our modern day human condition. Though technologies bring convenience to people, it also alienates human relation and depletes earthly resources. Air pollution and water pollution suffocate our environment and damage our natural ecosystem. No wonder Van R. Potter voiced that for human kind to survive the future, we must develop a new ethics that is responsible to the world and to nature.

Fritz Jahr, a philosopher, a pastor, an ecologist and a bioethicist called for a moral obligation in all human endeavors and voiced that we must be responsible toward all forms of life. His bioethical imperative that appeals for a "respect for every living being and life should be treated as an end in itself" has extended the human-centered relation to a broader scope of biological world. Based on his theory, we can assume that his ethics is not only physically oriented but goes deeper to the total being with biological, social, psychological and spiritual formation. Bioethics needs to be integrated and holistic.

In 'Bio-Ethics' Jahr wrote: "From Bio-Psychik it is only a step to Bio-Ethics, i.e. the assumption of moral obligation not only towards human, but towards all forms of life. In reality, bio-ethics is not just a discovery of modern times. An especially attractive example from the past is the figure of St. Francis of Assisi (1182-1226) with his great love towards animal, his warm sympathy for all forma of life, centuries before Rousseau's romanticism for the entire nature" [3]. He concludes: "The guiding rule for our actions may be the bio-ethical demand: respect every living being on principle as an end in itself and treat it if possible as such "[3].

Mutuality as the basis of bioethical effort

Jahr argued if we have a compassionate heart towards animal, then we will not withhold our compassion and help towards suffering humans. If someone's love is great enough to go beyond the border of human-only and sees the sanctity even in the most miserable creature, he or she will find this sanctity as well in the most poor lowest fellow human, will hold it high and will not reduce it to class of society, interest group, one party or what else may be considered. On the other hand, senseless cruelty towards animals is an indication of an unrefined character dangerous towards the human environment as well [3]. What Jahr tries to say is that unless we can care one among the least of all, we cannot care about others at all. This view underlies the foundation for an integrated bioethics because an integrated bioethics will make sure that all related factors in life must be brought in and considered so as to find a mutual respect among all related things, beginning with the relationship between patient and physician, husband and wife, among social interactions, earthly living environment and the world. "Do to others as you like others do to you" [5: Analects 13:19] is a golden teaching in China and in Christianity. This mutual respect and enhancing should become the basis of an integrated bioethics. In his article 'Three Studies on the 5th Commandment', 1944, he pointed out: if one protects oneself in this regard against harm, one does at the same time, good to one's neighbor, actually to one's entire country [3]. Three aspects of mutuality must be considered in clinical sense:

(1) Mutual communication: the gap between patient and physician must be bridged to ensure the line of communication is smoothly open and comprehended. Without this candidness, no trust can be established. Mutual communication fortifies the four basic rights of a human kind, right to be heard, the right to be informed, the right to safety and the right to choose. When the line of communication runs uninterruptedly without any hindrance and hesitation, can a good patient physician relationship be established. Mutuality ensures that the two parties are in equal position and function as partners rather than the master and the servant. Patients do not exist for physicians nor physicians for patients. They establish and fulfill each other.

(2) Mutual subjection: patients should be aware that physicians have the knowledge and skill that the patients do not have. Patient comes because he needs help and here physician offers his expertise. Patient, in a sense, is subject to physician. But physician cannot correctly diagnose the problem of patient unless patient honestly and thoroughly provides information of his condition. Though medical technology can detect something that cannot be discovered through dialogue, still, this mutual subjection plays an important

role in patient-physician relationship. They need each other and serve for each other.

(3) Mutual respect: the principle of autonomy is an expression of respect but it is to be based on justice. Justice refers to treating the like cases alike and treating every patient with similar earnest without showing favoritism due to patients' social status, gender, color or religion. The dignity of patient as a human person is undeniable and physician must show due respect in caring attitude. So is vice versa.

D. Interdisciplinary nature of bioethics

The tremendous development of medical technology in the last few decades has changed the face of medicine. Medical sciences can no longer isolate itself from the rest of academic world but must act in absolute solidarity with other disciplines by initiating a sincere dialogue with behavioral sciences, philosophy and theology. Without this new attempt, medicine may lock itself in an ivory tower seeing man only from a narrow diagnostic-prognostic angle. Because of the broad scope of medicine, prophylactic, therapeutic and planning for the future, medical scientists must seek a holistic vision of man. The modern physician can no longer approach biological and medical decisions without being ready to face the complicated social factors that effect human wellbeing. Medical scientists must take a serious look into the social responsibility they bear for the whole humanity when pursuing a breakthrough of new medical horizon.

In facing new realities and changes in genetic biotechnology, medical educators need also to reconsider the content of medical curriculum. Is the goal of medical education simply to train physicians skilled in treating diseases yet unaware of human predicaments and the complicated social fabric that make a person what he is?

The main concern of medicine used to be human physical condition, as it was believed that a person's illness was due to physical problems. But this understanding has been challenged. The role of medicine has been expanded from merely a bio-medically-oriented science to a holistic bio-psycho-social-environmental emphasis. The physicians of tomorrow must know not only how to treat physical diseases with empathy but also to take his responsibility seriously as healers of whole person to the whole Human society.

An emphasis on medical humanity is one of the moves we must take to respond to the new reality of medicine. Medical humanity, literally speaking, refers to those humanity courses having to do with medicine based on a bio-psycho-social understanding such as medical sociology, biomedical ethics, medical psychology, history of medicine, communication between physicians and patients ... etc. In reality, medical humanities, when properly planned and offered, will enable medical students to develop a new understanding of life and a concern to social justice. It is a design to help students develop a new value theory so that they will become people-oriented physicians. But in order for students to understand the basic human nature, more courses must be considered as social customs, religious beliefs, cultural traditions, – they all effect our attitudes toward health, life, and even death. Thus, a philosophy of life, religion and cultural anthropology must also be introduced. Medical humanities cover a wide spectrum of social sciences and humanities. The purposes of introducing these courses are no other than: cultivate a holistic understanding of personhood – enable medical students to know the human side of medicine and the impact of environment and life-style on health – provide students opportunities for continuing reflections on themselves as they grow in medical knowledge – foster a sense of "Medicine as Vocation" – study ethical theories and their applications to the clinical decision making – promote a consciousness of physicians' social responsibility. In one word: medical humanities attempts to equip students with a better understanding of human being so as to produce physicians of tomorrow who are not only effective healers of diseases, but also restorer of human wholeness (a person of harmonious combination of body, mind and spirit) and promoter of social justice.

E. Empathy for right and fairness, compassion and love

Jahr discussed the tensed interaction of 'Egoism and Altruism', 1929, and pointed out that these two are not necessarily contrasting against each other, and "not necessarily incompatible ethical adversaries" as there is altruism in egocentrism and egocentrism in altruism. In his words, "egoism is not without altruistic aspects so is altruism not thinkable with any egoism" [3]. Egocentric attitude can be instinctual for survival and it can become "a conscientious reasoning form of thinking in regard to a potentially justified struggle for life attitude and the best methods in such a struggle" [3]. He however felt that Altruism is naturally present in human and should carry a much higher recognition in public. Jahr listed justice, compassion, pity, love, - as expressions of this altruism.

Interestingly enough, Mencius, the Second Sage after Confucius in Confucianism, has the similar observation of human kind. He says "all men have a mind which cannot bear to see the suffering of others … if now men suddenly see a child is about to fall into a well, they will without exception experience a feeling of alarm and distress" [6: Mencius II A:6]. This inner sense of concern echoes Jahr's description. The nature of egocentrism and altruism exist simultaneously within human kind and at times conflict with each other, we must excel for a higher value. Even Apostle Paul experienced the similar conflict within himself. He said "the good that I desire I don't do, instead, I do the evil that I hate". Paul, however, admonished himself to overcome the desire for selfishness. In order to guide people to make sure the higher values are observed and precepts, laws, and principles have been given.

The question of "do we need principles or guidelines?" has been debated ever since Georgetown scholars published the principles of medical ethics. The opposition to principlism seems to be louder in Europe and Asia. For example, Prof. Sakamoto of Japan argued that we must do away with "reasonable, well organized and principle-oriented "bioethics and Prof. Carl B. Becker, an American teaching in Japan, explained: "perspectives from cultures like Japan, which do not subscribe to absolute value structures, cast doubts on the universality and applicability of principlism to modern medical ethics" [7] Still, principles provide a good guide to those who are not sure which way to turn in a dilemma.

For an integrated bioethics, principles can indeed help. Here are some suggestions from Asian perspectives, or shall I say, the Asian principles for an integrated bioethics.

Compassion

Compassion is central to Confucian teaching. Mahayana Buddhism also sees it as a fundamental virtue. Mencius believes that all men are endowed with a concern toward others, in his own words, "a man without the feeling of commiseration is not a man. The feeling of commiseration is the beginning of humanity" [6: Mencius II A:6] Buddhism has taught compassion for all living creatures, including animals.

Compassion cannot be directed only toward humankind, it has an impersonal goodwill to all living beings. This reflects Jahr's basic concern to extend ethics to bioethics including all forms of life. From the perspective of integrated bioethics, compassion is not based on sympathetic feelings, it is more empathy than sympathy, as Jahr said: Empathy for right and fairness, compassion and love.

Ahimsa

Asians have always upheld the idea of non-violence, especially in the

Buddhist and Hindu traditions. Ahimsa goes beyond the concept of non-maleficence to extend "Do-no-harm" to all living creatures. Ahimsa, from the Sanskrit, is normally translated as nonviolence and reverence for life. In practice, it means abstaining from animal food, relinquishing war, rejecting all thought of taking life, regarding all living being akin. This noble thought reflects the idea of non-maleficence to the point that we should not harm our fellow humankind, or any living beings. The loving care should be extended to all life forms. – In the Confucian tradition, Ahimsa is expressed through filial piety. Our body is a gift from our parents and we must safeguard and cherish it.

Respect

Respect is based on the spirit of filial piety. According to the Confucian tradition, children are required to pay due respect to their parents and elders. Furthermore, this piety is to be extended to a larger scope of family, where brotherly love is emphasized along with mutual respect required for social dealings. Mencius was once asked if men love their neighbors' children in the same way as they love their brothers' children, he answered: "treat the aged in your family as they should be treated and extended this treatment to the aged of other peoples' families. Treat the young in your family as they should be treated and extended this treatment to the young of other people's family". [6: Mencius II A:6] This notion of respect to all living is surely the foundation of integrated bioethics.

Dharma--responsibility

Dharma is a Hindu concept meaning "pattern of right living". Every one in his/her social station is endowed with duties which must be carried out. Without following this, disgrace descends on all who flee their duties as illustrated in the Hindu epic, the Bhagavadgita. In one passage the Hindu diety Krshna answered Arjuna, a warrior who was afraid to fight as a soldier in fear of acquisiting karma. Krshna said that real duties of life must not be abandoned. Performing duties of station in life is a service to God thus no karma will be created. On the contrary if a person refuses to carry out his duties, disgrace descends upon him. [8]

Applying this dharma to bioethics means that we have to care not only our own body but our environment and the whole nature including all life forms.
This principle of duty is also expressed in Confucianism through the concept of the Rectification of Names. Names refer to social stations of each person. When asked what is the rectification of names, the master answered: "let the ruler be ruler, the minister be minister, the father father and the son son"

[Analects 12:11] meaning that each person should act according to what a person is expected to do in his profession and station in life. Every name contains certain implications which constitute the essence of that class of things to which this name implies. Doing should agree with names. The ruler must bring wellbeing to his subjects, minister loyal to his master, father care to his young and son filial to his superior.

When God created the world he charged humankind to be a steward of the universe taking care of the earth. Integrated bioethics will emphasize that human kind has a responsibility to fulfill his duty as a good steward not only to himself, his fellow men, his environment but also to the whole universe. We must be responsible, therefore the Bioethics Imperative: "respect every living being on principle as an end in itself and treat it, if possible, as such".

Conclusion

Integrated bioethics respects all forms of life and takes all things into consideration, it is open to all possibility and calls for justice, compassion, love, tolerance and responsibility in all its endeavor.

References

1. Potter VR: Bioethics Bridge to the Future. Englewood Cliffs, NJ: Prentice-Hall International, Inc. 1971: 1-2
2. Sass HM: Fritz Jahr's Early 1927 Concept of Bio-Ethics. Kennedy Inst. Ethics J 2007, 17(4):279-295
3. Fritz Jahr: Essays in Bioethics and Ethics, in this volume
4. Tai MC: The Way of Asian Bioethics. Taipei: Princeton International Co. 2008: 46-48
5. Confucius: Analects
6. Mencius: Mencius I and II
7. Becker CB: Problems of Principlism in WASP Bioethics in Bioethics in Asia, ed. Fujiki N, Macer D; Eubio Ethics Institute, Tzukuba, Japan 1998: 77
8. The Bhagavadgita, Bolle, Kees trans. Berkeley, University Of California Press, 1979: 2:5-22
9. Sass HM: The many faces of the Bioethical Imperative, in this volume

FRITZ JAHR AND LOVE OF LIFE

Darryl Macer[1]

Bioethics is both a word and a concept. The word comes to us only from a German paper of 1927 (Jahr, 1927), although in English it was amplified by Potter (1970), yet the concept comes from human heritage thousands of years old (Macer, 1994). It is the concept of love, balancing benefits and risks of choices and decisions. This heritage can be seen in all cultures, religions, and in ancient writings from around the world. We in fact cannot trace the origin of bioethics back to their beginning, as the relationships between human beings within their society, within the biological community, and with nature and God, are formed at an earlier stage then our history would tell us.

We can find various definitions of bioethics, the simplest would be consideration of the ethical issues raised by questions involving life ("bio"). I would include all of the above issues of medical ethics, as well as questions I face each day, like "What food should I eat?", "How is the food grown?", "Where should I live and how much disturbance of nature should I make?", "What relationships should I have with fellow organisms including human beings?", "How do I balance the quality of my life with development of love of my life, other's lives and the community?", and so many more you can think of. The history of bioethical reasoning is influenced by our genes, and the forces that shaped and continue to shape these genes into the people, society and cultures that we have. We now have the power to change not only our own genes, but the genes of every organism, and the power to remodel whole ecosystems of the planet, which has made many focus on biotechnology applications however the key questions are more basic. We have the power to remodel whole ecosystems of the planets. New technology has nevertheless been a catalyst for our thinking about bioethics, which have been stimuli for research into bioethics in the last few decades.

A fundamental way of reasoning that people have when making decisions is to balance doing good against doing harm. We could group these ideals under the idea of love, though the question of benefits for whom and harm to whom is central to deciding whether an action is one of love or not. One of the underlying philosophical ideas of society is to pursue progress. The most common justification for this is the pursuit of improved medicines and health, which is doing good. A failure to attempt to do good, is a form of doing harm,

[1] Darryl Macer, Ph.D., d.macer@unesco.org, Regional Adviser in Social and Human Sciences for Asia and the Pacific,Regional Unit for Social and Human Sciences in Asia and the Pacific (RUSHSAP), UNESCO Bangkok, 920 Sukhumvit Road, Prakanong, Bangkok 10110, THAILAND, www.unescobkk.org/rushsap

the sin of omission. This is the principle of beneficence. This is a powerful impetus for further research into ways of improving health and agriculture, and living standards.

Fritz Jahr (1927) wrote in the tradition of many sources of wisdom found in Asia and the Pacific. In his call for a review of the ethical relations of humans towards animals and plants, he followed the traditions of millennia in Asia and Pacific traditions, and those of numerous indigenous persons across every continent. If one has ever walked in a jungle, or swum in a coral reef, the words, "The strict distinction between animal and human being [Mensch], dominant in our European culture up to the end of the 18th century, cannot be supported anymore." would have not only resonated, but made us wonder what sort of world tried to distinguish human beings from other animals.

Jahr introduced some form of Buddhist views to European readers, in a shift to a biocentric or ecocentric world view, influenced also by the increased understanding of biological evolution. The respect for life has always been expressed by some in society, in each religion. Jahr (1927, p. 2) described reincarnation through other animals as being more beautiful than the Christian tradition of St. Francis of Assisi. Compassion has been proposed as common ground for bioethics in a Thai Buddhist context (Boyd, 1998), and compassion can be considered as love as is discussed in the writings of Buddha. Generally compassion is focused on a particular context in suffering whereas social justice looks at inequality. In Theravada Buddhism the four prime virtues are mindfulness of friendliness (*metta*), compassion, joy and equanimity. In Buddhism in general the self-less love is the bodhisattva in Mahayana Buddhism. The bodhisattva, by virtue of his decision to decline the rewards of nirvana, *"until the last blade of grass has been liberated"* is revered in Buddhism as the infinite and inexhaustible reservoir of compassion (*karuna*). The bodhisattva could have passed himself into eternal bliss, but remains in the mortal while other beings are bound by the triple evils of greed, hatred, and delusionment. The six Perfections are giving, morality, patience, vigor, meditation and wisdom) and other virtues include friendliness, good will, loving kindness, benevolence and sympathetic joy. Benevolence and altruism have broader meanings, not being restricted to suffering.

The adaption of the Kantian imperative to treat all beings follows Hindu roots, "*bio-ethical Imperative: Respect every living being, including animals, as an end in itself, and treat it, if possible, as such!*" (Jahr, 1928). He discussed whether this bioethical imperative could extend the Fifth Commandment of Moses, not to kill. Let us consider a few other quotations of philosophers that relate to this concept (Macer, 1998). We can find words for love in every language, though the use of these words vary widely within each language and between them. A few examples: Confucius, *Analects* (China, 6th Century B.C.): "To love a thing means wanting it to live." – Erich Fromm, *The Art of Loving* (Germany, 1956): "If I truly love one person I love all persons, I love the world, I love life." – Pierre Teilhard de Chardin, *The Phenomenon of Man* (France1959): "Love alone is capable of uniting living beings in such a way as to complete and fulfill them, for it alone takes them and joins them by what is deepest in themselves." – Jean-Paul Satre, *The Words* (France, 1964): "When we love animals and children too much, we love them at the expense of men."

There are several basic theories of bioethics, and the simplest distinction that can be made is whether they focus on the *action*, the *consequences*, or the *motives*. Another separation that is used is deontological theories, which examine the concepts of rights and duties, and teleological ones, which are based on effects and consequences. If we use the image of walking along the path of life, a teleologist tries to look where decisions lead, whereas a deontologist follows a planned direction.

Aristotle in *Nicomachean Ethics* wrote that morality is the pursuit of a "final good" or "supreme good". This may be accepted, but the question is what this "good" is? The final good was often interpreted as happiness, which leads us to one of the main teleological theories, utilitarianism. Utilitarianism looks at the consequences of an action, and is based on the work of Jeremy Bentham (1748-1832) and John Stuart Mill (1806-1873). They could have been rediscovering what Mo Tzu had taught in China in the 6th century B.C. The principle of utility asserts that we ought always to produce the maximal balance of happiness/pleasure over pain, or good over harm, or positive value over disvalue. Initially they focused on the value of happiness, however recently other intrinsic values including friendship, knowledge, health, beauty, autonomy, achievement and success, understanding, enjoyment and deep personal relationships have been included (Beauchamp and Childress, 1994). Utilitarianism may appear cold and calculating, but it has been said by many, including its founders, to be an expression of brotherly love.

Albert Schweitzer (1966) with *Respect for Life* could be said to have used the idea of love of life as a ground for respect of life, but he did not overtly use the term love - preferring respect or reverence. He argued for a reverence for all life. This approach makes no distinction between higher and lower life forms, saying that we can not judge other lifeforms in relation to ourselves. It makes the point that it is very difficult for us to understand or judge the importance of other living organisms in the natural order. The only reason for harming life he sees is necessity. However, what is "necessary" can vary widely between cultures.

In 1993 a Parliament of the World Religions was convened. It included peoples from many religions and cultures, who came together and made a *Declaration toward a Global Ethic* (Küng, 1996). They wrote that an ethic already exists in the religious teachings of the world which can counter the global distress. They included four irrevocable directives common to religions: (1) Commitment to a culture of non-violence and respect for life; (2) Commitment to a culture of solidarity and a just economic order; (3) Commitment to a culture of tolerance and a life of truthfulness; and (4) Commitment to a culture of equal rights and a partnership between men and women. - They pointed out several directives that are found in all religions, including: have respect for life. They extend this principle of respect to the lives of animals and plants. Another was to deal honestly and fairly; to speak and act truthfully, and to respect and love one another.

All living organisms are biological beings, and share a common and intertwined biological heritage. Humans are members of the species *Homo sapiens,* one of the millions of species alive on the planet Earth. Fundamentally we must ask whether humans are a special form of life, different from other living creatures? We must also compare humans with other species and see where differences may be. We may also look at individual humans and ask whether there is any significant difference between individual members of the human species that could influence the ethical duties we have to them.

Human beings are created in the midst of an intricate biodiversity, which is yet to be comprehended. The process or time scale over which all life was made is not so remarkable as the species and ecosystems that we have today, or those that we can see from the fossils. Biological data tells us that all human beings have the same basic set of genes, the variation found in any one population covers almost all of the total variation, and that humans share a common male and female African ancestor. Changes in DNA sequences have also been used to trace the way that different organisms evolve, called phylogenetic trees. We can compare the DNA of species alive today, and investigate trends in the sequence change, and we can also look at DNA from past organisms, which is a more direct measure of the change over time.

The concept of love is applied to love of animals and to love of nature and this is found in many cultures. Bioethics has origins in exploring human relationships with animals and with nature, and this chapter explores the former and the following the later. In most religions there are verses implying some animals may have a soul, and at least a way of worship. The Prophet Mohammed divided creatures into angels, men and beasts, *"God the Most High created the angels and placed within them the intellect, He created the beasts and placed within them sensuality, and He created the children of Adam and placed within them both intellect and sensuality. So he whose intellect dominates his sensuality is higher than the angels, and he whose sensuality dominates his intellect is lower than the beasts"* (M IV between 1496 and 97). This verse also shows the distinction between angelic men, "ordinary" men, and bestial men (Jalal al-Din Rumi, 1983). Animals are not given religious prescriptions, although they are widely used as images in religion.

Whether the soul of a chimpanzee is different to a human soul has been debated for millennia, and is a question already found in many religions. It is a question only God can answer, not humans. It is a non-scientific question, like many other important questions of bioethics, the value of life, the value of love, and the meaning of existence. Scientific questions are those we can disprove by experiment, and there are many that we cannot. The concept of evolution means we see humans as living creatures derived from other living forms, and we can look at the evolution of all behavioural traits including love.

Many people accept that all humans are equal in moral status, and all humans are of superior moral status to nonhuman animals. From these two moral principles they put human welfare ahead of animal suffering, Peter Singer argues that these two moral principles cannot be defended within the terms of a nonreligious approach to ethics. He concludes that there is no rational ethical justification for always putting human suffering ahead of that of nonhuman animals. He argues that "if we are considering public policy in a pluralistic society, we should not take a particular religious outlook as the basis for our laws" (Singer, 1990).

In the conclusion of *Good Natured*, de Waal (1996) makes a poignant comment, "Animals are no moral philosophers. But then, how many people are?" We find few people who use deep moral principles to consider the consequences before they act. We can also think of how many of our daily activities we actually make moral judgments about. A chimpanzee sharing food with a hungry companion cannot be distinguished from a person working in a soup kitchen. de Waal makes a list of the following capacities found in other species that are necessary for human morality: (1) Sympathy-related traits: Attachment, succorance, and emotional contagion. Learned adjustment to and special treatment of the disabled and injured. Ability to trade places mentally with others (cognitive empathy). – (2) Norm-related characteristics: Prescriptive social rules. Internalization of rules and anticipation of punishment. – (3) Reciprocity: A concept of giving, trading, and revenge. Moralistic aggression against violators or reciprocity rules. – (4) Getting along: Peacemaking and avoidance of conflict. Community concern and maintenance of good relationships: Accommodation of conflicting interests through negotiation.

There are many species which show different levels of the above characters, and we should treat them after considering the level of autonomy that they have. Rather than expecting animals to speak the same language as humans, we should examine the actions and behaviour – the same criteria we use to examine true love from mere words in human beings. Many human persons have difficulty expressing the language of ethics and principles, but it does not mean that they lack the values used for balancing and making decisions. The same can be asked of members of other species. One of the categories that is central to this ethical approach based on the love of life is whether animals are also capable of loving others. In evolution it is assumed that selfishness is required for selection. Natural selection means survival of the fittest, and selfish behaviour allows an individual to leave more offspring. At the genetic level it means a selfish gene will try to replicate itself and leave more progeny (Dawkins, 1976). If a gene does not do this it will not last. When we look at animals we see that some animals exhibit non-selfish behaviour, called altruism. Some even give when there is no hope to receive any genetic benefit, helping unrelated individuals. We must therefore ask the question is altruism the basis for love?

Altruism is found in many species, and this led to the development of sociobiology – the study of animal behaviour from an evolutionary perspective. If love is the basis for ethics – we have several options. We love everything, or we love things we like (partiality), or should we love something more if it can show love (reciprocity)? In terms of evolution the most likely origin of love would be as a reciprocal tit-for-tat system of mutual reward that promotes selective advantage to those who cooperate sometimes compared to those who never do. As Wright (1994) reviewed theories of evolution for a gene for brotherly love, we can also see advantages to survival if a community has love for each other in all social animals. However, there are limits to this love. J. B. S. Haldane is said to have said that he would never give his life for a brother – but, rather, for "two brothers or eight cousins", this equally the proportion of genes that they share with us.

We can look at some interesting examples among animals of altruism, and analyze this from the viewpoint of natural selection, asking why that behaviour would survive. One example is warning calls of individual birds, which may put them at higher danger of being targeted but warns the group of a predator, so they can all take flight or prepare for defense. Vampire bats are not our image of a loving animal, but unrelated bats in the same colony may feed each other. If one bat is successful and another unsuccessful, the successful one may feed the unsuccessful hungry bat. This is called reciprocal altruism, because the bats may reciprocate on another night. This behaviour can be mathematically modeled in such a way to show that it could be a selective advantage for these bats, because they meet frequently, and by helping each other they help themselves leave more offspring. We have numerous examples in the animal kingdom of reciprocal altruism within species, but few between species.

Altruism within a species is supported by kin selection, the genes may be promoted through survival of any member of the species that has the genes, not only by the individual or the immediate family. But between species there is no specific reason to risk life for another, or even to spend vital energy and resources on saving another species at the cost of decreased promotion of the genes of the species itself.

Those species which favour their own species over any other individual of another species could be called speciesist. In the same way that racism or sexism are biases, so is specisism. Human beings are generally specisist also. Peter Singer (1976) argues that "pains of the same intensity and duration are equally bad whether felt by humans or animals", and we should not be prepared to inflict pain on other animals that we would not bear ourselves, unless there is some overwhelming justification for it.

Some of the examples of inter-species altruism are stories of dolphins saving drowning humans in the ocean; they are found throughout history and throughout the world (Kellert & Wilson, 1993). Cases have been documented on video of dolphins saving people who lie face down in the water.

The Jewish-Christian-Islamic religions say that humans were made in the "image of God" (Genesis 1:26-7). What is this image? Part of this image is what we could call the "human soul", as discussed above. Another approach to investigating the soul is to look for characteristics that have been claimed to make humans higher than animals, and to which some people say reveal the soul. These include intelligence and intellect, and language-ability. It is very difficult to estimate the language ability of animals who use different communication systems. One way is to examine the complexity of the language, which would eliminate most animals from being close to humans, but there would still be doubts over some.

It has been found that chimpanzees and gorillas can be taught sign language (with names such as Koko, Kanzi, Washoe, Ai), and talk to humans in it, but only to the extent of composing two or three word sentence replies. This sentence structure represents a stage equivalent to a human baby learning to talk, before they can start to actually make longer sentences. However, they can communicate their needs and desires through this medium, and have even gone on Internet for conversations with human beings. This ability should give them at least certain rights, and together with social evidence some claim we should give the higher primates, chimpanzees, gorillas, orangutans, and humans, equal rights in what is called the Great Ape Project (1991).

Confucius in the *Analects* wrote that the presence of *jen* (human-heartedness, the extension of acts of affection, patience, and understanding) designates a human being as opposed to an animal. *Jen* is an embodiment of goodness, wisdom, and courage in a descending order of importance (*Analects*, 14:30), and in the widest sense it refers to a person who possesses the virtues of kindness, gentleness, humanness and unselfishness (6:28). If we can find these characters in animals then we could say that they too possess *jen*, a similar concept to altruism and love. James Rachels (1990) in *Created from Animals* advances the view that in secular philosophy we should not be speciesists, but judge animals as individuals with differing moral worth which may be similar to humans. The sensitivity to pain and capacity for intelligent behaviour are the main qualities we should use to judge whether to use an organism for human ends, as is discussed later in this chapter in the section on animals. The book claims that Darwinism has undermined theological underpinning's of human superiority over animals, which is something many people will refute, especially those who are religious. At the other extreme, the views of Aquinas, who thought animals had no rights, are incompatible with what we know about the biological continuum between animals, and humans. We can argue for a religious difference between the point at which we violate the rights of humans and animals, or use the language of duties, but we certainly have duties to some animals based on their characteristics which we are increasingly becoming aware of. Our ethics must be built on the knowledge that we have, and change when that knowledge informs us of new ethically important qualities of animals such as pain, self awareness and rationality.

There are dangers to having respect for life dependent upon revealed characters of behaviour, or an act-centered definition, because not all individuals are able to show acts or even have the potential for future acts. In conclusion we must broaden our horizons to look at the autonomy of all organisms. Frans de Waal (1990) looked at the origins of right and wrong in different animals. Sympathy is a character at least seen in dolphins and whales. Dolphins have been videoed saving companions by biting through harpoon lines and hauling then out of fishing nets. The sympathy shown by whales to other members of their pod once injured is used by whalers, so that once one sperm whale is harpooned, other members of the pod will encircle the boat trying to help the injured companion, while the whalers will find it easy to kill many more. Sympathy in this case means both recognizing someone else's pain, empathy, and doing something about it. We can also note the cases of dolphins saving human lives.

The inter-relatedness of all living organisms can be readily seen in most ecosystems. All organisms need water, all organisms have the same genetic code and share similar genes. All creatures appear, at first sight at least, to be temporal, they live and they die. This relatedness is expressed by the idea that they are all alive. They share something – life. There is also continuity between inorganic and organic, ecology refers to the relationship of every organism with the environment. Is there anything which distinguishes living organisms from non-living materials beyond that they use information to replicate, and that information is non-random?

Ed Wilson (1984) proposed the theory *Biophilia*, saying that human beings inherently have a love for nature. He defined it as *"the innate tendency to focus on life and life-like processes"*, noting that *"to the degree that we come to understand other organisms, we will place greater value on them, and on ourselves."* It is still a matter of debate whether it is real or not (Kellert & Wilson, 1993). Included in the hypothesis is that the human inclination to affiliate with life is inherent (biologically based), part of our species evolutionary heritage, associated with human competitive advantage and genetic fitness, likely to increase the possibility for achieving individual meaning and personal fulfillment, and that it has a self-interested basis for a human ethic of care and conservation of nature. Although many in modern society seem to destroy nature, it is suggested that this may merely be a result of an unnatural estrangement with nature.

One of the aspects of nature which people seem to love is a diversity of living organisms. People put high value on biodiversity. The United Nations World Charter for Nature (1982) declared, *"Every form of life is unique, warranting respect regardless of its worth to man". This type of valuation is extrinsic. We need to ask whether there is intrinsic value to nature and life? We could reconsider the term selfishness as the conservation of intrinsic value, but we left with a fact that "The planet loves life and so do we"* (Rolston, 1994). We may not know where the value of life comes from, but every organism loves its own life, and at least tries to reproduce its own kind. Some organisms can picture others as themselves, giving extrinsic value to the other, but at least in their actions to love life, all try to survive.

An extension of love to other species could be considered under the concept of stewardship. Stewardship can apply to both the way people use other humans and the rest of nature. It has often been neglected, but has a long history in many religions, being central to a Judeo-Christian doctrine of creation (Macer, 1998). Usually people prefer to ignore it and to think of dominion of humans over the earth, treating the earth with little value, however we see what problems this has caused. There are numerous pollution problems that we can readily see, which affect humans and other species. Throughout time many have considered nature has intrinsic value, but usually these calls have been neglected. Alfred North Whitehead (1925) in *Science and the Modern World* said, *"The western world is now suffering from the limited moral outlook of the three previous generations ... The two evils are: one, the ignoration of the true relation of each organism to its environment; and the other, the habit of ignoring the intrinsic worth of the environment which must be allowed its weight in any consideration of final ends."* The intrinsic value of nature can be argued by Christian and Buddhist values, as shown by Schumacher (1968, 1974). These values were shared by Jahr.

Biodiversity is a word used to picture the great diversity of living organisms on the planet. Just as the individual processes of life are dynamic, so is the composite of the lifeforms. The idea of dynamism also implies a balance. This is illustrated by the words biosphere, foodwebs or ecosystem, with the largest ecosystem being nature itself. The dynamic nature is implied in both science – the second law of thermodynamics, and religion – in the religious doctrines, both Judeo-Christian creation and preservation; and other Asian religions with "harmony". Nature changes with time; some organisms die and others thrive, and nature has done so in dramatic ways in the past. There are various religious stories to support the preservation of biological diversity, the most famous of which is the story of Noah, which is shared by the Judeo-Christian-Islamic traditions. Noah preserved all the domestic and wild animals from environmental catastrophe, a catastrophe that it says was caused by the actions of humans.

If we consider a complete bioethics we must include the duties we have to human beings as well as to nature. In the midst of growing awareness of environmental change and damage we should be aware of the need for sustainable living. We not only have to view the environment in its role as essential to human existence, but we should value the environment itself. The destruction of the environment and disregard for other beings, ignores love. The destruction of the environment was something already clearly noticed at the end of the nineteenth century in industrialised Europe, as witnessed also by Jahr.

Love has more claims to be the principle ethical ideal than desire coming from autonomy. Bioethics does involve all of life, if we do not love all of life we cannot love other people. We need to seek ways to balance need and desire, and just distribution of the freedom that everyone is allowed. As response to the bioethical concerns about our environment that grew through the twentieth century we can also see international treaties on environmental protection outlining some of the limits of damage to the common environment that will be tolerated by other countries, such as the convention on ozone-damaging chemicals, and on deep sea dumping. We should not only aim to give the greatest good for the most, nor the greatest freedom for the most, but the greatest love for all.

Another environmental problem is greenhouse warming, which results mainly from energy use. This problem however can only be solved by individual action, to reduce energy use. We could do this by turning off lights, turning down heaters and air conditioners, building more energy efficient buildings, shutting doors, and driving with a light foot. These are all simple actions which everyone must do if we are concerned about our planet, like being tidy with litter and recycling, yet not many do so. Energy consumption could be reduced 50-80% by lifestyle change with current technology if people wanted to (Macer, 1998). After the Fukushima nuclear accident in Japan, many reactions were taken off line for maintenance and routine checks, and through the Summer of 2011, business men took off their jackets and ties, as an example of lifestyle change. By reducing electricity consumption from air conditioning less energy was used. New technology for energy production, such as solar or wind power, may help, but lifestyle change can have much more immediate affect.

We cannot leave it to governments to look after the planet, the actions of individual members of the human community are required. Some types of environmental improvement can be brought about already by individuals. That is the important call of bioethics and responsibility in the tradition of Fritz Jahr, and others before and after. The broadest concept of the human family is the entire world, and the term human family has been used in United Nations declarations. It has ancient roots, whether it be in Christian concepts of the world or of Mo Tzu in China. Mo Tzu argued that practicing universal love was in one's long term interests not only because other human beings tend to respond in kind to benefits and harms received, but also because heaven wills those practicing the doctrine shall ultimately benefit. As a biologist I see the development of the value of love of life quite consistent with a holistic view of life. The word "Eu-bios" means good-life, and such a life must be sustainable. The pursuit of a bioethical life, a good life for all, is something we can see across the world, and is a core of the bioethical imperative.

References

1. Beauchamp TL and Childress JF (1994) *Principles of Biomedical Ethics.* Fourth Edition. New York: Oxford University Press.
2. Boyd A, Ratanakul P and Deepudong A (1998) Compassion as common ground, *Eubios Journal of Asian and International Bioethics* 8(2) 34-37.
3. Dawkins R (1976) *The Selfish Gene.* Oxford University Press.
4. de Waal F (1996) *Good Natured. The Origins of Right and Wrong in Humans and Other Animals.* Cambridge, MA: Harvard University Press.
5. The Great Ape Project, P.O. Box 12838, A'Beckett Street, Melbourne, Victoria 3000, Australia.
6. Jahr F (1927) Bio-Ethik. Eine Umschau ueber die ethichen Beziehungen des Menschen zu Tier und Pflanze, *Kosmos. Handweiser für Naturfreunde* 24(1) 2-4.
7. Jahr F (1928) Tierschutz und Ethik in ihren Beziehungen zueinander. *Ethik. Sexual- und Gesellschaftsethik. Organ des, Ethikbundes'* 4(6/7) 100-102.
8. Kellert SR and Wilson EO eds (1993) *The Biophilia Hypothesis.* Washington D.C.: Island Press.
9. Kung H ed (1996) *Yes to a Global Ethic.* London: SCM Press.
10. Lovelock JE (1979) *Gaia - A New Look at Life on Earth.* Oxford University Press.
11. Macer DRJ (1994) *Bioethics for the People by the People.* Christchurch: Eubios Ethics Institute.
12. Macer DRJ (1998) *Bioethics is Love of Life: An Alternative textbook.* Christchurch: Eubios Ethics Institute.
13. Maulana, J al-Din R (1983) *The Sufi Path of Love.* Albany, N.Y.: State University of New York Press.
14. Rachels J (1990) *Created from Animals.* Oxford University Press.
15. Rolston, Holmes III. (1994) *Conserving Natural Value.* New York: Columbia University Press.
16. Schumacher EF (1996a) The age of plenty: A Christian view, pp. 159-172 in Daly HE & Townsend KN eds, *Valuing the Earth. Economics, Ecology, Ethics.* Cambridge, Ma: MIT Press.
17. Schumacher EF (1996b) Buddhist economics, pp. 173-181 in Daly HE & Townsend KN eds, *Valuing the Earth. Economics, Ecology, Ethics.* Cambridge, Ma: MIT Press.
18. Schweitzer A (1966) *The Teaching of the Reverence of Life.* London: Peter Owen.
19. Singer P (1976) *Animal Liberation.* London: Jonathan Cape.
20. Singer, P (1990) The significance of animal suffering. *Brain and Behavioral Science* 13: 9-12.
21. Whitehead AN (1925) *Science and the Modern World.* Cambridge: Cambridge University Press.
22. Wilson EO (1984) *Biophilia.* Cambridge: Harvard University Press.
23. Wright R (1994) *The Moral Animal. Why We Are The Way We Are. The New Science of Evolutionary Psychology.* New York: Vintage Books.

NEUROSCIENCE AND THE IMPORTANCE OF A NEURO*BIO*ETHICS: A REFLECTION UPON FRITZ JAHR

James Giordano[1]
Roland Benedikter[2]
Nikola Boris Kohls[3]

> "*...one has tried and is still trying – the longer the more – to support bioethical thought by biological and biopsychological arguments - and not without success...*"
> Fritz Jahr, 1928

Introduction

We offer the premise that by virtue of its history and canon, neuroscience is not necessarily a science of "reduction". It may allow some construct of inter-theoretical reductionism, so as to take psychology and explain it, in part, through neurobiology. But to contextualize the reality of an organism in the world, psychobiology must be framed within the ecology of systems-in-systems. Within this schema, the function of parts can never be fully disembedded from the whole, nor can the whole be depicted or understood absent its parts. In other words, there is reciprocity of the entirety of systems and their component elements. Thus, any realistic approach to understanding a system such as the embodied brain, and its role in consciousness, psychology and the ecological interactions of the organism-in-its-environment – requires insight to both the parts and the whole of the system. In this light, we posit that at its most fundamental level, neuroscience can be seen as a study of relationships. It seeks to understand, for example, the relationship between the internal and external environment of an organism. It seeks to elucidate the structural and functional relationships of molecules and cells, working as networked systems within organisms, enabling interactions within and between groups of organisms that are nested in environments in and across time. Simply put, it attempts to explain neural systems *qua* systems. Neural systems are complex and dynamical, and are embodied and

[1] James Giordano (corresponding author), jgiordano@potomacinstitute.org, Center for Neurotechnology Studies, Potomac Institute for Policy Studies, Arlington, VA, USA; Department of Electrical and Computational Engineering, University of New Mexico, Albuquerque, NM, USA; Oxford Centre for Neuroethics and Oxford-Uehiro Centre for Practical Ethics, University of Oxford, Oxford, UK.
[2] Roland Benedikter, The Europe Center, Stanford University, Palo Alto, CA, USA.
[3] Nikola Boris Kohls, Generation Research Program, Human Science Center-Bad Tölz, Ludwig Maximilians Universität, Munich, Germany.

embedded in complex dynamical systems (i.e.′- organisms, and environments, respectively). One cannot extricate one from the other and still get a full and meaningful explanation of how they interact (Kelso, 1995; Kelso and Engstrøm, 2006; Giordano, 2008). Nor should one dare to try, for that would ignore the current corpus of information about the ways that nervous systems function, and there is something disingenuous, if not fallacious, about not embracing and applying the knowledge amassed to date. Fidelity to the tenets of scientific philosophy compel the use knowledge in ways that are not anachronistic or dogmatic, but instead employ the most current fund of knowledge as a human endeavor *in* human endeavor (Dewey, 1916; Popper, 1934; Kuhn, 1962; Nagel 1961).

The inter-dependence of human endeavor (and the human condition) and the environment, and sensitivity to this ecology should foster imperatives to use current knowledge and capabilities in ways that are creative but none the less responsible (Arendt, 1964). Such perspectives (and hopes) are certainly forward-looking. Yet we assert that in many ways they stand upon and pay homage to the much of the work of Fritz Jahr (1895-1953), a Protestant pastor and philosopher who argued that scientific progress fostered a more naturalistic philosophy that is based upon an iterative body of knowledge, and in this way requires an equally current – and revised – moral code and ethics (Jahr, 1927; Sass, 2007). Indeed, we hold many of Jahr's views to be prescient, as they anticipated a neuro-biocentric world view that has only now come to the fore (Giordano, 2010a). To be sure, neuroscience has become a social force, and while the past decades have been instrumental in formalizing the field and its importance, historical events of the 1800s were important to establishing 1) studies of the brain~mind[1], 2) Jahr's views, philosophy and ethics, and 3) the stage for modern neuroscience and neuroethics, or perhaps more appropriately, neuro*bio*ethics.

Historical trends, impetus and effects

"The natural scientist accesses his observations with an unshakable belief in the veracity of the objects he is studying. [...] He cannot be deceived by nature as there is neither caprice nor randomness within the natural things"

Wilhelm Wundt

The Age of Enlightenment initiated a socio-cultural transformation that advanced trends of individualization, secularization and scientification, and fostered an outgrowth of new (and radical) scientific concepts and philosophical thought (Kohls, 2001; Benedikter, 2001; Benedikter, 2005). From this followed the rise of academic psychology in the 1880s, which was developed as a new and independent field of inquiry that employed the scientific method to approach the study of consciousness. This represented a significant change in and upon the explicit and implicit interpretational frameworks of both the natural sciences and the philosophy of mind. Important to this shift was that in the early to middle 19th century, the paradigm of biology and medicine became more focused upon the integration of anatomy and physiology. From such a structural and functional perspective, it became possible and philosophically and culturally acceptable – if not preferable – to attempt to define and describe concepts such as "life", and "mind" in physiological terms. The application of empirical, physiological approaches to the study of "mental" phenomena such as sensation and perception established the methodological and epistemological groundwork for physiological psychology. In this way, the Cartesian differentiation of body (*res extensa*) and mind (*res cogitans*), and the corresponding idea of a *primum movens homunculus* were gradually replaced by psychophysiological theories of cognition, emotion and actions.

Physiological and experimental psychology can be considered to have been "formalized" as institutionally valid academic disciplines (i.e. – distinct from philosophy) in 1879 by the establishment of Wilhelm Wundt's laboratory at the University of Leipzig. However, some of the circumstances surrounding this event are also relevant to the current status, and possible trajectories of cognitive neuroscience, neurophilosophy, and neuroethics. Following training in Helmholtz`s laboratory, Wundt published a two-volume book on the principles of physiological psychology in 1874 (Wundt, 1874) that was to significantly contribute to a building schism in German academia in which older, more "romantic" ideas about nature were confronted with, and displaced by empirical findings. Wundt focused in that domain of psychophysiology where exact and objective physiological mechanisms were more seemingly clouded by psychological phenomena (Krüger, 1994). Wundts works at the interface of brain and mind met considerable intellectual resistance, and from 1877, Wundt, still a junior academic at the University of Leipzig – struggled to establish experimental psychology as an independent academic field (Bringmann and Tweney, 1980). Through these efforts were gained the greater effect of Wundt's work upon the overall philosophical Zeitgeist. The predominant philosophical construct of the time was *Naturphilosophie* (philosophy of nature), a reflection of earlier romanticist ideals that postulated that all forms of life are imbued with a spiritual power that can only be appreciated through transcendental speculation.

This was challenged by the new and rapidly developing natural sciences that were grounded to an empirical, rational approach that was substantiated through experimentation (Paul, 1984). The decline of *Naturphilosophie* in the latter 19th century spawned the growth of idealism and empiricism (and its core construct of materialism), two very distinct philosophical perspectives that were to influence, and be influenced by increasing nationalistic ideas and tensions. This schism between idealist and materialist thought was not limited to the academic sphere, but rather, was to influence socio-cultural attitudes toward science, philosophy, and ultimately moral conduct and ethics (Treitel, 2004). At its most fundamental level the distinction can be seen as reflective of, and consequential to attempts to define and convert spiritual and religious (i.e. – more dogmatic) ideas about the soul into a more methodologically doctrinal scientific theory of mind. The scientific movement sought to replace older, supernatural ideas in order to obtain a theory of mind that was free of un-testable metaphysical assumptions.

Yet, attempts at scientific naturalism also manifested ethical dimensions and implications regarding the psychobiological nature of the self, and the relevance of such information to attitudes and conduct. Thus, experimental psychology, the new discipline of the mind and behaviour – and apparent wellspring of modern neurosciences – fostered an importance of, or at least implicit concern with ethical questions from its earliest days. This most certainly set the stage for Fritz Jahr's reflections and ideas about the role of biology, and the natural sciences in general, in formulating moral constructs, and developing a system of ethics that can address the profound questions that are generated by an appreciation for life and the capacity for sentience and suffering (Jahr, 1927; 1928). To be sure, many of these questions remain unresolved, and are at the forefront of contemporary neurosciences and the quest for a valid and meaningful form of neuroethics, amidst renewed controversy between (post)-idealistic and (post)-empiricist positions (Churchland, 1981; Churchland, 1986; Giordano, 2009a; Giordano, 2009b; Racine, 2010; Harris, 2011).

The epistemological breach between holism, empiricism, and positivism became evident by the latter 19th century, in what Lamont has referred to as "the crisis of evidence" (most notably apparent in spiritists' claims of experimental "proof" justifying their beliefs) (Lamont, 2004). It is important to view Wundt as a guardian of the humanistic tradition, as he (and others) recognized the need to both overcome the speculative approach of holistic science so as to achieve progress, and still maintain certain humanistic values that preserved society's spiritual needs and demands. Wundt viewed these values to be corrupted by spiritism, and thus opposed spiritism on scientific and moral grounds as antithetical to the achievement of meaningful intellectual and social progress. Wundt's rejection of a supernaturally metaphysical source of consciousness not only established the naturalistic basis of psychology, but also instantiated a tradition of secular inquiry into human consciousness and its implications through the use of experimental methods. According to his autobiography, while Wundt did not consider himself to be a reductive materialist, he could not conceptualize consciousness as an event of the natural world to be a result of non-causal, supernatural factors (Wundt, 1920). In eschewing reductive materialism and Cartesian dualism, Wundt described brain and mind states as independent, yet synchronized functions that are both necessary for consciousness: claiming that while conscious processes are associated with brain functions, both appear to have a "causality" of their own – a form of token physicalism that remains pertinent to this day. It is noteworthy that much of the debate that characterized the academic, intellectual and social climate of Leipzig in the later 19th century still undergirds the current scientific and cultural discussion about consciousness and its implications for morality and ethics (i.e. – the landscape of neuroethics). Indeed, many of today's debates about the "easy and the hard problems" of neuroscience (i.e. √ the relationship of the physical brain to consciousness, subjectivity of experience, and the "self"), and the ethico-legal and social issues and questions that stem from such problems, arise from empirical and idealistic viewpoints and hypotheses (Chalmers 1995; Giordano 2010; Giordano and Olds, 2010; Giordano, 2012).

The desire – if not necessity – to explain, interpret, and manipulate consciousness has been an unquestionable constant throughout modernity. Hermann Ebbinghaus' (Ebbinghaus, 1908) claim that psychology *"...has had a long past and only a short history"* is noteworthy in that while authentic studies of consciousness really only began at the end of the Enlightenment, the roots of such speculation and inquiry are much older, and this is meaningful to the tacit ethical and societal values and implications generated by the mind-body problem. Thus, the "tools-to-theory" heuristics of the early days of Wundtian psychology led to the development of new techniques and technologies that were then used to further shape both science and society. Martin Heidegger has described how technology transforms human understanding of, and orientation towards the world. Heidegger claimed that through a process termed "enframing", human beings use technology to define levels of order in the natural world, and then engage technological tools to order and control their environment (Heidegger, 1953). The advancement of Wundtian psychology to its present iteration as neuroscience has enabled the development of technical methods and technology that have increasingly been used as both metaphorical and literal models for interpreting consciousness. A theory of mind is a powerful tool, not only for explaining the nature and mechanisms of thought, but also for establishing the moral, ethical and legal constructs that are based upon distinctions of what and who has a "mind", and what having a mind "means" for the ontologies and categorical distinctions upon which to derive and base social attitudes and conduct.

Fritz Jahr, bioethics and modern neuroscience

It is upon the historical backdrop of the late 19th century that we view the work of Fritz Jahr. Writing as something of a contemporary of Albert Schweitzer, and coming from a similar philosophical tradition, Jahr acknowledged the scholarship of Schleiermacher and Krause in advocating that an understanding of nature compels respect for living beings (Jahr, 1927). Like Schweitzer, Jahr's position was that organisms' embodied life (what Schweitzer, after Goethe, referred to as *Beseelt*) must be appreciated as a basis for respect that reflects a sense of reverence for the natural world (viz. *Ehrfurcht;* Schweitzer, 1923). Indeed, Jahr's take on the empirical trend of the mid- to late-1800s and the technological turn of the 20th century was that the ethos of science is not merely to understand nature so as to impart ability for control, but rather that it should foster a sense of humility: "...scientific triumphs...have taken away the dominant position of the human being in the world in general. Philosophy, formerly prescribing leading ideals for the natural sciences, now has to build her systems on the basis of scientific knowledge *from* the natural sciences..." (Jahr, 1927).

Jahr saw such knowledge as complementary to a broadened Kantian ethic of responsibility to self – as a biopsychological organism – that extended to others, both as biopsychological organisms, and in appreciation of the asymmetries of any and all relationships (Kant, 1785). To Jahr, such a stance toward other living beings was possible "...only...when we are somewhat knowledgeable about their physiological and psychological properties and life conditions" (Jahr, 1928). Thus, physiological and psychological studies (i.e.- what is today entailed in and obtained through the contemporary biological, cognitive, behavioural and social neurosciences) are fundamental to developing 1) an understanding of biological organisms, and 2) a basis for the way(s) that we morally regard and treat humans and other species.

Indubitably, modern neuroscience conveys great epistemological and physical power in its techniques and technological tools. Given such power, it is important to consider what *should* be done with this capability, acknowledging the potential to use any scientific and technological tool to evoke good or harm (Heidegger, 1953; Habermas, 1968; Jonas, 1973; Lenk, 1998). Here we must return to a view and appreciation of neuroscience as a human endeavor. In this light, humanity bears responsibility for the ways that neuroscientific knowledge and its outcomes are employed. Modern neuroscience has extended the boundaries of self-understanding, and we posit, pro Jahr, that this knowledge is crucial to the ways that both humans and non-human beings (e.g., animals, and perhaps one day, even sentient machines) are regarded and treated. But neuroscience does not merely afford means to understanding; the techniques and technologies of brain science can be used to control cognition, emotion, and behavior. Nor does neuroscience exist in a social vacuum, and thus, its tools and outcomes can both affect, and be subject to social ideals, values, norms, and mores and the economic and political forces that contribute to them. Therefore, any consideration of neuroscientific knowledge and outcomes must be seen as viable resources to influence the nature and scope of the human condition – if not the world at-large – and hence it is important to reflect upon and address how such information and resources will affect individuals, groups, society and nature.

The role, value, and importance of a neuro*bio*ethics

The field and practice of neuroethics is dedicated to such tasks. In accordance with the definition originally provided by Adina Roskies (Roskies, 2002), neuroethics entails two "traditions": the first encompasses multi-disciplinary studies of putative neural bases of moral cognition and behavior (viz. 'neuromorality' or, we believe a better description might be 'neuroecology'). Neuroethics in the second tradition is a form of bioethics, in that it addresses ethical issues arising in and from neuroscientific research and its applications in medicine, public life, and national defense. We are fond of the term "neuro*bio*ethics", believing that it links the field to its Jahrian roots, and may afford a more accurate description of what the discipline and practices are – and should be (Giordano and Olds, 2010; Giordano, 2011a; Giordano, 2011b; Giordano, 2012b). Neurobioethics obtains the inter-disciplinarity of bioethics, and axiomatically grounds its discourse and actions to the naturalistic philosophy of biology. Moreover, the term is intended to convey a specialized – and perhaps unique – bioethics that reflects the rapidly changing epistemic capital, persistent "hard problems," and far-reaching implications of both the neurobiological substrates of morality, and neurocentric considerations for the moral concern and treatment of living organisms, that views consciousness, painience, and sentience as principal qualities upon which to base, initiate, and enact moral regard (Giordano, 2009a; Giordano, 2009b). But it's important to go beyond purely philosophical musings, and advance this construct into the practical realm.

Neurobioethics in the first tradition (i.e. – "neuroecology") may be a form of meta-ethics, in that it enables an understanding of those neural processes that function in moral cognition, decisions, and actions (Giordano, 2012; Giordano, 2011b). This could provide a more accurate lens through which to perceive the ways that humans (and perhaps other species) obtain, develop, and manifest moral capabilities as part of their interactions with others. But, we contend that the validity of this ecological approach necessitates framing within a bio-psychosocial systems model, given that humans (and other species) likely possess certain proto-ecological sensitivities that are then inculcated and reinforced through environmental experience(s). Thus, neuroecology describes those internal functions that are operative in, and contributory to feelings of "rightness" and/or "wrongness" as derived from and responsive to environmental influences that affect an organism's cognitions, and activities (if not overall existence). To re-state our initial assertion, we do not view this as overtly reductive, but instead offer that this conceptualization establishes a naturalistic framework for moral decision-making that comports well with the philosopher Owen Flanagan's notion of ethics as "human ecology" enacted in the psycho-social domain (Flanagan 2002; Flanagan 2007).

Neuroscience has enabled deeper and wider insight to *putative* substrates and mechanisms of consciousness, mind, self and personhood (Perry, 1995; Dvorsky 2011; Van Gulick, 2011) Despite (genuine recognition of) current limitations in the type and extent of such information, the knowledge gained to date has initiated moves from longstanding, dogmatic notions of self and person, to a broader construct, that's more inclusive of the possibility – if not probability – of animal "persons" and machine "selves" (Uttal 2003; Farah and Heberlein, 2007; Tallis, 2011). Of course, differing viewpoints exist, not only within the field of brain-mind studies (including its disciplines of neuroscience, psychology, philosophy, etc); but also between various camps within the sciences and humanities, and even within the public sphere (Metzinger, 2003) In the main, these differences reflect and\or stem from various epistemological and anthropological positions that continue to pose questions for both scientific inquiry and social conduct. Transformations in the construct of self and personhood are certain to impact ethico-legal considerations, policy decisions, ecological trends, if not the human condition at-large. Indeed, Jahr stated that knowledge and a better understanding of nature serve to awaken, broaden, and deepen end ethical constructs and practices (Jahr, 1928).

Working in our group, Nicholas Fitz and Daniel Howlader are focusing upon the ways that the increasing advancement, and societal reliance and role of neuroscience and neurotechnology may change current and longstanding ideas of self and personhood, and foster re-examination of more neurocentrically-oriented views of animals, embryos and fetuses, the obtunded, senile, hybrid human-animal and human-machine beings, and even so-called intelligent or sentient computers. Working from the premise that it's not so much a question of *if* such epistemic shifts will occur, but *when* this will happen, Fitz and Howlader are questioning what society will do with this new information and its potential implications for policy and law (Fitz, Howlader and Giordano, 2011).

Our general position is that a deepening understanding of the ways that nervous systems and brains are involved in (or evoke) those characteristics that we value as individuals, groups, and a species should compel and sustain the ways that we regard and treat the organisms that possess nervous systems that give rise to such characteristics. Moreover, neuroscience has – and will likely continue to – demonstrate that despite a wide array of individual differences, there are features that are common to nervous systems, and to the organisms in which they are embodied. Simply put, we must ask whether and in what ways neuroscience might demonstrate the ways that humans- and other species – are alike and differ. Is it possible that neuroscience might afford both purchase and leverage to reconcile such apparent differences? On some level, we believe (or at least hope) so; but we hold that a more critical question is whether humanity - as individuals, groups, cultures and a species - will be willing and able to embrace such knowledge to prompt positive change in the ways that things that "have a brain and are a mind" are viewed, valued, and treated.

This prompts questions about the nature and implications of neurocentric criteria of normality and diversity, ontological status, and the ways that beliefs and systems of governance are formed and formulated (Blume, 1998; Hildt, Franke, and Huber, 2008). In the main, we ask 1) whether (and how) insight(s) to the neuroscience of painience and sentience (or the translation of neuroscientific information and technology to create organisms that are sentient and\or painient) could provide a metric for moral and social regard and behavior (Giordano, 2009b), 2) whether we will be sufficiently sensitive to, and wise enough to appropriately weigh and act upon such knowledge, and 3) if and how such information can – and should be used to inform – and perhaps revise – standards and practices of morality and ethics.

We argue that if neuroscience is to have any value as a human endeavor, then the information it yields must be leveraged in both understanding and action. The discovery that an entity is painient and sentient is not esoteric, but rather *means* something both about that organism, and the ways that it should be considered (Ryder 2001; Giordano, 2009a; Giordano, 2009b). Neurocentric criteria, namely, whether a being manifests the ability for pain/suffering, some form of emotion, awareness of self, and the type and extent of these properties, are arguably important for moral regard – and the ethical, legal and social treatment of other beings (Giordano, 2009b). For example, together with researcher Sherry Loveless, we have posited that neuroscientific studies should sustain an anti-speciesist "Jahrian" neurobioethics, that is applicable to guidelines, polices and laws that direct and govern the use, care, and treatment of non-human animals (Loveless, 2011).

Yet, there is much that is still not known about the brain, consciousness, and how biological, psychological and social domains interact. It's folly not to critically assess what this science holds for the present, foolhardy not to recognize the promise – and perils – that such science and technology may incur, and frighteningly dangerous not to devote time, effort and resources to studying, and developing ways to prudently guide each and every step ahead. (Habermas, 1968; Jonas, 1973; Lenk, 1998; Janicaud , 2000; Giordano, 2010b; Giordano, 2011a; Giordano,2012a; Giordano,2012b; Giordano, 2012c). We call for frank, pragmatic assessment of neuroscientific and neurotechnological capability and limitations, and an openness to revising scientific facts, philosophical doctrine, and social constructs in preparation for and recognition of the potential proximate, intermediate, and distal effects that such new knowledge – and values – are likely to incur.

We maintain that neurobioethics could serve to guide the scope and tenor of current neuroscientific research and applications, and ensure preparedness for the contingencies and consequences of neuroscientific advancements in the future. Working with philosopher John Shook, we are examining if current ethico-legal concepts and criteria are adequate to deal with the contingencies posed by today's neuroscientific and neurotechnological challenges, or if ethical and legal concepts and systems need to be adapted, or even developed anew to sufficiently account for and meet the epistemological, anthropological and socio-cultural (and economic) changes that neuroscience fosters (Shook and Giordano, 2011; Shook and Giordano, in press; Palchik and Giordano, 2011).).As Jahr noted some 80 years ago, new science and technology add to the palette of philosophy, and require new forms of ethical reflection, terminologies, and concepts (Sass, 2010). Given the reciprocal relationship of knowledge, technology, and culture it will be critical to develop ethical, legal, and political systems that appropriately reflect scientific advancements, apprehend the realities of social effect(s), and aptly guide, if not govern the use and manifestations of neuroscience in the public sphere (Benedikter and Giordano, 2011; Benedikter and Fitz, 2011).

Conclusions

So if neuroscience – and neurobioethics – are to impart a sense of practical wisdom in the spirit of Jahr, then perhaps the Thomistic sentiment, *recta ratio speculabilium, recta ratio agibilium* - reflecting upon the Aristotelian notion of the right measure of knowledge to compel the right measure of actions, is fitting. We opine that neuroscience should serve as both a lens upon the natural world, and as a mirror upon the human brain~mind and humanity's place and conduct in nature. This mirror of introspection should not only be focused upon the brain, but must also reflect upon the ways that neuroscience and its techniques are used.

We posit that Jahr's legacy doesn't necessarily lie in the value of understanding the natural world, but rather the value of what is done with such knowledge. Biologies may differ, yet individuals and species share certain things in common. The psychologies that arise from and interact with these biologies also differ, yet these too share particular commonalities. And in moving toward a more pluralized and globalized society that is a result of what human brain~minds can conceive and achieve, we argue that there is a necessity to seek unity amidst plurality, and hope that neurobioethics can be of value in this regard (Giordano 2011a). Indeed, Jahr's legacy may be humanity's, for the ways we use the knowledge and power of neuroscience and neurobioethics will likely determine the human condition, natural ecology, and the world humanity creates for the future.

Notes

1. To date, the primary "hard problem" of neuroscience remains how consciousness or "mind" occurs in the physical substrate brain. As matter of fact, any and all postulates on this relationship remain hypothetical, if not wholly speculative, and thus the enigmatic relationship of brain and mind persists. Current evidence suggests that this relationship is something of a complementarity, in which brain affects "mind" and the process or properties of "mind" influence the physical substrate of brain in some ways. To preserve the ambiguity of this relationship of structure and function, we refer to the brain~mind, by using the (~) or tilde symbol, to infer complementarity throughout the body of this paper.

Acknowledgments

This work was supported in part by grants from the J. W. Fulbright Foundation, Nour Foundation, William H. and Ruth Crane Schaefer Endowment, and funding from the Center for Neurotechnology Studies of the Potomac Institute for Policy Studies (JG); Europe Center of Stanford University (RB); and Brain-Mind and Healing Research Program of the Samueli Institute, and Peter Schilffarth Institut für Sozialtechnololgie (NBK). This essay was completed while the senior author (JG) was Fulbright Visiting Professor of Neuroscience, Neurotechnology, and Neuroethics at the Human Science Center-Bad Tölz of Ludwig Maximilians Universität, Munich, Germany, and the author is grateful for the support and collegiality provided by this institution. The authors thank Daniel Howlader, Nicholas Fitz, John Shook, Hans-Martin Sass, and Sherry Loveless for their contribution to this manuscript.

References

1. Arendt H. (1964). Labor, Work, Love. Presented at the Christianity and Economic Man: Moral Decisions in an Affluent Society conference, November 10, Chicago.
2. Benedikter R. (2001). *Einführung in das postmaterialistische Denken*, Vol. 1. Vienna: Passagen Verlag.
3. Benedikter R. (2005). *Perspektiven postmaterialistischen Denkens*, Vol. 7. Vienna: Passagen Verlag.
4. Benedikter R, and Fitz N. (2011). Technophilia and the new media: Contemporary questions of responsible cultural consumption. A call for public debate. *Synesis* 2(1): G62-G68.
5. Benedikter R, and Giordano, J. (2011). The outer and inner transformation of the global sphere through technology: The state of two fields in transition. *New Global Studies* 5(2).
6. Blume H. (1998). Neurodiversity: On the neurological underpinnings of geekdom. *The Atlantic*.http://www.theatlantic.com/magazine/archive/1998/09/neurodiversity/5909/.
7. Bringmann WG, and Tweney RD. (1980). *Wundt Studies*. Toronto: Hogrefe.
8. Churchland PM. (1981). Eliminative materialism and the propositional attitudes. *Journal of Philosophy* 78(2): 67-90.
9. Churchland PS. (1986): *Neurophilosophy: Toward a Unified Science of the Mind-Brain*. Cambridge, Massachusetts: Massachusetts Institute of Technology Press.
10. Dewey J. (1916). *Democracy and Education: An Introduction to the Philosophy of Education*. New York: Macmillan.
11. Dvorsky G. (2011). Rights of Non-Human Persons. Institute for Ethics & Emerging Technologies. http://ieet.org/index.php/IEET/RNHP.
12. Ebbinghaus H. (1908). *A Summary of Psychology*. Leipzig: Veit.
13. Farah MJ, and Heberlein AS. (2007). Personhood and neuroscience: Naturalizing or nihilating? *American Journal of Bioethics* 7(1): 37-48.
14. Flanagan O. (2002). T*he Problem of the Soul: Two Visions of Mind and How to Reconcile Them*. New York: Basic Books.
15. Flanagan O. (2007): *The Really Hard Problem: Meaning in a Material World*. Cambridge, Massachusetts: Massachusetts Institute of Technology Press.
16. Fitz N, Howlader D, Giordano J. (2011). The Self and personhood: Neuroethical implications of a shifting paradigm. Poster presented at the International Neuroethics Conference, November 10-11, Washington, DC.
17. Giordano J. (2008). Complementarity, brain-mind, and pain. *Forsch. Komplementarmed* 15(2): 2-6.
18. Giordano J. (2009a). *Pain: Mind, Meaning and Medicine*. Glen Falls, Pennsylvania: PPM Press.
19. Giordano J (2009b). The neuroscience of pain, and the neuroethics of pain care. *Neuroethics* 3(1): 89-94.
20. Giordano J. (2010a). Neuroethics- coming of age and facing the future. In: Giordano J, Gordijn B. (eds.) *Scientific and Philosophical Perspectives in Neuroethics*, ix-xxi. Cambridge: Cambridge University Press.
21. Giordano J. (2010b). The mechanistic paradox: The relationship of science, technology, ethics and policy. *Synesis* 1(1): G1-4
22. Giordano J. (2011a). Neuroethics - two interacting traditions as a viable meta-ethics? *American Journal of Bioethics-Neuroscience* 3(1): 23-25.

23. Giordano J. (2011b). Neuroethics: Traditions, tasks and values. *Human Prospect*, 1(1): 2-8.
24. Giordano J. (2012a) Integrative convergence in neuroscience: trajectories, problems and the need for a progressive neurobioethics. In Vaseashta A, Braman E, and Sussman P. (Eds.), *Technological Innovation in Sensing and Detecting Chemical, Biological, Radiological, Nuclear Threats and Ecological Terrorism (NATO Science for Peace and Security Series)*. New York: Springer.
25. Giordano J. (2012b). *Advances in Neurotechnology: Premises, Potential and Problems*. Boca Raton, Florida: CRC Press.
26. Giordano J. (2012c). Icarus folly: On the capabilities and limitations of neuroscience and neurotechnology. In Giordano J (Ed.) *Advances in Neurotechnology: Premises, Potential and Problems*. Boca Raton, Florida: CRC Press.
27. Giordano J, and Olds J. (2010). On the interfluence of neuroscience, neuroethics and legal and social issues: The need for (N)ELSI. *American Journal of Bioethics-Neuroscience* 2(2): 13-15.
28. Habermas J. (1968). *Technik und Wissenschaft als Ideologie*. Frankfurt am Main: Suhrkamp.
29. Harris S. (2011). *The Moral Landscape: How Science Can Determine Human Values*. New York: Free Press.
30. Heidegger M. (1953). Die Frage nach der Technik. In *Vorträge und Aufsätze*. Pfullingen, Germany: Neske.
31. Hildt E, Franke A, and Huber L. (2008). *Normality, Normalization and Enhancement in the Neurosciences: Ethical, Sociocultural and Neuropsychiatric Aspects of Cognitive Enhancement*. Johannes Gutenberg-Universität Mainz. http://www.philosophie.unimainz.de/1889_ENG_ HTML.php.
32. Jahr F. (1927). Bio-Ethik: Eine Umschau über die ethischen Beziehungen des Menschen zu Tier und Pflanze. *Kosmos. Handweiser für Naturfreunde*, 24(1): 2-4.
33. Jahr F. (1928). Tierschutz und Ethik in ihren Beziehungen zueinander. *Ethik. Sexual- und Gesellschaftsethik. Organ des Ethikbundes*, 4(6/7): 100-102.
34. Janicaud D. (2000). *Phenomenology and the Theological Turn: The French Debate*. New York: Fordham University Press.
35. Jonas H. (1973). Technology and responsibility: Reflections on the new tasks of ethics. *Social Research* 40(1): 31-54.
36. Kant I. (1785/2004). *Grundlegung zur Metaphysik der Sitten*. Göttingen, Germany: Vandenhoeck & Ruprecht.
37. Kelso JAS. (1995). *Dynamic Patterns: The Self-Organization of Brain and Behavior*. Cambridge, Massachusetts: Massachusetts Institute of Technology Press.
38. Kelso JAS, and Engstrøm DA. (2006). *The Complementary Nature*. Cambridge, Massachusetts: Massachusetts Institute of Technology Press.
39. Kohls N. (2004). *Aussergewöhnliche Erfahrungen - Blinder Fleck der Psychologie? Eine Auseinandersetzung mit aussergewöhnlichen Erfahrungen und ihrem Zusammenhang mit geistiger Gesundheit*. Münster: Lit-Verlag.
40. Krüger L. (1994). *Universalgenie Helmholtz. Rückblick nach 100 Jahren*. Berlin: Akademie.
41. Kuhn TS. (1962). *The Structure of Scientific Revolutions*. Chicago: University of Chicago Press.
42. Lamont P. (2004). Spiritualism and a mid-Victorian crisis of evidence. *The Historical Journal* 47(4): 897-920.
43. Lenk H. (1998). Distributability problems and challenges to the future resolution of responsibility conflicts. *Society for Philosophy and Technology* 3(4): 68-93.

44. Loveless S. (2011). Welfare & Ethics in Veterinary Medicine. *Today's Veterinary Practice* July/August: 50-55.
45. Metzinger T. (2003). *Being No One: the Self-model Theory of Subjectivity*. Cambridge, Massachusetts: Massachusetts Institute of Technology Press.
46. Nagel E. (1961). *The Structure of Science*. New York: Harcourt, Brace and Company.
47. Palchik G, and Giordano J. (2011). A new lexicon for neuroscience and neuroethics. Poster presented at the International Neuroethics Society Conference, November 10-11, Washington, DC.
48. Paul R. (1984). German academic science and the Mandarin ethos, 1850-1880. *British Journal for the History of Science* 17(1): 1-29.
49. Perry J. (1995). The Self. In *The Stanford Encyclopedia of Philosophy*.
50. Popper KR. (1934). *Logik der Forschung*. Tuebingen, Germany: Mohr Siebeck
51. Racine E. (2010): *Pragmatic Neuroethics: Improving Treatment and Understanding of the Mind-Brain*. Cambridge, Massachusetts: Massachusetts Institute of Technology Press.
52. Roskies A. (2002). Neuroethics for the New Millenium. *Neuron* 35(1): 21-23.
53. Ryder RD. (2001). *Painism: A Modern Morality*. London: Open Gate Press.
54. Sass HM. (2007). Fritz Jahr's Bioethischer Imperativ. 80 Jahre Bioethik in Deutschland von 1927 bis 2007. *Medizinethische Materialien*, Heft 175 (Juli): 1-33.
55. Sass HM. (2010). Selected essays in bioethics 1927-1934 of Fritz Jahr; postscriptum. *Medizinethische Materialien*, Heft 186 (November): 24-32. See also in this volume.
56. Schweitzer A. (1923). *Kulturphilosophie II: Kultur und Ethik*. Bern, Germany: Paul Haupt.
57. Shook J, and Giordano J. (2011). Neuroethics: Out of the toolbox and into the future. Poster presented at the International Neuroethics Society Conference, November 10-11, Washington, DC.
58. Shook J, and Giordano J. (in press). Toward a cosmopolitan neuroethics: Considerations for international relevance. *Bioethics*.
59. Tallis R. (2011). *Aping Mankind: Neuromania, Darwinitis and the Misrepresentation of Humanity*. Durham, United Kingdom: Acumen Publishing.
60. Treitel C. (2004). *A Science for the Soul. Occultism and the Genesis of the German Modern*. Baltimore & London: Johns Hopkins University Press.
61. Uttal WR. (2003). *The New Phrenology: The Limits of Localizing Cognitive Processes in the Brain*. Cambridge, Massachusetts: Massachusetts Institute of Technology Press.
62. Van Gulick R. (2011). Consciousness. In Zalta ED (ed.) *The Stanford Encyclopedia of Philosophy*. http://plato.stanford.edu/archives/sum2011/entries/consciousness
63. Wundt W. (1874). *Grundzüge der physiologischen Psychologie*. Leipzig: Engelmann.
64. Wundt W. (1920). *Erlebtes und Erkanntes*. Leipzig: Kröner.

THE MANY FACES AND COLORS OF THE BIOETHICS IMPERATIVE

Hans-Martin Sass[1]

Jahr's concept of the *Bioethical Imperative* extends Kant's Categorical Imperative to all forms of life. Kant, in his 'Groundwork of the Metaphysic of Morals' 1785, had defined as the foundational moral Imperative: 'Act in such as way, that you treat humanity, whether in your own person or in the person of any other, never merely as a means to an end, but always at the same time as an end.' Jahr replaces the categorical 'always' by a hypothetical and situational 'if possible'; thus modifying the inflexible categorical structure of Kant's model into a pragmatic and situational model of balancing moral principles, obligations, rights, and visions in a Bioethical Imperative: 'Respect every living being on principle as an end in itself and treat it, if possible, as such!' The vision of integrative bioethics as a new transdisciplinary academic field and as a new personal ethics calls for a new culture towards all forms of life, towards the land and environment, and for a virtue of civility in social, professional, and political environments. He develops his original concept of a new integrative academic discipline and a content-rich Bioethical Imperative replacing Kant's formal Categorical Imperative subsequently in slightly more than a dozen articles in various educational and religious journals.

Jahr also underlines, that – if not the term – the concept, the culture and the mission of bioethics had already been with humankind most likely since prehistoric times and has not been a heritage of one culture or one continent alone: Respect for the entire world of bios, not just for humans, for plants and animals as well, for the environment, natural and social, for the earth. Taoist reverence for nature, Buddhist compassion with all forms of suffering life, Francesco Assisi's call for sisterhood and brotherhood with plants and animals, also Albert Schweitzer's philosophy of respect for all forms of life in support of his medical missions in Africa and Aldo Leopold's land ethic, are prime examples of deep human compassion with non-human life and examples of a human commitment in respect to many other forms of life and environment.

---

[1] Hans-Martin Sass, hsass@potomacinstitute.org, Academic Fellow at the Potomac Institute of Policy Studies, in Arlington, Virginia, USA, and Founding Member of the Bochum Zentrum fuer Medizinische Ethik, is em. Prof fuer Philosophie at Ruhr University, Bochum, Germany, em. Senior Research Scholar at the Kennedy Institute of Ethics at Georgetown University, Washington DC, and Honorary Professor at Peking Union Medical College, Beijing, China.

In reviewing new physiological knowledge of his times and moral challenges associated with the development of secular and pluralistic societies, he redefines moral obligations towards human and nonhuman forms of life and outlines the concept of bioethics as an academic discipline, as a foundational cultural and moral principle, and as a virtue. He argues that new science and new technology require new ethical and philosophical reflection and resolve. He also calls for the development of a new and clear terminologies and defined fields in the humanities, and for normative and practical visions of bioethics and her sub disciplines.

The original vision, its differentiation and explication.

The article *Bioethics. Reviewing the Ethical Relations of Humans towards Animals and Plants,* 1927 [Bio-Ethik. Eine Umschau über die ethischen Beziehungen des Menschen zu Tier und Pflanze] served as an Editorial in the leading German science journal ‚Kosmos. Handweise für Naturfreunde und Zentralblatt für das naturwissenschaftliche Bildungs- und Sammelwesen', Stuttgart 1927, 24: 2-4. He discusses most recent results of neurophysiology and psychological studies on plants and animals and presents Bio-Ethics as a new academic discipline and a new moral attitude corresponding with the term Bio-Psychik, coined by philosopher and psychologist Rudolf Eisler as the newly developing soul science of all forms of life. Thus, the origin of the term and the concept of Bioethics are closely related to 19th century progress in the life sciences, in particular in experimental physiology and psychology. Traditionally European thinkers had followed Aristotle who had distinguished between physics, ethics and logic as the three parts of philosophy. Eisler and Wundt felt it necessary to make a distinction between physics as covering unorganic matter and psychics as organic matter such as plants and animals having the capacity for growth, reproduction, functional activity and continual change preceding death. In 1878 Wilhelm Wundt had founded the first research Institute for Experimental Psychology in Leipzig. Wundt's 3-volume book ‘Grundzüge der physiologischen Psychologie', which documented similar neural and physiological reactions and goal- and survival-oriented activities in humans, animals and plants, had its 6th edition in 1908-1911. Wundt's ‘Vorlesungen über die Menschen- und Tierseele' 1863 [6th ed. 1919], Fechner's ‘Nana oder das Seelenleben der Pflanzen', 1848, and ‘Elemente der Psychophysik', 1860, were bestsellers. Thus, methods and arguments were sharpened to overcome the Cartesian mechanical body-soul dualism and to research more directly life- and survival-oriented wills and interactions between living and sensing environments and living and sensing beings. Only recently has neuroscience, subsequently also neuroethics, become new and fascinating frontiers. In the interest of methodological clarity, Rudolf Eisler in ‘Das Wirken der Seele.

Ideen zu einer organischen Psychologie' [1909: 32] suggested a new term employing recent research methods for 'psychological facts as biological factors': 'If we want to hold on to the unity of natural causality also in the realm of organic matter, we have to add (not replace) Biopsychics to Biophysics and Biochemistry and to recognize that psychic movements of lower or higher form, simple and complicated wills, tendencies to protect organic unity and drives and willings [Wollungen] ... are means to the highest goal – ruling or modifying directly or indirectly actions of life. ... Far away from describing the will as a product of mechanical reflexes, reflexes are better understood as residuals of original will processes'.

Eisler included the term 'Biopsychik' in 'Wörterbuch der philosophischen Begriffe' (1910, 3. ed., Vol. 1, p. 192). But 'Bioethics' has become a dictionary reference only recently in an article by E. M. Engels in 'Metzler Lexikon Religion' (1999, p. 159-164). Jahr's Bioethical Imperative is not inflexible and categorical like Kant's, nor does it enjoy the luxury of content-free formality. The Bioethical Imperative is content-rich and balances values and life goals of living entities in their struggle for life and in their need for food, space and development. Jahr, on the other side, takes issue with, what he terms, 'Buddhist fanatics', who do not kill even deadly snakes, because they 'also are our brothers and sisters'. Of course, we make ethically sound decisions in killing rats and microbes in our hospitals and homes; for some of us the gift of cut flowers to certain occasions is a part of our culture but the careless picking of flowers by rascals who then throw them away anyway is considered to be uncivilized and an immoral violation of the Bioethical Imperative. To strive and struggle for survival is an essential part of live, thus the obligation and drive to live has to be balanced with the respect towards the life and strive of others.

Death and the Animals [Der Tod und die Tiere], 1928, presents references from the bible and from philosophical thinkers that animals alike humans may not end their lives with cessation of their biological existence, rather continue to live in a spiritual world with each other and with us. His position, however, is more influenced by biblical sources than by Asian and Hindu concepts of reincarnation into other forms of life.

The Relationship of Animal Protection and Ethics, 1928 [Tierschutz und Ethik] underlines the importance of ethical sentiment, empathy, compassion and support for animals and plants as a part of moral and social obligations humans owe each other. Animal ethics and ethics among humans are not in conflict, on the contrary: they complement each other. While Kant reserved his moral imperative to humans only and kept it formal, Jahr's imperative encompasses all living beings and their interactions in order to present a flexible and pragmatic imperative, which takes into account the struggle for survival as an essential part of all life.

Social and Sexual Ethics in the Daily Press, 1928 [Soziale und sexuelle Ethik in der Tagespresse] calls on educators and ethicists to not just publish in academic journals or debate in their close inner circles, but to use modern media, in particular mass media, to form moral thoughts and attitudes and to contribute to ethics education. Long before we had our contemporary academic and political discussions about the origin and course of public opinion, but at a time when secular and fascist ideologies started to take public opinion hostage in the way the Fascists did, Jahr recognizes very clearly this new battle ground over culture, ethics and civilization in the coming information age. Ethics and bioethics are not private matters, but belong into the public realm. Therefore the active use of modern mass media is an obligation and a mission for ethicists and bioethicists. Who in the 21th century would disagree?

Ways to Sexual Ethics, 1928 [Wege zum sexuellen Ethos] strongly suggests to not only inform the young generation about biological and physiological facts of sexual activity, but also to take a broader approach in discussing other important ethical and cultural issues related to love and lovemaking.

Egoism and Altruism. Two Basic Moral Problems: Opposition and Alliance, 1928 [Zwei ethische Grundprobleme in ihrem Gegensatz und in ihrer Vereinigung im sozialen Leben] concentrates on one of the least addressed issue in ethics: the conflict between egotism and altruism. In consequence of his Bioethics Imperative to recognize and to care for all forms of life, he has to recognize the struggle for life and survival among various forms of life, humans, plants, animals, also within and between biotopes and natural and social environments, including nations, businesses, communities, institutions, and parties, – all life forms are in competition, each fighting for themselves but needing others as well to form networks, to build reciprocity, and to express compassion. Individuals need to support each other; humans do it conscientiously or compassionately, other forms of life in the best interest of coexistence in symbioses, biotopes and environments. Jahr values both – altruism and egotism – as natural properties in all struggles for life. He calls for reasonable and pragmatic use of integrating and complementing mutual aid, of supporting each other in a compassionate way of reasoning, feeling and acting. He comes close to review social organizations as living beings as well, each having their own vision and goal for their life, success, and survival; he hints at new methods to analyze personal, social, professional and cultural interrelations in the workplace such as in hospital wards and offices, in clubs, parties, neighborhoods and churches. He thus finds a normative basis for integrative dialectic interplay between egotism and altruism in the empirical facts of good and successful natural and social lives.

Character Dictate or Freedom of Thought? 1930 [Gesinnungsdiktatur oder Gedankenfreiheit? Gedanken über eine liberale Gestaltung des Gesinnungsunterrichts] Jahr involves himself quite bravely into a growing debate about modes and models of character formation and education. The old models of paternalistic indoctrination of the Imperial times were also used by Fascist right-wing and left-wing groups for indoctrination in the Weimar Republik. He supports an interactive pedagogical partnership model of information and discourse in the classroom, respect for individual values and a nondirective interactive moral discourse [Arbeitsunterricht]. The journal 'Die neue Erziehung. Monatsschrift für entschiedene Schulreform und freiheitliche Schulpolitik' had to stop publication in 1934, when Nazi politics and ideology became even more dominant in all areas of life. Jahr's ten points of advice in liberal value-based interactive teaching are not less valid today than they were 80 years ago. Recognizing cultural and individual differences in thought, conviction and attitudes, an interactive discourse and teaching approach is even more warranted today in the wider international and intercultural dimensions of bioethics and ethics.

Our Doubts about God, 1933 [Unsere Zweifel an Gott] is a response to an earlier article in the 'Ethik' journal and stresses the practical relevance of a 'religion of love' and compassion and the relativity of scientific and religious insights, both standing on less firm grounds than their followers like to believe.

Three Studies on the Fifth Commandment, 1934 [Drei Studien zum fünften Gebot] not only present the Bioethical Imperative in a broad perspective including individual health responsibility and public health obligations, but position it into concrete actual challenges of his times and into a wider vision of global, universal, and integrated bioethics. Methodically, Jahr uses a 2500 year old rule given to one group of nomadic tribes in the deserts of the Arabian Peninsula to hermeneutically delineate three modern and actual challenges of the Bioethical Imperative in his times. Dealing with tradition the way he does, may alert contemporary teachers and writers in bioethics to the hidden wealth of values, virtues and principles in our traditions. Those traditions can be brought to good use today even more so, when bioethics literature rarely quotes references older than 5 or 10 years, thus imitating publication attitudes in natural sciences which are necessarily to the benefit of bioethical reasoning and application. While some challenges in bioethics and ethics might be new and specific, the vision for the future can very well be laid out in the hermeneutics of classical old texts of various traditions and cultures, as exemplified by Jahr.

Faith in a Hereafter and Ethics in Christianity, 1934 [Jenseitsglaube und Ethik im Christentum. Eine nachösterliche Betrachtung] delineates the interrelationship between active deeds and faithful believing, hinting at the fact that moral conviction and moral works may be a way to also understand and eventually accept the belief in a world to come, a life integrating physical and spiritual existence.

The Ethical-Social Importance of the Sunday [Die sittlich-soziale Bedeutung des Sonntags], 1934, expands the biblical notion of a God-given day of rest to a broader cultural and social concept of holiday and rest for all forms of life, of recreation and meditation for humans regardless of their religion.

Doubts about Jesus? [Zweifel an Jesus?], 1934, is a reflection and meditation on Richard Wagner's ‚Parsifal' opera. Compassion is the dominant and leading virtue and principle to understand, to gain knowledge and to live and survive, the means and the goal of learning.

Ethical Reflections on inner-church Quarrels [Ethische Betrachtungen zu innerkirchlichen Glaubenskämpfen], 1935, is a strong critique of ideological fights within churches and between different denominations and religions. Jahr argues, that compassion and moral deeds are most fundamental and practical expressions of a true faith, and that inner-church fighting is not compatible with Christian ethics at all. Jahr would find examples for his criticism in the 21th century as well. In 1935 the fight between the 'German Christians' [Deutsche Christen], who were close to the Nazis and the Fascist movement, and the 'Confessing Church' [Bekennende Kirche], standing on traditional grounds of belief and work, had reached a first high stage of confrontion and fight; Jahr trying to gain a higher ground above both parties, emphasizing common grounds and the shortcoming of everything human.

Faith and Works. Opposition and Alliance [Glaube und Werke in ihrem Gegensatz und in ihrer Vereinigung], 1935, presents a situational and utilitarian interpretation of Kant's Categorical Imperative. Jahr describes a similar dialectical interdependence of believing and doing as he has done in his article on egoism and altruism. He values Schleiermacher as the most important modern theologian and expresses hope that both, Catholics and Protestants, are moving into the direction of a 'Christianity of life and work', surpassing previous contradictions and quarrels.

Three Stages in Life [Drei Abschnitte des Lebens], 1938, demonstrates the influence of theological and spiritual tradition (Comenius) and of modern science (Fechner) on Jahr's concept of life as a continuum, integrating and modifying. A few years earlier, in ‚Life after Death' (Vom Leben nach dem Tode. Aus J. A. Comenius ‚Didactica Magna' zusammengestellt. Ethik. Sexual- und Gesellschaftsethik. 1933, 10: 50-51) Jahr presented without much interpretation a long quote from visionary Comenius main paedagogic work ‚Didactica Magna', exemplifying this integrative spiritual understanding of life in quoting from Comenius: 'Threefold are life and house of life for each of us prepared: the mother's womb, the earth, the heaven … first we are just given life with movement and conscience beginning, secondly life, movement and knowledge in its beginnings, thirdly the limitless abundance of all'. It is not only in Christianity and other monotheistic religions, but even more so in Asian and other spiritual traditions, that the concept of life is much wider and more integrative and comprehensive than individual life between birth and death. Such a comprehensive and inclusive concept of bios, life, and bioethics does not represent the mainstream of contemporary bioethics, but a majority of future bioethicists might see this differently and cherish Pastor Jahr's integrative and broad inclusion of invisible and spiritual forms of life. Cultural and spiritual history has a life of its own, just as we as individuals have our ages, maturities, and changing understandings of interactions and environments. Most recent Quantum mechanics and multiverse models in astrophysics and biology today support Comenius' and Jahr's methodological exercise.

The Sunday – a Secular Holiday [Der Sonntag – ein weltlicher Feiertag], 1947, picks up the earlier article from 1934 on the ethical-social relevance of the Sunday as a holiday for all forms of live, but Jahr now emphasizes even more the transreligious and transcultural importance of a free holiday for recovery, recreation and reflection. The change of focus, not of content, in both articles demonstrates Pastor Jahr's bioethical emphasis far and beyond any specific religious tradition.

The Heritage and Future of the Bioethics Imperative

Jahr's vision of an integrated new and global ethics and culture centers on the concept of a Bioethics Imperative. He reviews the impact of science and technology on human ethics and it may guide the systematic study of human conduct in the areas of the life sciences and of personal, professional and public moral commitment and conduct towards all forms of life, and examines new knowledge and related conduct and commitment in the light of traditional moral values and principles. Therefore, Bioethics is a new discipline, a principle, and a virtue, – all in close discussion with Kant –

extending and modifying the formal Categorical Imperative towards a more encompassing and accommodating integrative content-based Bioethical Imperative. The Bioethical Imperative takes the struggle for life and survival seriously and as a central issue of life. The struggle for survival and life is an essential expression of all forms of life. Ten different areas in academics, politics, and personal and cultural life will need to be influenced and modified by the new concept of the Bioethical Imperative:

(1) A new Integrative Academic Discipline: The Bioethical Imperative is a necessary result of moral analysis and reasoning in the Humanities, based on empirical physiology and psychology of humans, plants, and animals. As such it needs to develop, to educate and to steward personal and collective cultural and moral attitudes and calls for new respect and responsibilities towards all forms of life. The 'Sanctity of Life' is the foundation of Jahr's 1927 Bioethical Imperative, while 1788 Kant named the 'Sanctity of the Moral Law' as the foundation of the Categorical Imperative: 'The moral law is sacred (inviolable). The person is not sacred, but humankind in his person must be recognized as sacred. Everything in the entire creation, if one wants and has power over it, can be used as a means only; only the human person and with it every intelligent being is an end in himself. He is the subject of the moral law, which is sacred, based on the autonomy of his will' [Kant: A156]. It is interesting to see, how well Jahr quotes new scientific publications in developing an ethical response to most recent scientific information.

(2) A new Foudation of an Integrative Virtue Ethics: The Bioethical Imperative is based on historical and other evidence that 'compassion is an empirical established phenomenon of the human soul'. There is 'wrong love' and 'true love', however. The old lady fattens her poodle while letting her servants suffer, displays false love and compassion similar to those, who practice corruption, favoritism, and unfair dealings with fellow humans. Also, there is no conflict between compassion to all forms of life and compassion towards fellow humans. Jahr argues that animal protection has a positive effect on ethical behavior towards humans, popular education and public education, and that even those, who do not accept bioethical reasoning, should accept animal protection as part of a culture of civilized and moral behavior among humans.

(3) A new Integrative Golden Rule Principle: The Bioethical Imperative strengthens and complements moral recognition and duties towards fellow humans in the Kantian context and should be followed in respect of human culture and mutual moral obligations among humans. However, the Bioethical Imperative, based on compassion and love, cannot allow itself the

Kantian luxury of just being formal. The Bioethical Imperative nevertheless is rigorous and categorical in its requirement to make deliberate pragmatic, situational, and prudent moral choices in the respect of all forms of life. The golden rule does not to promote one single principle over the others, rather balances and integrates principles and virtues, depending on situations and parties involved. If the other formulation of the Categorical Imperative by Kant 'Act only according to that maxim whereby you can at the same time will that it should become a universal law' can be considered a Golden Rule, as some experts do and as Jahr understood Kant, then the Biblical 'love your neighbor' command and Kant's imperative together with other religious or philosophical Golden Rules can be summarized in Jahr's bioethical imperative.

(4) A new Personal Health Care Rule and Public Health Ethics: The Bioethical Imperative includes obligations towards one's own body and soul as a living being. Jahr is primarily interested in the wider aspects of recognizing and teaching bioethical virtues and principles. Thus, moral duties towards one's own body and soul provide a bridge to biomedical ethics and public health ethics in the modern contemporary sense, towards interactive and interrelated goals and visions in personal and public health and hygiene, also in personal and public morality. Even better knowledge in modern predictive and preventive health care today has not yet found a successful and efficient way into health care ethics and lifestyle culture.

(5) A new Public Health Care Rule and Ethics: Pastor Jahr voices a critical and quite conservative view towards public health issues associated with moral and cultural changes during the 20ties and 30ties of the last century. Going strong against the zeitgeist, he argues that fulfilling obligations towards oneself is also a duty towards others and towards public health, underlining the close interaction of personal care for health and public health care.

(6) A new Global Stewardship Rule and Ethics: Jahr broadens the 5th Commandment 'thou shall not kill' into a universal rule and ethics of positively and proactively caring for the health and life of the land, of the environment, and of this globe as a part of a living cosmos: 'This all shows the universal importance of the 5th Commandment, which needs to be employed in regard to all life. Rewriting the 5th Commandment results in the Bio-ethical Imperative: Respect every living being on principle as a goal in itself and treat it, if possible, as such!'

(7) A new Management Rule and Corporate Ethics: The Bioethical Imperative has to recognize, to steward, and to cultivate the struggle for life

and survival among forms of life and of natural and cultural living environments. Thus, social environment need similar attention as natural biotopes. Jahr uses the then popular term biozoemose (Wörterbuch der philosophischen Begriffe, ed. Eisler, 4. ed., 1927, vol. 1, 226). The term was defined by R. H. France for living environments. In Jahr's bioethical model of interacting forms of life in living environments – including natural, social, economic, political environments – it includes social institutions, such as those in the care for health care, as well. Jahr has not elaborated in detail on these types of life forms, which consists of individuals and groups and which have their own interest in life, in success, recognition and survival. They will have to develop their own personal identity, corporate or institutional ethics and attitude as a good or bad corporate neighbor within their respective neighborhood and their social and natural environment. The universal Bioethical Imperative can also be detailed into specialized fields of ethics of care, such as for the internal metabolisms of hospital wards and similar institutions of service and production. Institutions in the various fields of health care interact, serve, compete and struggle with other institutional life and have individually their own internal rules and real individuals, who in various capacities belong into these institutions. For Jahr, bioethics and environmental ethics, land ethics, corporate and institutional ethics, social and sexual ethics have to follow the same principles and virtues of responsibility and respect towards natural forms of life and forms of life created by humans.

(8) A new Terminology Rule and Terminological Ethics: There is another insight we can gain from Jahr's reasoning: the need to have a clear and precise terminology. He coined the term Bioethics in order to provide for clear and distinct reasoning and resolve in our dealings with living forms of reality as different to nonliving forms, and in stewarding most modern science and technology and their applications into a morally responsible way. Is it professionally and logically not correct to use such a wide term as Bioethics for more precise and distinct issues such as in medical ethics without any further distinction and differentiation. Therefore we must define more precisely what we mean in clinical ethics, ethics of medical research, ethics in the care for the demented, ethics in law, ethics in politics, – just to name a few specialized fields? Spinoza in his Ethics once said 'omne esse verum quod valde clare et distincte percipio' [only that is true which we perceive clearly and distinctly] and Wittgenstein would add 'whereof one cannot speak, one must be silent'. Unclear terminology leads to unclear investigations, goals, and actions, – not only in the Science but in the Humanities, including ethics, as well. If ethics and every-day attitudes can learn anything from science, then the fact that precision in definition is a priority and a precondition for clear conceptual and practical work, for communication and for cooperation and for further development.

(9) A new Rule and Ethics of Differentiation: Jahr, when coining the term Bioethics followed the differentiation in the terminology of the most modern sciences, psychology and physiology in particular, which had developed a term bio-psychik – not in use anymore today – for analyzing forms of living nature from other forms of non-living nature. Unclear terminology leads to unclear reasoning and acting; it is an expression of unclear thinking itself. There are different terms available for different subjects, fields, and issues: bioethics, medical ethics, hospice ethics, health policy ethics, hospital ethics, biomedical ethics, medical research ethics, physician's ethics, nursing ethics, health care ethics, nursing ethics, public health ethics, prevention ethics, gene ethics, consultation ethics, environmental ethics, animal ethics, global ethics – just to name a few.

(10) A new Interaction and Integration Rule in Ethics: For Jahr animal ethics and social ethics are different fields, but they interact and integrate, bringing different shapes and shades of the Bioethical Imperative together and describing the multitude of ethical obligations in the 21th century, some reciprocal among humans, some more or less paternalistic / maternalistic in compassionately and professionally caring for the weak, frail, and incompetent, some in stewarding plants and animals as co-creatures and partners-in-life, others in mitigating or creating natural and social environments as human habitats and in an ethics of the land, even in geos as a living entity herself. Those moral obligations and opportunities will and must overlap and interact in different ethical, philosophical and cultural models of personal, professional or institutional ethics, – medical ethics including prevention, treatment, care, research, – bioethics covering respect for and duties to all forms of life, environmental ethics accepting responsibility for natural and man-manipulated environment and their sustained survival and health. A new field of geo-ethics is already visible, applying reason and responsibility in response to global natural or man-made changes in the protection and promotion of human lives and human cultures in regard to environments and institutions. Recent earthquakes, tsunamis, and nuclear disasters have reminded us that the earth, indeed, is a living being and needs to be treated as such, i.e. respecting her, if not for herself but then at least in our own interest. Jahr has already an understanding of individual, cooperating, conflicting, and collective models of the 'will to live and to survive', including resistance to or eagerness for change and suggested new methods in the study, design and stewarding of environments and organizations. Nothing else is needed for the expertise and ethos of Land Ethics, and the ethics of global responsibility in personal and in corporate and institutional ethics than a universal and integrative prudent and reasonable application of the Bioethical Imperative.

Section 4

Branches of Integrative Bioethics

PATH-MAKER IN BIOETHICS - REV. FRITZ JAHR

Jayapaul Azariah[1]

> Right (Dharma) and wrong (Adharma)
> Do not go about saying: "Here we are!"
> Nor do Gods, Centaurs, Ancestors say:
> "This is Right and that is Wrong"
> (Apastamba Dharma Sutra 3.7.20.6)

The case

A path-maker, in any civilization, is the one who brings about, for the very first time, a paradigm shift in an ideology. In every geographical region, a foot-path, devoid of grass, does exit to reach a remote village. But who is the person who walked for the very first time to make that foot-path? In short nobody can name that person. In reality, when that foot-path becomes an established way when more and more people tread on that foot-path. Then some distinguished person may give a name for that foot path. In a particular sense a pioneer is the path-maker. In the context of bioethics it may be pointed out that the development of objective moral values constitutes that early foot-path which involves the collective approach of a community to make rules, regulations and norms to uphold social values. The thesis of the present paper emphasizes the fact that in earlier civilizations the essence of bioethics was practiced in day to day life but without the label "bioethics" (Azariah, 2009). But who christen that foot-path – the way of life as "bioethics"?

There is a biological father of bioethics. For the past four decades. Prof. Potter of the University of Wisconsin had been considered as the person who first coined the new word "Bioethics" in 1970. But that was wrong. It was a German by the name of Rev. Fritz Jahr who first coined the word 'Bio-Ethik' in 1927. He was a Protestant pastor, philosopher, and educator hailing from Halle an der Saale of Germany. Hence, he is the real biological father of bioethics. The rest of the Americans who nurtured this bioethical theme are surrogate fathers – thanks be unto them for their untiring work. It is time that the discipline of bioethics is reunited to its real and rightful biological father. Rev. Jahr wrote an article entitled "Bio-Ethics: A Review of the Ethical Relationships of Humans to Animals and Plants" and proposed a "Bioethical Imperative, extending Kant's moral imperative to all forms of life" (Sass,

[1] Professor Dr. Jayapaul Azariah, Founding President, All India Bioethics Association, Indiranagir, Chennai 600020, India; hildaazariah@yahoo.co.in

2008). Rev. Jahr's coverage of bioethics was comprehensive whereas American bioethics is usually restricted to medical ethics. But Fritz Jahr's view of bioethics being holistic is similar to Asian bioethics. It covered all of life, including environmental ethics as well as all life-forms. It is noteworthy that Fritz used the phrase "living being" instead of "living things". The latter phrase was introduced by evolutionists which is a contradiction to its meaning (Azariah, 1994). Philosophically the term 'living' is used to designate those organisms with 'life' and non living materials are considered as 'things". For Fritz wrote "So that the rule for our actions may be the bio-ethical demand: Respect *every living being* on principle as a goal in itself and treat it, if possible, as such!" (Jahr, 1927, also in this book). When bioethicists use the phrase 'living being' it addresses many core bioethical issues.

In order to establish the priority of Rev. Fritz Jahr as the one who first coined the word "bioethics" it is very necessary to provide an overall review on the perception of earlier bygone civilizations which practiced bioethics in their daily life but without the label of 'bioethics'. One such civilization is the eastern 'Vedic' civilization. Secondly, life is one single entity. But humans do not know and cannot answer the question "What is life?" since the word 'life' is indefinable. Plants, animals and humans have 'life'. Is 'life' in all these three categories similar in consistency or made up of same or similar components? In any case the one who recognized the rights of plants, animals and humans is the rightful custodian of the discipline of bioethics. Therefore any ideology that recognizes the "rights" of plants, animals as well as human rights has the right to be designated as 'bioethics". And the person who first employed the word "bioethics" is the one who is the pioneer and a path-maker in bioethics. In this light the case of Rev. Fritz Jahr will be argued.

ELSI (Ethical, legal and social issues)

Whenever Science and Technology advances with rapid straits there are moral issues. An issue is both a problem and a question which is indicative of the moral attitude of a given society to a particular issue that it faces in its techno-scientific-domain. These issues that arise due to rapid advancement of scientific knowledge and technical application are usually multidimensional in nature. Some issues originate at a particular time of a given society and linger on to the future generations. Such a situation demands the creation of a discipline which can act like a bridge between two generations or among many disciplines as well as a bridge to the future. These lingering issues may be Ethical, Legal and Social Issues, commonly referred to as ELSI. When there is an issue, the process of decision-making is invoked to solve the problem.

The process of decision making in daily life needs a FRAME WORK which could be in the domain of (i) human moral reasoning, (ii) rational mind (iii) philosophical outlook in life (iv) spiritual perception (v) a religious bend of mind or (vi) a combination of all of these mindsets. When ELSI are complicated, then there exists an ethical dilemma which is a state of mind wherein a person is unable to make a moral judgment or make a moral choice. The decision that is made may be right or wrong depending upon the store of available information. Ethics deals with humans making a wrong or a right choice. But it becomes 'bioethics' when decisions are addressed to life situation. Hence, the term "Bioethics" embraces all of life-related issues that emanate from the complex science and complicated technological based lifestyle. Therefore bioethics is essentially an art of making a choice between two or more alternatives that makes life a worth living reality. In order to make a decision between a set of issues there are always norms and principles. The question remains then - "From where do we get these norms and principles?"

ELSI are solved from two different end-benefactors in view. Accordingly there are two different approaches to solve problematic questionable and life related-issues. These are (i) anthropocentric and (ii) non-anthropocentric approaches. The former aims to achieve benefits only for human life. In this category the two newly developing centric way of life are (i) web-centric-mobile lifestyle and (ii) outer space ethics. On the other hand, the non-anthropocentric position has a comprehensive agenda to accrue benefit to all forms of life, plants, animals and humans. Non-anthropocentric approach is further divided into (i) Cosmocentric (ii) Theocentric (iii) Biocentric (iv) Ecocentric.

The West versus the rest

The Western Society is predominantly anthropocentric in their way of life. The anthropocentric view started with the teachings of the Greek philosopher Aristotle. He wrote "plants exist to give food to animals and animals to give food to man – domestic animals for their use and food, wild ones, in most cases, if not in all, furnish food and other conveniences, such as clothing and various tools. Since nature makes nothing purposeless or in vain, all animals must have been made by nature for the sake of men" (Politics 1256). Therefore it is obvious that the Greek culture did not simply recognize the 'rights' of plants and animals. In their view, plants and animals were subservient to humans.

The alternate view is that of a non-anthropocentric approach – an approach which is essentially 'biocentric". In reality to make a choice between a biocentric way of life and an anthropocentric way of life is very difficult. When human life is subsistent on plants and animals then it is a matter of conscience i.e. how a human conscience is educated to live as a vegetarian or as a non-vegetarian! It is interesting to note that both Rev. Fritz and the Indian scientist Sir Jegadesh Chandra Bose recognized that plants have life. At the same time it should be recognized that it is impossible for humans live without 'killing' plants. The Vedic culture recognized such a difficulty and 'solved' the conflict of interest by recognizing the rights of plants from one angle while from another view point it recognized the humans as the benefactors. The following quotation cited in the next section may vividly portray the point under consideration.

On the other hand, eastern societies and specially the Vedic society, culturally exhibited a Cosmocentric way of life. They recognized the existence of Cosmos (beauty) in the universe as an integral part of their daily life. They also recognized the existence of "*Aanantham"* (Sanskrit) (eternity) and "*Suuneeyam*" (outer space) and "*Aath-man*" (inner space-being) and "*paramaathman*" (the Supreme Being). When they lived their life with reference to a larger cosmic scale which is eternity then their value system changes in the sense that it is not geocentric. But they led a happy life on earth. They did not make a distinction between human life and other life-forms. They also practiced a "theocentric" approach by naming the rivers and the mountains with various names of gods and goddesses. They recognized the various varieties of plants and vegetation in their locality and planted the most dominant tree species as the "temple tree" (*stalla veericham*) and enforced a ban on cutting those trees.

There are references in Tamil literature that the kings practiced a "biocentric" approach to life. A king named *Paari Val-lal*, on seeing a tender plant-climber withering without a supporting structure he then left his chariot as a support for the climber plant and walked back to his palace. Another emperor King *Chivee Sakkaravathy* did not hesitate to give his own flesh in order to secure the release of a bird from the hunter's cage. The Buddhist King Asoka was the first Indian king who introduced "Medical Humanities" on a national basis. He built hospitals for both humans and animals.

Environmentally there was a sound policy to keep ecobalance. The word 'sewage' was non-existent in *Tamil Cungakala* literature since they converted the waste water into a resource. It is surprising that both Tamil and German languages do not have a native word for "natural resources." Every natural entity was important and enjoyed an equal statues and rights from their view point. They also practiced a cyclic 'ecocentric' way of life. Like all ecological cycles their way of life was cyclic including cyclic "rebirth". With globalization and liberalization these time tested way of life is being replaced with western anthropocentric way of life. As a result the whole globe, both western and eastern societies, is being influenced by a "www.com-web centric" way of life.

Holistic eastern-Vedic view

The Vedic culture prepared medicinal concoctions from plant extracts. They also offered their first fruits of harvest to the fire goddess "Agni". It is written: "he enlightened seers looked towards the plants with great love, awe and respect. The plants were regarded as having divine, curative and healing powers. The seers seek humble permission of a plant before harvesting or cutting even a leaf from it for reemployment in herbal combinations". Therefore it is evident that it was a common practice among the Vedic way of life to treat plants as intelligent creatures and plants were given equal rights and importance in their daily lives! Their feeling of ahimsa (non violence) is revealed in another prayer. They prayed: "Oh! Ye Plants, shoot forth in hundredfold leaf as I cut for curing human diseases. If you prosper, then we men will grow in thousand fold". They were very clever people in the sense, in trying to be biocentric they were really anthropocentric. That they were anthropocentric is revealed in differential allotting the blessings. They gave a blessing of 'one hundred fold" to plants whereas of the blessing was "one thousand fold" to all humans (Rajimwale, 1994). The Vedic culture solved it by assigning differential but just-rights to plants and humans. In short they practiced a type of distributive justice recognizing their (plants) due rights. They proclaimed their moral obligation to act on the basis of *fair distribution – Distributive justice* – between competing claims.

In short, the eastern view of the earth, rivers, animals and plants was very unique. Ancient India the rights of the plants were recognized since they considered herbs as divine: "O herb thou are born divine on this divine earth". Sages prayed for permission from plants to pluck their leaves for use in health care practices. The earth was divine and so they named the rivers with the names of goddesses. Water was sacred akin to life. There was a beautiful harmony between the community and the nature. They led a happy and a disease free life by administering medicinal extracts prepared from herbal plants.

India is an ancient land. Its antiquity extends far back in history to the Vedic civilization. Ancient India practiced rich bioethical principles in science, medicine and cosmology. In medical science and technology, ancient classical treatises like *Charaka* and *Sushrata* are standing testimony that there was a well established health care system with well qualified surgeons and physicians like *Charaka* who was the official physician of the kings like *Kanishka* in the first century AD. Scientific knowledge was so well advanced that medical interventions like lithotomy (operation for the removal of bladder stones) and repair of damaged nose were of common occurrence. Advanced knowledge on internal medicine was at their finger tips as they were well versed in botanical knowledge of medicinally important herbs and plants (1008 in number), 25 types of salts, 64 types of arsenic, 9 different metals and about 120 varieties of metallic essences. They were expert physicians in alleviating pain and curing diseases. Physicians were prescribed a balanced personal bioethical behavior (Azariah, 2003). Human health was their prime interest.

The foetus is a person

Both Hindu scriptures (Vedic period) and Christian scriptures (AD 1) affirmatively confirm that the foetus is conscious person. Prof Peter Singer has characterized cardinal characteristics of the human person. Consciousness is a chief ethical parameter, which is the capacity for physical, social and mental interaction with other beings, having conscious preferences for continued life, and having enjoyable experiences. In both Hindu and Christian scriptures references to the foetal quality of life can be found

The Indian Vedic classical Epic *Mahabharaatha* outlines the episode of *Krishna's* conversation with his pregnant sister *Subharhtra*. This Hindu Scriptures tell the story of *Abhimanyu* who was conceived in the womb of *Subharthra*. The conceptus *Abhimanyu*, as a foetus responded to *Krishna's* narration on the complete military strategy of a specialized military "Chackkra-uha". The foetus heard the secret of braking such a complex military formation and how to get into it. *Abhimanyu* responded to the narration which angered Krishna. Hence Krishna did not complete it as to how to get out of the formation after entry. At the age of 16, *Abhimanyu* broke the military formation, but was killed since he did not know how to get of the complex military formation and the listening capacity of the foetus *Abhimanyu* can throw light on the bioethical concept of foetus as a human person which stands in contradiction to the western bioethical position (Azariah, 2006).

That they had knowledge of multiple cloning is evident from the narration of Queen *Kandhari*, the wife of King *Dhritrashtra*, from whose womb 100 bits of the developing *conceptus* (embryo) developed into boy-children. Such expert knowledge about early human development enabled them to formulate laws governing foetal extraction and regulation of abortion (Azariah, 2003). In this context, it may be worth mentioning that a very ancient Japanese

medical system also recognized the medical importance of foetus (Shinagawa, 2000).
During the time of Rev. Jahr both the disciplines of Psychology and Neurophysiology were well advanced. Jahr observes that psychology is not limited to humans any more today, it applies the same methods to animal life: and as there is comparative anatomic-zoological research, there are highly educative comparisons between human and animal souls. Indeed also the beginnings of plants psychology can be recognized. Even beginnings of plant psychology appear… Under these circumstances it is only consequent E. Eisler in summarizing uses the term Bio-Psychik (soul science of all life forms). It is only a small step from here to Bio-ethics (Jahr, Bio-Ethik, 1927 highlighted by Jahr) i.e. the assumption of moral duties not only towards humans but to all living beings as well (Jahr, 1927, in this volume). Interestingly Rev. Jahr introduced the term 'plant-soul' for the first time. In sum the guidance for our moral actions has to be the Bio-Ethical Imperative: Respect every living being, including animals, as an end in itself and treat it, if possible, as such! And if someone does not want to accept the absolute validity of such a rule in as far as animals and plants are concerned, he should – repeating what was said before- in recognition of the moral obligation.

Human values and value systems

Ethics is a branch of philosophy. As an applied and normative philosophy it covers and advocates moral Philosophy with its prescribed norms and principles. Rev. Fritz Jahr 'biologized" applied ethics and coined the word "Bioethik". And his newly coined includes the ethical principles of American philosophers like Beauchamp and Childress who advocated the principle of (i) 'respect to autonomy' = Self love (ii) 'justice' = Love the neighbor (iii) 'beneficence' and = Do Good and (iv) 'non-maleficence'. = Do no harm. The Principle of autonomy (self-being – a person) deals with Self dominion (Rule). In Jahr's notion of Bioethik justice is of fundamental value. Jahr as a Christian pastor may have based his notion of justice on (Bible) Proverbs 3:27: "What is due to a person – Those who deserve it" Justice is 'fairness'. Justice is interpreted as "fair, equitable and appropriate treatment in the light of what is due or owed to persons." Justice is suggested to be the *first virtue* of social institutions, as truth is of systems of thought. Rev. Jahr's concept of Bioethik is based on (i) decision making (ii) making a choice between right and wrong concepts/guidelines/principles and actions (iii) calculating the benefits and risks. Thus establishing a just and healthy society was his ultimate aim. Hence it was a brand new academic discipline.

The Western notion of beneficence and autonomy were inbuilt into his Bioethik concept. Beneficence generally means doing good to life-forms or doing acts of kindness. Over and above refraining from doing harm to others, the principle of beneficence requires us to be concerned and promote the welfare of others. The term beneficence is inclusive of acts of mercy, love/kindness, charity, altruism, and humility. It is the central theme of ethical theories such as utilitarianism (the principle of utility) and common morality theories. In these theories, beneficence is conceived as the aspect of human nature that motivates us to act in the interest of others and is considered to be the goal of morality itself.

Coining new words

This rediscovery of Jahr's bioethical work (Sass, 2008) falls in line with the coining of other scientific new words by other German scientists. It is interesting to note that the word "biology"; was in usage among German scientists. It appeared in a book title written by a German physician called Gottfried Reinhold Treviranus, who titled his book (1802) as "*Biologie oder Philosphie der lebenden Natur fur Naturforscher und Aerzte* – Biology or philosophy of the living Nature for scientists and physicians. (Azariah, 2009). The word "ecology" was coined by Ernest Haeckel in 1869. It is he who popularized Darwin's idea of Origin of Species through natural selection (1859) in Germany. Francis Galton was the cousin of Charles Darwin. Galton disproved Darwin's idea of 'pangenes' and 'pangenesis' and coined the word "Eugenics" around 1869. (Wikipedia, 2011).

The term "germ plasm" was first used by the German biologist, Weismann August Friedrich Leopold (1834-1914) in 1883. Weismann carried out a series of experiments in which he cut the tails of mice and allowed them to breed. He found out conclusively that acquired physical traits cannot be inherited. He proposed the Germplasm theory of inheritance in 1883 and the continuity of Germplasm in 1885 (EC 2008). New scientific knowledge generated by Weismann laid the foundation for the understanding the scientific work of Mendel, which provided a key to the modern understanding of genetic inheritance. Mendel's contribution is the foundation of the discipline of genetics The Danish botanist, Wilhelm Johannsen, in 1909, coined the word 'gene' and the study of genes became a disciple of science, which came to be known as "genetics" (EB, 2011). The word *Biotechnology* was *coined in 1917* by a Hungarian Agriculture Engineer Karl Ereky. He was studying the process of fermentation using yeast cells. This new discipline has many beneficial applications to human welfare Since the word 'Biotechnology' is an umbrella term to embrace many other science disciplines, it may indicate that in doing science, moralist found evidence for the introduction of ethical issues. This may have laid the foundation for the creation of a new technical term namely "Bioethics" in 1927.

The time gap between 1917 (genesis of Biotechnology era) and 1927 (genesis of Bioethics era) is very important. It needs to be documented as to what experimental scientific studies were carried out which may have introduced by the need to coin the word "Bioethics". This is the time when the Theory of Evolution (TOE) received a fatal blow through the works of Mendel which emphasized the important role played by the genome in the heredity of traits that are of survival value and at the same time under estimating the importance of natural selection in the Origin of Species. Concurrent with the two fundamental and revolutionary eras, namely biotechnology and bioethics, is the resurgence of the almost "knocked out" theory of evolution which suffered a severe setback due to the advancement in the field of genetics. As a result the Theory of Evolution was modified to be rechristened as the "New Synthesis". Such a paradigm shift in ideology resulted in creating a challenge to reinforce the established moral values. It is a well known fact that TOE is fundamentally atheistic in its approach which might have come in direct conflict with the Christian beliefs of Rev. Fritz Jahr.

Facts from supporting disciplines

Rev. Fritz Jahr was aware of and recognized that "there are highly educative comparisons between human and animal souls. Indeed also the beginnings of plants psychology can be recognized. Even beginnings of plant psychology appear.... Under these circumstances it is only consequent when R. Eisler in summarizing uses the term Bio-Psychik (soul science of all life forms). It is only a small step from here to Bio-ethics (Bio- Ethik, highlighted by Jahr) i.e. the assumption of moral duties not only towards humans but to all living beings as well." (Sass, 2009)

It may be good to recall that August Friedrich Leopold Weissmann (1822-1911) was an experimental scientist. His lecture entitled "The Duration of Life" marked a significant step in the area of human desire for long length of life. In this lecture, given in 1881, before the *Deutsche Naturforscher Gesammlung*, he exposed the loss of an immortal property of "immortality" of single celled organisms like Amoebae through the evolution of multi cellular organisms where there is a division of labor (EC, 2008). Amoeba is indeed immortal. That the property of potential immortality is lost in higher organisms but still retained in all somatic cells of multi cellular organisms. This fact was scientifically proved by Brody (1945). In his classical text book entitled *Bioenergetics and Growth* (1945), results of an interesting experiment have been cited. The experimental results reported in the book referred to above, relates to the tissue-culture of living cells taken from a living human. A few body cells were cultured on a nutrient medium and soon the cells grew and divided by binary fission producing cells of the same age and genetic makeup. The process went on. The research scientist kept changing the tissue culture medium and sub cultured the cells by transferring a sample of the growing and dividing cells that were isolated cells from a donor body. These living cell-cultures will live even after the death of the host body from which these cells were obtained. There were no signs of aging and death. Hence Brody (1945) concluded that this isolated human bodily materials (cells) are "*potentially immortal*", a confirmatory experimental analyses which endorsed the core idea of Weismann's lecture on "The Duration of Life".

The massive surge of population genetics between 1918 and 1932 posed a very difficult question whether or not Mendelian genetics could be reconciled with natural selection and with the gradual gene mutation in evolution. Evolutionary biology confronted the issue of the conflict between a spurt in broad-scale- changes found in fossil species and small mutational changes in a local population. The contrast between macroevolution and microevolution became dramatically evident. Therefore, a well coordinated effort was made between 1936 and 1947 to contain the damage caused to TOE, which is also the active period in Rev. Jahr's life! An evolutionary synthesis began to take place to unify the contrasting differences among the disciplines of biology, systematics (Taxonomy), invertebrate and vertebrate morphology, genetics, cytology, ecology and paleontology, The neo Darwinian players in this synthesis were: R. A. Fisher, Theodosius Dobzhansky, J. B. S. Haldane, Sewall Wright, E. B. Ford, Ernst Mayr, Bernhard Rensch, Sergei Chetverikov, George Gaylord Simpson, and G. Ledyard Stebbins. The term "Modern Synthesis" (a union of ideas) was introduced by Julian Huxley in his new book *Evolution: The Modern Synthesis* (1942) which later came to be known as the "new synthesis, the modern synthesis, the evolutionary synthesis, millennium synthesis and the neo-Darwinian synthesis" (Wiki, 2011). Evolutionary biology is atheistic at its core doctrine which may have confronted the belief system of Rev. Fritz Jahr who functioned as a Christian pastor. In 1930 another well reputed group of scientist, Wells H. G, Julian Huxley and G. P. Wells (1930) brought out a book on the "The Science of Life". Life became a subject of research and reverence and life was in danger of losing its sanctity and reverence. The importance of Charles Darwin's theory for Fritz Jahr's conception of bioethics has been discussed in a detailed paper by Engels (2011).

Death as an alternative to social problems?

"Western man as firmly held to the supportive principles as promoted by Hippocrates from 450 BC until the turn of the century. Since the mid 1800's certain groups and individuals have been promoting death as a solution to social problems. The belief; that every human life is of equal worth and is worthy of protection and loving care has come under increased attack. Such a sanctity-of-life-ethic- approach is in danger of being replaced by a cheap 'quality of life- ethic' thinking! Advocates of this new ethic maintain that some persons should be 'terminated' because they are unwanted, unplanned, imperfect or perhaps not properly productive (Bergel 1998). Thomas Jefferson the founder of spiritual America considered "The care of human life and not its destruction is the first and only legitimate object of good government. The State comes into existence that man may live – Aristotle.

In Hinduism death is positive since life after death is accompanied by 'rebirth' which is governed by one's own karma. On the other hand a Buddhist considers pain and suffering are necessary agents of purification to attain 'nirvana'. The following real life incidence may illustrate the point in question. Two bright young IT professionals intelligent, rich and happy and well accomplished, based in Goa, India, lost further interest in life and ended their life together by saying "We've lived a full life, now we choose to end it" (Bombay Mirror, 2011) . The question is then raised "Is it really our own choice to end our lives when we feel it is the right time? " (Bombay Mirror, 2011). On this report Dr. S. S. Narayanan of the Indian IRB network group raised a question "shaven heads were they inclined towards Buddhism?" Mr. Dzongsar Khyentse Rinpoche is a Buddhist monk and a film-maker in the city of Pune, India. According to him "to be a Buddhist is not important, to be Buddha (enlightened) is". Dr Narges Mahaluxmivala a Clinical Development Researcher highlighted the Tibetan Buddhist notion as outlined in the book "The Tibetan Book of Living and Dying" authored by Sogyal Rinpoche. According Buddhism a person achieves inner peace by continued practice of spiritualism which provides a confidence within oneself so that there is no fear of death. Further, pain and suffering, are the agents of a natural process of purification. Renunciation of worldly assets and social isolation ending in shaving of the head are tenants of Buddhist faith. May be Rev Jahr anticipated that people's empty lives may drive them to such a tragic end (Mahaluxmivala, 2011). Dr. Lalit Vaya, Clinical Research Consultant & Chief Coordinator, PGCCR, Apollo Hospital, Ahmedabad drew a parallel in the Hindi novel by Ms. Late Shila Chandra Joshi, titled '*Rhituchakra'* wherein the author comments "I have lived a full life and nothing is left to be achieved" (IRB Network, 2011).

That death is not the solution to social problems is depicted in a recent movie entitle 'Somewhere" (December 22, 2010). A. O. *Scott,* a film reviewer, wrote in The *New York Times.* "Films are starting to reflect the economy's impact on some battered *Americans***.** Sofia Coppola's new film presents a star, who has everything, including emptiness”. A US group, the Final Exit Network (FEN), has put up billboards with the slogan "My Life, My Death, My Choice", to spread their message that people have the "right to die". The final exit is death. The solution is to lead a purposeful life on earth. Unfortunately evolution does not have the word ‘purpose’ or ‘aim’ in its agenda.

A contemporary bioethical problem – abortion

In USA, unethical practices in abortion bioethics were present as early as 1929. However, information on 1929's abortions was unearthed now only for reporting. About eight years ago the North Carolina State’s (NC in USA) lawmakers woke up to unearth such a past prevalence of forced sterilization of teens, soon after the child birth. Such involuntarily coerced sterilization in NC was inflicted between 1929 and 1974. During that time period forced sterilization was carried on 7,600 women, children (85%) and black people (40%). “It began as a way to control welfare spending on poor white women and men, but over time, North Carolina shifted focus, targeting more women and more blacks than whites. A third of the sterilizations performed in North Carolina were done on girls under the age of 18. Some were as young as nine years old” The procedure is as follows: In 1967 a 13 year old girl Elaine Riddick got pregnant after being raped by a neighbor in Winfall, N. C.“ The State ordered that immediately after giving birth, she should be sterilized. Doctors cut and tied off her fallopian tubes”. She is rendered infertile for the rest of my life and she feels “I have to carry these scars with me” (Kessel and Hopper, 2011).

The fight against abortion in USA continues. A constitutional amendment was filed in Mississippi to declare a fertilized human egg to be a legal person, effectively branding abortion and some forms of birth control as murder. Such a move is also on the anvil in some more states in US (Eckholm, 2011). About 57% of the Mississippi voters voted against Initiative 26 – a ballot to amend the state's constitution to redefine “person” as “every human being from the moment of fertilization, cloning, or the equivalent thereof”. If this amendment is brought about then it would effectively criminalized abortion in that State. It has been argued that such an initiative jeopardize child birth through IVF as well as other contraception. It will also bring in legal implications in human embryonic stem-cell research.

When abortion was legalized by the States then it opened the flood gates of a new campaign which promoted death of foetus as a social solution for social problems. A godless doctrine in evolution and a surge in doing experiments on “life” coupled with a threat to terminate human life by prescribing a “fixed period" may have disturbed Rev. Jahr". It should be highlighted that 1800s is a turning point in human rights and justice for all life-forms. Justice is rooted in making right choice recognizing the rights of people and plants and animals. In the case of humans the right to privacy has been emphasized. The "Right to position" includes the (i) Pro-choice (ii) pro-life: as in the case of abortion or no abortion. It will be of interest to study who are those "certain groups and individuals have been promoting death as a solution to social problems" and link it with the crimes against humanity administered by then scientific community! Societies in industrialized nations are aging. The aged are considered redundant to society and there even is a move to legalize compulsory euthanasia at the age of 80 years.

A singular case of unethical precedence

On May 3, 2011 the world's greatest evolutionary super power took revenge on its Number One terrorist enemy by shooting him at a point blank range. This is a singular case of revenge. Hence, it raises bioethical problems and questions such as "Is targeted assassination unethical?" Why America, the citadel of bioethics, did not exercise forgiveness but resorted to revenge?

Journalists in Bioethics raised a pertinent question "But what if the killings were something that the US government is neurophysiologically programmed to do? Is there an evolutionary value to take revenge on another person?" Cook (2011) considers that the killing of Number One enemy of America could possibly be described as biologically-determined revenge. Whatever it is or it is not, "justice" has given room for further debate. And it neither has fundamental bioethical value. According to McCullough evolutionary value of revenge is an evolved instinct "designed" to deter others from imposing harms after they have already harmed us. But this does not bear any characteristic features of objective moral values. This is considered not a good bioethical approach in teaching Bioethics since it does not carry with it any objective moral value towards the sanctity of human life. Taking revenge cannot also be taught as a core bioethical value in bioethics education. The main reason for McCullough to dispense practical advice for making the world a more forgiving place is because there is no living example of an evolutionist who has richly exhibited the Spirit of Forgiveness. Moreover, US also did not practice sound bioethics in breaking the political code of bioethics when it transgressed a country's territorial sovereignty nor in conducting a fake immunization program to obtain blood samples of close relative of its Number One enemy.

Ethics and bioethics

Bioethics, as an applied and normative philosophy, places an emphasis on objective moral values so as to serve as a bridge to the future. It also functions as a bridge among sciences and social sciences. Bioethics is a comprehensive umbrella term. It includes environmental ethics, medical ethics, nursing ethics, processional ethics, computer ethics, information ethics, engineering ethics, and space ethics. Due to the rapid advancement of science (knowledge) and technology (skill), one is currently faced with the interdisciplinary dimensions of the current global problems in bioethics. Therefore, bioethics, as a trans-disciplinary subject, is a bridge between sciences/art subjects and a bridge to the future. It is essentially a decision making process and it addresses ethical, legal, and social issues (ELSI), which are "theoretical and practical moral issues, raised by medicine and the life sciences as applied to human beings and humanity's relationship with the biosphere".

On this basis, André E. Hellegers used the term bioethics in the area of public policy making to resolve ELSI Issues. These ELSI have a techno-science basis such as medicine and biotechnology. Hence approaches taken to resolve these issues are important. Therefore, UNESCO defined bioethics as "the systematic, pluralistic and interdisciplinary study and resolution of the ethical issues raised by medicine and the life and social sciences as applied to human beings and their relationship with the biosphere, including issues relating to the availability and accessibility of scientific and technological developments and their applications" (UNESCO, 2005). The cementing factor and coupling link that make the system work on a universal or global bioethical scale is love (Macer, 1998). Since 'love is universal' it is human love that is the prime mover in human social development. Therefore, "universal cross-cultural ethics should be developed to allow diverse views to be maintained even with a singly community, as well as throughout the world in the global community."(p. 82).

Conclusion

It is evident that Rev. Jahr was aware of the social, legal and ethical issues of his time and therefore, he felt the need to evolve a brand new discipline of "Bioethics". His work on bioethical prospects and concepts should form a fundamental basis for designing any curricula in bioethics – either regional or global.

References

1. Azariah, J, 1994 Part 1. Ecology and Religion: Spirituality Mode – A keystone in ecobalance. IN Bioethics for the People by the People. Ed. D.R.J. Macer. Eubios Ethics Institute. NZ, Pp 98 – 104
2. Azariah, J, 2003 The Long History of India Medical System and Current Perspectives in Health Care Bioethics. IN The Annals of Bioethics. Regional Perspectives in Bioethics. Eds. Peppin J. F. and M. J. Cherry. Swets & Zeitlinger Publishers. USSE The Netherlands. 285-300 (Pp 1-380)
3. Azariah, J, 2006 Emerging Biblical Insights in Bioethics. Dharma Deepika.(An Indian Missiological Journal) January 10: 5-23
4. Azariah, J, 2009 Bioethics Science: Is it? J. Med. Ethics Hist. Med. 2: 18 October. Online electronic version.
5. Bergel, G, 1998 Abortion in America. Intercessors for America USA pp ii-16
6. Bombay Mirror 2011 http://www.mumbaimirror.com/article/ 15/2011112020111200218504045ef859db/We%E2%80%99ve-lived-a-full-life-now-we-choose-to-end-it.html
7. Brody, S, 1945 Bioenergetics and Growth. Reinhold, New York. Pp 1033 (Online) pubs.acs.org/doi/abs/10.1021/j150446a008
8. Cook, M, 2011 Bioethics of assassination and Bioethics of revenge. Editorial. BioEdge, May 7, 2011. Online
9. EB, 2011 Genetics – ARTICLE Encycloaedia Britannica. Online http://www.britannica.com/EBchecked/topic/228936/genetics/48730/Historical-background
10. EC, 2008 Weismann, "August Friedrich Leopold." Complete Dictionary of Scientific Biography. 2008. Encyclopedia.com. 19 Aug. 2011 <http://www.encyclopedia.com>.
11. Eckholm, E, 2011 Push for 'personhood' Amemdment Represents new tack n Abortioon Fight. New York Times. October 25, Online
12. Engels, E-M, 2011 The importance of Charles Darwin's theory for Fritz Jahr's conception of bioethics. IN Jahr Annul, University of Rijeka, Faculty of Medicine Croatia. 2 (4): 475-504
13. IRB Network 2011 Comments on Bombay Mirror article. IRB network Online
14. Jahr, F, 2011 Essays in Bioethics and Ethics 1927 – 1947 [IN Fritz Jahr: Essays in bioethics and ethics 1927-1947, Eds. and translation Miller, I M and H-M Sass, 2011. Zentrum fur Medizinishe Ethik. Heft 188. pp 1-56] See also in this volume.
15. Kessel, M and Hopper, J, 2011 Victims speak out about North Carolina sterilization program, which targeted women, young girls and blacks [Mon Nov 7, 2011, Online Rock Center with Brian Williams]
16. Macer, D, 1998 Bioethics is Love of Life. Eubios Ethics Institute. Christchurch, NZ, pp 158
17. Mahaluxmivala, N, 2011 Comments on Bombay Mirror article. IRB network Online
18. Rajimwale, S G, 1994 Agnihotra: Farming Methods. Institute for Studies of Vedic Sciences Pp. 16 Stapur. India, p 1-3
19. Sass H-M, 2008 Fritz Jahr's 1927 Concept of Bioethics. Kennedy Institute of Ethics Journal 17 (4) 279–295. The Johns Hopkins University Press
20. Sass, H-M 2009 Asian and European Roots of Bioethics: Fritz Jahr's 1927 Definition and Vision of Bioethics. Asian Bioethics Review 1(3): 185-197
21. Shinagawa, S N, 2000 Tradition, Ethics and Medicine in Japan. Berliner Medizinethische Scheiften – Beitrage zu ethischen und rechtlicnhen Fragen der

Medizin. Heft 40/41 1-73.
22. UNESCO 2005 Explanatory memorandum on the elaboration of the preliminary draft declaration on universal norms on bioethics: First Intergovernmental Meeting of Experts Aimed at Finalizing a Draft Declaration on Universal Norms on Bioethics. Paris April 2005.
23. Wells H G, Huxley, J and Wells GP, 1930 The Science of Life. London vol 2. p. 505 This section (The ABC of genetics) was written by Huxley (Quoted from Wiki, 2011).
24. Wiki, 2011 Modern evolutionary synthesis. From Wikipedia, the free encyclopedia. Accessed on Oct. 2011
25. Wikipedea, 2011 Francis Galton. Wikipedia Free Encyclopedia. Online

CLINICAL ETHICS COMMITTEES AS LIVING ENTITIES

Arnd T. May[1]

"Our entire life and activity in politics, in business, in administration, in the laboratory, in the workshop, in the fields is … in its reasoning and goals not focusing on love in the first place, quite often rather focused on struggle with some sort of fellow competitors. Quite often we don't recognize it, as long as such a struggle is without hate and in an open and legally accepted way. As much as we cannot avoid the struggle with fellow humans, similarly the struggle for life with other living entities is unavoidable. Nevertheless, neither in the first nor in the second case, we will lose the idea of moral obligations as a principle" Fritz Jahr[2]

Clinical Ethics Consultation in Germany

Ethics committees for clinical research have been established since the late 1970s as institutional review boards (IRBs) at the request of the federal government. The German Medical Association (Bundesärztekammer) issued guidelines for establishing IRBs in 1985. Today, we find IRBs also in Medical Schools and in the State Chambers of Physicians for research conducted outside universities. Hospital ethics committees (HECs) dealing with individual cases in clinical ethics are not yet found at every hospital. Local initiatives have increased their numbers, as well as consensus conferences (Konsensuskonferenzen), such as at the Intensive Care Unit at the Catholic St. Josef Hospital in Bochum since the early 1990s. A joint recommendation of the German Association of Protestant Hospitals (DEKV) and the German Catholic Hospital Association (KKVD) focused on establishing HECs in hospitals in 1997 with three main objectives: education, case consultation and policy development. The organization Cooperation for Quality and Transparency in Healthcare (KTQ), offered certification procedures to hospitals requesting information from HEC or other ethics consultation practices in 2001 and many hospitals established HECs. A new Regulation for Medical Licensing (Approbationsordnung für Ärzte – ÄAppO) was issued in June 2002, requiring training in the history and theory of ethics in medicine. Case discussions are included in the ethics curriculum only since 2003/04. The annual meeting of the Academy for Ethics in Medicine (AEM) in Wittenberg in 2003 focused on the professional clinical ethics consultation.

---

[1] Arnd T. May, PhD, University of Halle-Wittenberg, Medical Faculty, Institute of History and Ethics of Medicine, Magdeburger Str. 8, 06112 Halle (Saale), arnd.may@medizin.uni-halle.de

[2] Jahr 'Animal Protection and Ethics' 1928, in this volume, p. 14

The task force Ethical Consulting in the Hospital (Arbeitsgruppe "Ethikberatung im Krankenhaus") resulted and a curriculum for ethics consultants drawing on their experiences was published in 2005[3]. It included the three functions of HECs, i.e., (1) education, (2) case consultation and (3) policy development, as well as the organization of ethics consultation. Courses were offered in Arnoldshain, Düsseldorf, Hannover, Mülheim/Ruhr, and Saarbrücken. The AEM task force also established a website [www.ethikkomitee.de] to provide information on ethics consultation and to establish a network.

In 2006 the Central Commission for the Safeguarding of Moral Principles in Medicine and Related Disciplines (Central Ethics Commission) of the German Medical Association published a statement on ethics consultation in clinical medicine (Ethikberatung in der klinischen Medizin), recommending the implementation of HECs or other structures of ethics consultation for improvement patient-centered care. It called on physicians to participate in ethical case consultation to improve the transparency in decisions of therapeutic options.[4] Few IRBs currently offer ethical case consultation. Originally begun among medical staff with some support from the institutions, a 1997 recommendation by the Catholic and Protestant hospital associations made Christian hospitals to leaders in established HECs, the preferred model being multidisciplinary. This approach to case consultation integrates a decision maker into the process and seeks to achieve consensus following deliberations of the arguments by the members of the HEC and its participants. Ethics consultation serves all caregivers, i.e. coworkers, teams, directors, patients, their legal advocates, relatives and companions. It also addresses moral conflicts of institutional unity with isolated participation, and it strives for transparency in the decision making process in the search of morally acceptable criteria; it aims at a "good" decision and "good" decision making processes, thereby strengthening the ethical competence of each individual. It contributes to the quality of care of patients and occupants.

Ethical case consultations undergird the difficult decision of treatment options and are accessible to all involved (i.e. varied specialists, patients and their relatives or advocates). In those case consultations ethical expertise and skillful moderation go hand in hand. The task of the consultant is to bring to light all case-appropiate details and to give equal chance for participation by all, as well as to clarify the ethical issues as they emerge. Theoretical aspects must be weighed against the given realities of each case, at the same time striving for a possible consensus. The resulting treatment decisions must be clearly documented in the medical record.

---

[3] Simon, May, Neitzke (2005)

[4] Central Ethics Commission of the German Medical Association (2006)

Clinical ethical consultation is concerned with everyday ethics of treatment and care, but also with interpersonal conflicts and rivalries, also with misunderstandings among stakeholders. Increasing numbers of hospitals have hired professional ethics consultants, some institutions have a single designated ethics consultant. Various working groups and forums have emerged as well. Most of the university hospitals have a clinical ethics committee. For quality assurance of these members, the workgroup "Ethikberatung im Krankenhaus" introduced a curriculum in 2005[5]. Various experiences in different hospital settings were used to form a basic knowledge and competence in the fields of ethics, organization and consultation for the ethics consultant. Guidelines were also established for eldercare in ambulatory or hospital settings in 2007[6]. The Akademie for Ethics in Medicine (AEM), a society of interdisciplinary and interprofessional medical ethics provided criteria for quality and basic demands for every kind of ethics consultation. In Germany this presented a new direction in betterment of the care of the ill and disabled. These consultants are members of ethics committees in hospitals and in-and out-patient medical settings. The concept of ethics consultation is limited when confined to single patient consultation in conflict. However it has gained broader applicability in the development and guidelines for the furthering of ethics in general. The question always arises of the specific knowledge needed for ethical case consultation. Without a doubt, knowledge in the areas of medicine, philosophy, theology, counseling and law are helpful. But how much knowledge in these areas would be deemed adequate? And how about compassion and competence in the art of communication and the ability to listen to each other and to inspire the same in others? How can one steer the group's reasoning and goal away from a struggle with some sort of fellow competitor to focus on love and caring? Professional ethics consultants in the healthcare setting need competence in communication and cooperation and thus should be trained very well in clinical ethics.

Clinical Ethics Committees and Consensus Formation

Clinical ethics consultation is widely performed as case consultation in multidisciplinary HEC's and integrates a legitimate decision maker into the process. The ethics consultant seeks to achieve consensus following deliberation on the arguments developed by the members and participants. It is common to have physicians and nurses as members of the HEC, bringing their individual and different professional values and beliefs. In ethics consultation members of an HEC are not representatives of their professions

[5] Simon et al. (2005).
[6] Bockenheimer, May (2007)

in the hospital. The widely accepted multidisciplinary approach of HECs in Germany helps to broaden the spectrum of discussion on ethical questions. Also, the variety of moral beliefs in secular German society may contribute to HECs having to address more controversial matters. They face issues, such as terminating tube feeding of incompetent patients, respecting advance directives and the expressed wish to forego treatment or surgeries when suffering from dementia or well, or respecting the refusal of blood by Jehovah's Witnesses. At Catholic hospitals, the priest will not be the one to define the correct interpretation of the religious normative grounds to which a Catholic hospital is bound. In ethical case consultation, deliberations are based on the patient's wishes and beliefs. The consultation may involve consideration of therapy options and substitute decision making by a surrogate for an incompetent patient.

HECs multidisciplinary focus on moral problems raises the question whether their deliberations can lead to consensus. The reflections on ethical principles can lead to a consensus among its members that finds its basis in an understanding of the Four Principles of Beauchamp/Childress[7]. For many HECs, discourse ethics (Apel, Habermas, et al.) is the normative conception and self-conception. Habermas's[8] discourse ethics attempt to distill the idealized moral point of view that accompanies a perfectly rational process of argumentation. Even this has its place in multidisciplinary HECs of the Catholic and Protestant hospital associations. On the other hand, participation in ethics consultation should sharpen the individual's understanding of his or her personal moral commitments. The values of an institution may influence the work of HECs. Finding consensus is not, however, the overriding goal of ethics consultation. Sometimes it may be more helpful to see through case-based moral reasoning the competing principles at issue, as reflected in the different solutions proposed by the individual members of a HEC. The members of an HEC should use a tool to structure the ethics case consultation.

Structured ethics case consultation

Case consultation in moral conflicts should be moderated and led by trained ethics consultants, who also have experience in understanding and solving interpersonal conflicts and misunderstandings. The different backgrounds of the persons in the profession hold different views on the relevant topics of a specific case. The Academy for Ethics in Medicine in Germany stresses that an ethics consultant must have competence in moderating a meetings, such as case consultations, and bring to it his own expertise in ethics, relying not only

[7] Beauchamp, Childress (2009)
[8] Apel, Kettner (1992)

on the knowledge of the four Principles by Beauchamp/Childress. Several publications for the structure of ethics consultation have been published. In 1987 the Bochum Working Paper[9] on the Practice of Medical Ethics (Sass/Viefhues) started the discussion about structured instruments for ethics consultation. It is an issue of team spirit and group mentality. As a member of an HEC I might clash with my fellow members, because of different moral believes and points of view. Am I stuck on my point of view? Is my blocking the flow of the meeting somewhat akin to a blockage of a healthy organism by one 'stuck' organ? Different values within the HEC do cause difficulties in reaching a consensus. The group-identity of a HEC depends on the variety of their members and a broad range of moral beliefs. Is it not like a living organism, needing eyes to see and feet to walk?

The team at a ward is a group or cluster of individuals with different values and beliefs. The common base of acting and decision making is defined by general principles like the Declaration of Helsinki or other codes specific to their profession such as medical or nursing ethics. A group of people and their different values will not be a homogenous mass with a consistent set of values. The expertise of an ethics consultant is to search for and to recognize relevant ethical principles in a particular case.

Clinical ethics consulting is case specific and therefore situational. A practical method must allow for "special cases".[10] Moderators for ethic case consultations are available to all participants. Central tasks include: 1. Structure, 2. Discourse, 3. Content, 4. Results. – 1: addresses the orderly frame and delivery of the consultation. – 2: sees to the participation of all involved and that everyone equally take part in the discourse. – 3: An understanding of the content of the consultation with its relevant ethical aspects. – 4: A summary and call to unite for the vote. Ideally, a vote or recommendation finds unanimous support. The vote also may demonstrate differing positions, when participants are unable to agree, but the reasons for disagreement should be spelled out should not be based in personal rivalries or misunderstandings.

A further consequence of working together on an ethics team is a molding of character, the so-called character formation: character [Gesinnung], in one way or the other, always based on morally grounded judgment. Different from character is science, which does not accept value judgment as a basis of work.[11]

---

[9] Viefhues, Sass 1987

[10] Drane, 2004, 296

[11] Jahr 'Character Dictate or Freedom of Thought?' 1930, in this volume

Organizational ethics and ethics consultation

Ethics consultation is an adventure, experienced not in homogenous groups, but in bundles of individuals, each with a philosophy of their own. The moderator has to be aware that each participant, who finds him/herself at that particular moment in this group, has needs of their own. These are not apparent in the normal sharing of these convictions, but they are at the very root of their personal thinking. No one may dominate in this chain of decision-making and an honest exchange of opinions is sought. This is well suited to Fritz Jahr's conviction, that the training for the conceptualization of one's own convictions must clearly address the plurality of convictions: no brake-in by dominance, but open discourse, respectively a 'democratization' of character formation. On a practical problem-solving level the moderator in ethics consultation must respect the metabolism within a team as a living entity with its own internal communication, not unlike the metabolism of living creatures. The characteristics and talents of its members are quite varied. Possibly, physicians are better trained in communication, but the less well presented concerns of others carry equal weight. The moderator becomes facilitator and consultant and will have to work on developing a common language and an a team sprit within such a living body. The struggle for dominance in ethics consultation may not overshadow or diminish compassion and caring for the patient or other team members. It seems that clinical ethics might increase with clinical experience and responsibility as a review among health care experts at Medicals school in Bochum and Magdeburg has shown.[12]

Nature does give varied expressions to all, some are tall, others small. Size does not determine speed. Some live at a slower pace, others hurry along. The ethics consultant must perceive these conscious or unconscious traits. They cannot be eliminated, but need to be incorporated. The consultant's task is to harmonize the different characters and different ways of communication. He is expert in the moral values and also moral diagnostitian. Sometimes the ethics consultant performs as a mediator and brings to the process reliable knowledge and facts. He sometimes is the bridge to authority and relays information to the hospital management. The management can then decide on other initiatives necessary to address certain topics. Ethics consultation can also help in conflict resolution of a team member with the management or within other members of the team. Different views of the patient treatment might also be solved by calling an ethics-expert from outside, because solving of internal problems within a ward or ICU can be limited for various personal and non-personal reasons. Thus, the common structure of decision making may be strengthened by conflict solution and communication.

[12] Baberg et al. (2004).

Conflict solution is a common phenomenon, well described and necessary in families, teams or other social structures with differences among leaders, in hierarchies, and in group mentality. Ethics consultation is a valid interaction for living entities. It places a high value on compassion and collaborative analysis. It fights the selfish struggle for dominance and egotism. Ethics consultation thus is normal and in most cases routine, but it also can become a battle ground in some rare cases. The overwhelming need for ethics consultation is fed by the need for better communication within a group of professionals who want better care for their patients. Fritz Jahr clearly has pointed out the dangers of moral scepticism and the challenges we all face in ethics consultation: "Only ethical skepticism seriously supports egocentric attitudes; but these are relatively few exemptions. Much larger is the number of those, who unconsciously make their own self the center of their interest. That this attitude is unavoidable and not necessarily to be criticized has been shown above. However, an exaggeration of egoism associated with endangering the ethos is obvious. These circumstances make it understandable that altruism carries a much higher recognition in the public eye. Even those, who call love a 'sentimentality', routinely refrain from admitting their psychological shortcomings (and, indeed, these they are). Most people, however, possess, depending on their predisposition, a serious will and confidence to be 'just', 'good', 'altruistic'".[13]

References

Apel K-O, Kettner M (Hrsg.): *Zur Anwendung der Diskursethik in Politik, Recht und Wissenschaft*. Suhrkamp, Frankfurt 1992

Baberg H, de Zeeuw J, Kielstein R, Sass HM: Behandlungsgebot und Behandlungsverzicht im Blick des Bochumer Inventars zur medizinischen Ethik (BIME), in: Sass H-M, May A (Hg.) *Behandlungsgebot oder Behandlungsverzicht*, Münster: LIT 2004, 171-212

Beauchamp TL, Childress JF: *Principles of Biomedical Ethics*, New York: Oxford University Press 6. th 2009

Bockenheimer G, May A: Ethikberatung – Ethik-Komitee in Einrichtungen der stationären Altenhilfe (EKA). Eckpunkte für ein Curriculum, *Ethik in der Medizin*, 2007, 331-339

Central Ethics Commission of the German Medical Association. Ethikberatung in der klinischen Medizin: Stellungnahme der Zentralen Kommission zur Wahrung ethischer Grundsätze in der Medizin und ihren Grenzgebieten (Zentrale Ethikkommission). *Deutsches Ärzteblatt*, 2006 103(24), p 1703-1707

Deutscher Evangelischer Krankenhausverband (DEKV) & Deutscher Katholischer Krankenhausverband (KKVD), *Ethik-Komitee im Krankenhaus*; Berlin/Freiburg 1997

Drane JF: Methoden klinischer Ethik, in: Sass H-M, May A (Hg.) *Behandlungsgebot oder Behandlungsverzicht*, Münster: LIT 2004, 296

[13] Jahr 'Egoism and Altruism. Oppositon and Alliance', 1929, in this volume

Frewer A, Bruns F, May A (Hrsg.) (2012) *Ethikberatung in der Medizin.* Heidelberg
May A: Modelle von Ethikberatung in: Heinemann W, Maio G (Hrsg.): *Ethik in Strukturen bringen Denkanstöße zur Ethikberatung im Gesundheitswesen*, Freiburg 2010, 80-102
May A: *Autonomie und Fremdbestimmung bei medizinischen Entscheidungen für Nichteinwilligungsfähige*, Münster: LIT-Verlag 3. Auflage 2005
May A, Brandenburg B: Akzeptanz von Instrumenten der Vorsorge von medizinischen Laien, in: Sass H-M, May A (Hg.): *Behandlungsgebot oder Behandlungsverzicht*, Münster: LIT-Verlag 2004, 213-237
Sass H-M, May A: Advance Directives: Balancing Patient's Self-Determination with Professional Paternalism, in: Helmchen H, Sartorius N (Ed.) Ethics in Psychiatry, Berlin 2010, 147-159
Simon A, May A, Neitzke G: Curriculum Ethikberatung im Krankenhaus. *Ethik in der Medizin* 2005; 17:322–326
Viefhues H, Sass H-M: *Der Bochumer Arbeitsbogen zur medizinethischen Praxis*, Bochum 1987: Medizinethische Materialien Nr. 2
Vorstand der Akademie für Ethik in der Medizin (AEM) e.V.: Standards für Ethikberatung in Einrichtungen des Gesundheitswesens, *Ethik in der Medizin* 2010, 149-1

THE ACTUALITY OF THOUGHTS OF FRITZ JAHR IN BIOETHICS EDUCATION OR WHY FRITZ JAHR ADVOCATES CHARACTER EDUCATION*

Nada Gosić[1]

Introduction

This work presents a continuation of research on character education published in the creation of doctoral thesis titled "Bioethics Education: Contents, Methods and Models" published in the book titled *Bioethics Education*[1]. This book presents thoughts on the course contents that can help to obtain status of subject and programme of character education[2]. Today, owing to Professor Hans-Martin Sass's research on Fritz Jahr, we can actualise Fritz Jahr's standpoint on character education, compare them to contemporary standpoints and determine Fritz Jahr's contribution to the creation of a concept of the European model of bioethics education.

Jahr's decorum in character education

In 1930s Fritz Jahr defined ten ways to influence moral reasoning of young people[3]. The aim of this work has been based on this idea and it tends to test the application of Jahr's approaches to moral education in medical-ethical and bioethical course contents of medical schools students. On the theoretical level we have decided to state Jahr's method, or (even better) the educational principles and explanation of their educational importance. On the practical level we consider the implementation of each of these principles in the concrete bioethics course content. This means that the application refers to consideration and analysis of concrete medical-ethical issues and bioethical problem situations that physicians confront. We cannot form the complete assessment of evaluation on the actualisation of Jahr's educational principles in this article. This can be done after their application in pre-clinical and clinical bioethics courses. As we, for the time being, realise bioethics courses in pre-clinical teaching, the application of Jahr's educational principles in this

---

*The paper was presented at 1st International Conference "Fritz Jahr and the Foundations of European Bioethics", Rijeka/Opatija, March 11-12, 2011, and published in Gosic N (2011) *Jahr* 2(4) 407-414.

[1] Nada Gosic, ngosic@medri.hr, University of Rijeka, Faculty of Medicine, Department of Social Sciences and Medical Humanities, Braće Branchetta 20, 51 000 Rijeka, Croatia.

[2] Lickona, Th. (1996). Article "Eleven Principles of Effective Character Education", in: Journal of Moral Education, 1, 93-100

[3] Jahr, F. (2011) Selected Essays in Bioethics 1927-1934 Fritz Jahr. Postscript and References by Hans-Martin Sass. Bochum: Zentrum für Medizinische Ethik (Medizinethiche Materialien 186). See also in this volume.

article is focused on the adoption of basic knowledge of medical ethics and bioethics, on bioethical theories and principles, paternalistic relationshop to patients and on informed consent, as well as privacy and confidentiallity.

According to pedagogical-methodical demand, we have devided educational principles inside Jahr's decorum into those ones which represent the condition of the character education in general (the first three principles), those ones which implement the collaborative teaching and learning as condition for the character education (principles 4, 5, 6 and 7) and finally those ones which form the teacher's character characteristics (principles 8, 9 and 10).

Analytical review of Jahr's educational principles

Principle no. 1 *Do not teach predetermined subjective disposition.*

This principle refers to indoctrination with teaching contents. It serves as a warning for teachers not to be partial in presenting and explaining of bioethical problems. Disrespect of this principle leads to indoctrination of students and make them form opinions which are the result of influence and personal preference of the teachers. The more complex is the case dilemma the higher is the degree of indoctrination.

Messages from this principle are :

- The teachers indoctrinate students with teaching contents in all cases in which they choose only one content and make it crucial in deciding.

- Without taking into consideration and observing the professional, scientific, personal, familial, social and other contents which influence the person who must take the decision, the teacher directs students to a solution produced by the dominance of one bioethical content.

- Students have no possibilities to choose bioethical contents within bioethical problem situation.

Our suggestion is to test this affirmation on the problem of abortion. Making a decision *For* or *Against* abortion is not simple nowadays. The decision should take into consideration the dilemma about the real beginning of human life, medical conditions, rights of a mother and rights of an unborn child, psychological factors which accompany the process of decision making and administrative-legal, political influences, as well as influences of media.

For instance, the teacher's intention may be to present only harmful medical consequences of abortion like sterility, body injuries, bleeding, infections etc. In that case these problems become the main contents in deciding and indoctrinate students. If the teacher presents abortion only from theological perspective, God's commandments, especially the commandment *Don't kill* ! can create the sense of guilt and the guilt will become the main content in deciding. It is similar if the teacher presents only psychological perspective. Depression, fear, loss of self respect can be the contents that the teacher will point out in the presentation.

Principle no. 2 *Strictly avoid the cover-up of a predetermined opinion with so-called objectivity and with wrongly called interactive teaching.*

This principle refers to indoctrination with teaching method.

Messages from this principle are :

- The methodological indoctrination pressumes the intrusion of one methodology to the student.

- This indoctrination promotes the standpoints and the importance of conclusion of only one methodology or science while other ones are ignored.

- The sum of scientific methodologies alone without their mutual interaction is not a true interaction.

Opposite to this, Jahr points out that each science has different methodological approaches and that each science itself can offer different solutions. Thus the teacher should refer the students to an interaction of the methodological approaches which are necessary in the solving of bioethical problems. This must be presented in the teaching concepts. With the application of different scientific methodologies students are in the position to find and emphasize the similarities and differences and their contribution to achieving solutions. The component part of this principle is also the application of the same methodology to different problems and the observation of consequences of solutions obtained in both ways. The message is that intention of interactivity in teaching itself is not sufficient and teaching really becomes interactive when we include those methodological approaches which contribute to the solution of the problem.

We can show this on the example of Euthanasia. God's commandments or some other methodologies of theological ethics have meaning and importance in the explanation of the problem of euthanasia. In a concrete case in which a person does not express his/her religious beliefs, these methodologies get different meaning. It is just the opposite in the cases of blood transfusion refusal of Jehovah's Witnesses. In the first case religious belief of a patient does not influence the decision and the standpoints about euthanasia. In the second case religious belief defines the patient's decision which can lead to euthanasia. Inclusion of the legal perspective which permits or prohibits euthanasia or determines patients' rights can help in both cases. The students also need to know that such decision can be brought under the influence of family values, as well as of community values and cultural conditions that the question about death and dying is considered in. Therefore the definition of interactivity in teaching directs to the inclusion of those methodological approaches which are necessary for the solution of a certain problem. It does not direct them to stating and listing all scientific disciplines. Interactivity tends to help students with orientation and open-mindedness to different approaches and to responsibility for consequences of their application.

Principle no. 3 *It can methodologically not be accepted to present only what is suitable and to suppress unsuitable facts, to deny or to manipulate them at will.*

This principle demands methodological originality of the teacher.

Messages from this principle are :

- Teaching content can be approached in different ways.

- Bioethical problem situation demands an approach which expresses the total fate of a person affected by a specific situation.

- In the solving of the problem we include science and professions of the same nature like the corresponding problem.

- Pluriperspectivity is a necessary methodological approach in the understanding of bioethical problem situations.

It is recognized that, if a combination of approaches is used to interpret the problem and state the situation, the students will, while listening to the teachers' lectures, acquire an insight into pluriperspectivity as a necessary methodological approach to the understanding of the bioethical problem situation. We showed this on the course *Life and culture of dialogue in medicine.* In the age of highly developed technics and technology, life and its characterisics gain technological meaning and are interpreted by some scientists with the help of numbers, percentages, diagrams and other statistical indicators in medicine. Opposite to them, humanistic science and scientists who belong to those scientific branches, find out dangers of technological approach to life, ilness and health. The first ones equalise the approach to the problem and its solution, the others show that cases from medical practice must not be solved by using a technical and routine approach. They demand an approach which expresses the total fate of the person affected by a specific situation. If we include both viewpoints we will send the message to the students that this teaching concept can be approached in different ways and that they are free to choose approaches to solve a certain problem. They must also accept professional and ethical responsibility for their choice.

Principles: no. 4 (*Always consider different character attitudes*), no. 5 (*The benefits and shortcomings of different opinions and attitudes must be discussed*), no. 6 (*When you present your personal opinion, it must be done in an impartial form. Also, one should forget to discuss problems associated with one's own position*) and no. 7 (*Instead of presenting biased characte formation students should be given the opportunity to form their own opinion, respectively objective information should be given, so they may form their own character at a later date*) within Jahr's strategy of education refer to the character of the teaching process. With these principles Fritz Jahr formed the second phase of the teaching process.
Messages from these principles are :

- Applying these principles the teachers can create the climate of a dialogue.
- They can motivate the students for active presentation and advocacy of their standpoints.
- They can define students' relationship towards the presented ideas.
- They point out to the students that a problem can be solved with knowledge, professional experience and contains cultural perspective, familial values and media and political influence.

The issue of organ and tissue transplatation is a teaching content that is used to test the application of this principle. It is very important to emphasize that this problem is used to point out that critical relation and critical opinion contain a personal attitude that arises from the knowledge of the problem, the consideration of standpoints different from our own and respect of different, often opposed standpoints. This is an opportunity for students to defend their personal opinion and to realise that standpoints of other participants in teaching are important, too. They also realise that their personal opinion can be changed if the change has followed new learnings and facts. In relation to this, the teacher's task is to point out to the students that the change of opinion is a product of the critical exchange of standpoints and as such is legitimate.

Principles no. 8 (*Reason and science, people's highest authority never shall be missed in the formation or review of an already existing character*), no. 9 (*One should not claim that the youth is only ready for authoritarian methods, not for methods of freedom, a position which might be contradicted by some. But let it be. Seeding is always earlier than harvesting*) and no. 10 (*And if a new expected character is not developing, we should not forget that had happened under the old method even more often*), relate to the teachers' character characteristics. According to them the teachers are character persons if they satisfy the following requirements of the following messages:

- It is important for the students to understand interpretations.
- The teachers use practical examples related to the future professions of the students.
- The teachers listen to the students carefully.
- The teachers are open for all students' questions and commentaries.
- The teachers respect initiatives and attitudes of the students.
- The teachers show the understanding for the mistakes.
- The teachers create the positive emotional climate for the students' standpoints.
- The teachers are ready for changing the parts of the curriculum the students have argued for.

All the teachers also need pedagogical knowledge. We can also claim that the teachers, while teaching the students, notice that they necessarily need to learn more and improve their knowledge in all aspects.

Conclusion

Jahr's decorum directs all three segments of the educational process of bioethics teaching, the teacher and the student towards character education. The arguments that follow prove the pronounced statement.

1. Social changes influence teaching and the teaching of bioethics must satisfy social and individual components of upbringing and education.

2. The teaching of bioethics is a didactical-methodical, but also a social-cultural and communicational process. Its task is to stimulate the adoption of knowledge and to emphasize that legitimate argumentation can be done only with knowledge and information of the problem. It is directed to critical opinion based on knowledge and on the interaction of different and opposed standpoints.

3. The orientational knowledge is methodologically based on foundations of integrative pluralism of perspectives (pluriperspectivism).

4. The University is an institution which provides the orientation in social life.

5. During the teaching process the students adopt competences for ethical decision making and solving of ethical dilemmas. That is why the solution of ethical problems is not the question of personal talent, but rather a constructive acquirement of knowledge necessary for the forming of arguments.

6. Bioethics teaching motivates understanding of intercultural observation of bioethics problems.

7. The bioethics teacher has clearly distinguished traditional and new didactical-methodological roles. In his/her traditional role he/she defines teaching goals; he/she is a source of information, an interpreter of knowledge and organiser of teaching.

8. According to contemporary pedagogical demands he is a professional who plans teaching course strategies, motivates the students for collaboration and self instruction and forms an evaluation of the achieved work together with the students after the coursework.

9. In his/her contemporary role he/she is an enthusiasic person with developed communicational skills and ability to solve problems.

10. He/she is an authority to the students. He/she has built his authority in a partnership relationship with the students, with respect to the students' personalities, through the creative application of methodological

approaches, in motivation for work and learning and critical valuation of the students' and his/her own standpoints.

11. During bioethics courses the students acquire knowledge and skills for ethical analysis and solution of bioethical dilemmas.

12. They unite knowledge and make effort to adopt it.

13. They recognize and respect different standpoints and decisions based on them.

14. They show solidarity and sensitivity for problems affronting a sick person.

15. They are open to other cultures and efforts in the repression of stereotypes and discrimination arising from the ignorance of and disregard for different cultural forms.

Everything stated here refers to the fact that Fritz Jahr advocates character education not with the intetion to change the character of the students, but to make them open and tolerant to different and various ideas and standpoints.

Finally, according to Jahr's decorum, bioethical courses have the role of improving, and not changing the character of the students. During the courses the students are respected persons who express the need to widen their knowledge and adopt skills for the solution of problems. Owing to the collaborative relationship with the teacher they build an internal motivation for the realisation of the tasks determined by their future profession. To this cognition and message arising from it, the teachers of bioethics must not remain indifferent.

FRITZ JAHR'S BIOETHICAL CONCEPTION: WHAT ARE THE CHALLENGES FOR CHRISTIANS TODAY?

Geni Maria Hoss[1]

Introduction

For Fritz Jahr many of the problems that are generating the current ecological crisis presented themselves, in part, as a reality of that time and, in part, as a prognosis of a somewhat near future. He suggests that the problem is one of relationships and, because of this, the Bioethical Imperative which he proposes is an interpellation based on the experience of otherness and interaction with a broader perspective for the relations with all living beings. To respect the other – animal and plants – implies recognizing it as an end in itself which does not depend on the subjective valorization of someone else nor on any external conditioning.

To understand Jahr within the Christian scope, it is necessary to reflect on the dimensions of the faith that are implicated with regard to the biosphere. That is, it is necessary to deepen the Theology of Creation and its relation with the current ecological issues.

The Judeo-Christian tradition was, for a long time, harshly criticized by specialists in issues of the Environment. Some authors suggested that the *dessacralization* of Nature, in Judeo-Christian thought, opened paths to the destruction of nature on a large scale which resulted in imminent concrete threats for all life forms on the planet. Among others, there is Prof. Lynn White, University of California, who suggests that an important cause of the current ecological crisis is the indiscriminate use of the power given by the Creator, as written in the known publication *The historical roots of our ecologic crisis. Science, New Series, Vol. 155, No. 3767 (Mar. 10, 1967), pp. 1203-1207*. Another name worth highlighting is Carl Amery, author of the book, *Das Ende der Vorsehung. Die gnadenlosen Folgen des Christentums, 1972*.

Such denunciations and incompatibilities between scientific knowledge and the beginning of the world according to the traditional interpretation of the

[1] Geni Maria Hoss, geni.maria@yahoo.com.br, Master's in Theology, member of the research group in Pastoral Counseling, Social Practices and Interculturality of the Faculdades EST, Specialist in Bioethics from the Pontifícia Universidade Católica of Paraná, Professor/Consultant in the areas of Health Care Ministry, Hospital Humanization, Pastoral Counseling and Spirituality, Brasil.

Creation have given rise to fierce debates on the issue. Thus the need for a Theology of Creation which is capable of dialoguing with the Human Sciences and the Natural Sciences of today, in sum, which can furnish plausible substrates for an ecological reflection based on the Christian faith. In this way, the Christian communities can place themselves in the world scenario not as the villains of the destruction of the different forms of life on the planet, but as defenders, with an individual and collective ethical posture in regard to the biosphere and an effective insertion in the programs and policies of protection of biodiversity.

The growing concern with the sustainability of the planet in the various fields of society has challenged the Christian communities to greater and more effective involvement in favor of life on the planet in light of the faith in the Creator God.

However, one does not expect from the Christian an effort moved only by the current crisis and uncertainty as to the future of humanity and of the planet, but also moved by the respect for and enchantment with the divine creation. This paper analyzes Fritz Jahr's contribution toward the current challenges as he proposes the Bioethical Imperative, which implies in a change of attitudes and a new way of being in the world, contemplating the complexity of the whole of creation.

1. The re-enchantment with God's Creation in times of ecological crisis

The advance of the Natural Sciences and the ecological emergencies which present themselves today lead to new challenges for the Theology of Creation as well. Through scientific research human beings are granted ever deeper knowledge of the complex system of life on the planet. The autonomy of the Natural Sciences and of Theology, the distinct ways they have of looking at the same object, are not necessarily motives for reciprocal estrangement, but they require open and respectful dialog.

> The question about the origins of the world and of man has been the object of many scientific studies which have splendidly enriched our knowledge of the age and dimensions of the cosmos, the development of life-forms and the appearance of man. These discoveries invite us to even greater admiration for the greatness of the Creator, prompting us to give him thanks for all his works and for the understanding and wisdom he gives to scholars and researchers. (4:84).

Knowledge is not a barrier to faith in the Creation. On the contrary, it magnifies even further the greatness of its Creator. This theme is not covered through just one point of view. It is completed and broadened through different approaches.

The creation in its totality

"God looked at everything he had made, and he found it very good." (Gen. 1:31). Created by the Word, the creatures – of the whole of nature – form a universal, cosmic harmony. The story of creation portrays a diversified, majestic world, chaos put in order, with time and space well defined, in sum, a consummated fact. (cf. Gen. 1:1-31) However, the experience affirms the creation also as something open, extremely dynamic. There is a constant creation, recreation, regeneration, therefore, a vital, intrinsic force which lends autonomy to the creation which, if left with the necessary conditions for such, that is, avoiding destructive impact, it renews itself constantly. An immeasurable life force, the continuity of which depends today on the positive intervention of the human being in the sense of leaving it untouched or facilitating its process of renewal through the knowledge acquired about Nature and its own dynamics. Therefore it is necessary to respect the continuous process of creation-recreation. "The protagonist of the creation, in the Bible, is, in fact, God. And creation possesses a life generating capacity. This capacity is, in itself, sacred and undeniable, because the Giver of life, par excellence, is always, in the end, the Spirit of God." (2:66).

It can be affirmed, correctly, that creation is not a finished act, but is in "a state of journey" – in 'statu viae' - (cf. 4:90), thus, the journey of the Alliance is not only of the People of God, but it is a journey which integrates all of creation, is destined to the glory of God and in it God is manifest. "Ever since the creation of the world, his invisible attributes of eternal power and divinity have been able to be understood and perceived in what he has made"(Rom. 1:20).

The charge that the Creator God places on the human beings with regard to the creation has given space for various interpretations, with some tendencies being incoherent within themselves. "God blessed them, saying 'Be fertile and multiply and fill the earth and subdue it. Have dominion over the fish of the sea, the birds of the air, and all the living things that move on the earth". (Gen. 1:28). How to understand God's mandate with regard to creation? Is it truly the human being that holds the power over the other creatures? How can the human being give him/her self the right to destroy what God created and what is good?

> Misunderstandings always happen when Biblical passages are taken out of their traditional-historical context and used to legitimate other

> interests. That is why we must also pay attention to the Jahvist version of creation: Gen. 2:15 talks of the "garden of Eden" which the humans should "cultivate and care for". The dominion of the human person over the land should thus correspond to the activity of a gardener cultivating and preserving. In no way does it speak of an exhaustive and exploitative cultivation. (10:55)

The human being shares the historical space and experience with the other creatures, that is, with all life forms. The *dominion* is a mission of *caring*, of *respect* in regard to the other creatures since what God created is good. Words such as *power* and *dominion* are historically related to situations of exploitation and domination. The lordship desired by God, however, is about a respectful relation and an effective contribution so that the human being and the biosphere may live "a life in abundance" (cf. John 10:10), promised by Christ. Therefore, there is no place for indiscriminate use of the goods of the Earth and destructive interventions of life, since care

> means renouncing the will for power which reduces everything to objects, disconnected from human subjectivity. It means imposing limits on the obsession for efficaciousness at any price. It means tearing down the dictatorship of the cold and abstract rationality to make room for caring. It means organizing work in harmony with nature and its indications. It means respecting the communion that all things have between them and with us. (...) It means capturing the presence of the Spirit beyond our human limits, in the universe, in the plants, in live organisms (3:102)

The human being and his/her place in creation

And who is the human being in this rich biodiversity of the planet, what is his/her place? The Holy Scriptures say: "You have made him little less than the angels, and crowned him with glory and honor." (Ps. 8:6). An immeasurable greatness! However, one reads in a different biblical text: "but I am a worm, not a man" (Ps. 22:7) On one side, made to be close to God, a free being, thinking and enabled for transcendency and, on the other side, in the condition of creature, shared even with the minuscule and vulnerable living beings, remitting to the universal solidarity with all creatures. "God created man in his image; in his own divine image he created them." (Gen. 1: 27). To be *image and likeness* of God does not withdraw from the human being his/her condition of being creature – a living being – a condition in common with the whole biosphere. The Creator granted to the human being the ability of transcending him/her self. Therefore, s/he is capable of relating to the Transcendence in an experiential and conscious way. Because of this peculiarity, in vast Christian circles, the central position of the human being in the order of Creation is admitted while at the same time it is admitted that the human being is integrated in a complex vital system, where things are reciprocally interconnected and ordered, each with its own end. Thus, slowly, the conviction that things are ordered exclusively for the human being is being abandoned. From this it is deduced that the human beings have the greater responsibility in the sense of preserving and respecting life on the planet, this work of the Creator God. This conception is still not a unanimous consciousness among the Christians, nor in the interpretation outside the Christian circles. Herein lies the challenge for the Theology of Creation.

> The Christian theological path to overcoming the ecological crisis would be, therefore, around asking about the God of the Judeo-Christian revelation, of which the human creature is image. A God who reveals God's self not through dominion, but through letting go of God's prerogatives and through the humble coming toward God's creation to reveal God's self there, make a dwelling there, get to know and be known. To know God in this sense is inseparable from loving. Get to know the creation without then dominating it, not reducing it, instrumentalizing it, but place oneself, as well, modestly and marveled, in the school of love which looks, sees with respect and enters into a relation. (2:65)

Therefore, the displacement in this new reading happens especially in the aspect of the authentic meaning of the divine mandate: be caretaker, but not exploiter! Caretaker, yes! A care motivated by the love, founded in the Creator God. In the last analysis, for the Christian, ecology would mean returning to the centrality of the faith. God, the creator, the creation, all creatures, therefore, all living beings, in respectful reciprocity, journeying toward the definitive Kingdom

1. The bioethical conception of Fritz Jahr: Affinities with the current ecological conception in the Christian sphere

It is not the intention of this study to give a complete explanation of Jahr's thought, but to highlight some important aspects which deserve special attention for their importance in the current debates about life on the planet and for their contribution to develop a culture of caring in the Christian communities.

In the explanation of the bioethical imperative, Jahr affirms the need for a new paradigm since the old ones – through pressure from the changes in course in various spheres and from new knowledge – are about to succumb.

The author reinforces the need to overcome a vision of the world where the boundaries between the various life forms are too accentuated. "The strict distinction between animal and human being (Mensch), dominant in our European culture until the end of the 18th century, cannot be supported anymore." (8:2). What Jahr points out as being essential for reaching a new relation, an ethical relation with all living beings, is a thought pattern which, still today, holds an important space in the reflections on the theme:

> The ecological ethic aims at overcoming the radical division between humanity and nature – with the first one dominating the environment through growing technicization, and the second separating itself in the direction of preservation without any human incidence, as if the human being was not part of nature. (9:183)

Human being and biosphere: Dominate or get to know?

The Sciences of Nature, mentioned by Jahr for their valuable and necessary contribution to an objective view of the world, made possible when they are permitted their own research system. Jahr also highlights the importance of the knowledge resulting from the work of the Sciences for an authentic relation of respect.

> But effective and successful animal protection is only possible, when enough knowledge and at least some understanding of nature are present. The reason is, that we only can protect animals in reality, when we are somewhat knowledgeable about their physiological and psychological properties and life conditions. Therefore one of the main goals of the animal protection movement is the promotion of such a kind of knowledge and a better understanding of nature, – awakening, broadening, and deepening it. (8:12)

Knowledge is also a primordial condition for the communities, how could they care for that which they do not know? Theology is not sufficient unto itself, it needs Philosophy, it also needs the Sciences. It also needs to know the biosphere so that, in a more coherent way, it can propose paths which can, in fact, lead the Christians to a "respect for laws inscribed in creation". (4:102). The Natural Sciences, as they make known the mysteries of life, permit Christians to nurture the enchantment for Creation and the praise of the Creator. "Let them praise the name of Yahweh at whose command they were made". (Ps. 148: 5) Praise and enchantment which should reverberate in the human community in the form of ethical relations of respect toward all life forms.

The different: a school of learning par excellence

Glancing at other ways of relating to living beings in the very different cultures of Europe, Jahr seeks a way to question the dominant paradigms without, however, intending to impose strange concepts or yet ignore his own Christian roots. It is about evaluating the paradigms themselves from a different perspective, which implies a complex process of reciprocal learning. To learn from another presupposes an intercultural relationship where each, while preserving, strengthening and consolidating their identity, opens themselves up to that which is different and learns from it, permitting that their own values, precepts and actions be questioned. It is in the permanent confrontation with ones' own convictions that paradigms are actualized and plausible answers are found for urgent and complex issues of the reality related to life on all spheres, as well as to a future perspective for humanity and the whole biosphere. Jahr's analysis, applicable to today's cultural and religious conceptions, is of capital importance for a new way of approaching ecological issues. The conceptual differences with regard to the origin and meaning of life do not impede the creation, in the field of life on the planet, of a place for global encounter. This is about issues which break the boundaries of religious confessions and their own philosophies. The religions and religious currents may each have their doctrine, their temple or spirituality center, but not a planet just for them. This is the house for all, their *ethos.*

> Ethos, in its original Greek, means an animal's den or haunt or the house of humans, that is to say, that part of the world which we reserve to organize, care for and make our habitat. We need to reconstruct the common human house – the Earth – so that we can all fit in it. It is urgent that we model it in such a way as to give it sustainability to nourish our new civilizational dream. Today's human house is no longer a nation-state, but the Earth as a common fatherland/motherland for all humanity. (3, 27)

The concern with this common house constitutes an important place of dialog, of acceptance of the different and of reciprocal learning. Here, the dogmatic differences are minimized and different religions and Christian denominations come together in view of a dignified life and of common survival. "The religions bring with them a type of eco-religious anthropology, since they insert the human being in the dialog between the sacred and nature, defining the human's place and responsibility among, with and before the creatures." (9:173).

Understanding the suffering of the other – comprehending the suffering of animals and plants as well!

Various authors are cited by Jahr for their valuable contribution to a new ethical relation of the human being with the other living beings. Among them, Schleiermacher (1768-1834) Krause, (1781-1832), Schopenhauer (1788-1860)... In sum, these highlight the concern with an ethical relation extended to all living beings, the compassion, the respect for their rights and the sensitivity to avoid unnecessary suffering of animals and plants.

The mention of various authors calling for sparing living beings *unnecessary, without just cause* suffering, suggests that this is not about a utopic vision of a world without suffering. Although not explicated, the suffering of the mere process of living, and equally, the suffering caused to recover life, as in the case of Human and Veterinary Science, and various procedures in the area of Agronomy, are admitted. The issue at stake is the suffering caused *without need.* This constitutes a relevant task of Bioethics as it questions the procedures which cause suffering without a viable explanation for such, or when necessary, that methods be applied which permit the most comfort possible.

With the advance of the Natural Sciences in the last years, today there should be no doubts as to the suffering of plants, as Jahr calls for. However, plants are submitted to hard suffering when they are violently destroyed, many times, without any reasons for such, as, for example, the devastation of huge areas by fires or other ways. Radical groups in favor of animals propose as an alternative to the extermination of animals, that vegetables be consumed, but they do not reflect likewise about an ethical attitude toward plants. "Thus, in regard to animals such a rule has become evident, at least as far as needless torture is concerned. With plants it is different, so. For some, it seems at first unreasonable to have certain ethical obligations towards plants." (8:3)

In this particular, the Christians, following the tendencies of all of society, have concerned themselves with human suffering, but have dedicated very little reflection to the suffering of animals and have practically forgotten the suffering of the plants.

The biosphere as a talent and a gift – to beyond the immediate threat

Jahr foresees a more complex relation with the biosphere, not only to guarantee our own survival, but due to the recognition of the respect of the fact itself that the other – also animals and plants – exists. Jahr's concern with regard to public policies which are reduced to actions which attend emergency demands to save this or that part of life is of fundamental importance to society as a whole. This is so that not only are the emergency demands of the biosphere attended but mechanisms of positive action be created, that is, which can contribute to a culture of respect for life as a personal and collective principal which does not depend on continuous external interpellations, especially in the form of threat of extinction.

> Our regulations by law and police protect specific plants and flowers in certain areas (such as plants in the Alps) are based on totally different assumptions. The police state (Polizeistaat) intends to protect those plants from becoming extinct in those areas, also to be enjoyed by other people in the future. Whenever there are plants abundant, the state does not intervene to protect them as an end in themselves. (8;3).

This view integrates the essence of the Christian faith which, by recognizing the creation as a divine work, naturally cultivates attitudes of respect and care. From this perspective, the Christian communities have as their primordial task in the world context an effective work in favor of all forms of life on the planet, not only of those which are threatened.

A personal interpellation

Jahr believes in a promising future, but at the same time can foresee complex ethical issues which would come upon the human being in regard to themselves as well as to the whole of the biosphere, resulting especially from the technological advances. Therefore, legal norms, as they are becoming defined, are not enough. There is also need for an individual ethical posture which considers the possible interventions in life. Although in very different proportions, today there is no doubt left about the reasons that Jahr occupied himself with the issue, especially considering the medical area.

From a broad view, considering the culture, the scientific field, the thinkers who dealt with similar themes, Jahr arrives at the people's daily lives. Concrete examples of current life leave no doubt about the practical aspect of Jahr's reflection. He understands that the issues of life cannot remain in the world of ideas. According to the author, daily gestures of disrespect should be reprimanded.

> But, we are on the road of progress and animal protection gets more and more support in wider circles, such as no decent human being (anständiger Mensch) will without criticism accept, that a thoughtless lout (Flegel) without any afterthought beheads flowers with a stick while on the hike or that children break flowers only to throw them away after a few steps. (8:4)

This and other references from Jahr as to daily practice confer a pedagogical character to his expositions, of fundamental importance in the formation of communities.
To finally expound on the Bioethical Imperative, Jahr resorts to Biopsychology, although in an incipient way in the beginning of the 20th century, but which, due to its proximity to the theme, becomes important for the elaboration of his Idea. After presenting his reasoning in an objective and succinct way, he concludes: "From Bio-Psychik it is only a step to Bio-Ethics, i.e. the assumption of moral obligations not only towards humans, but towards all forms of life." (8:1)

Jahr's reflection finally becomes a clear and personalized call: Our self-education, in this regard, already has made considerable progress, but we have to go further, so that the guiding rule for our actions may be the bio-ethical demand: *'Respect every living being on principle as an end in itself and treat it, if possible, as such!* (8:4).

Recognizing the other – animals and plants – as an end in itself, demands a form of respect without conditions or preferences, but a way of relating based on the fact that the other exists. Its dignity and integrity are inherent in the fact of being in the world. "All things have their own stability, truth, goodness, their inner law and coherence which man should respect" (13:60).
It is about a proposal for a bioethical imperative, which challenges in a personal and concrete way, questions attitudes to be adopted regarding living beings, that is, a commitment to ethical relations which include, above all, respect. Jahr involves all the dimensions of the human being. The change in behavior in the relations of the human being constitutes an ever greater urgency in society, therefore, also in Christian communities, due to the demands of the biosphere, but also to faith and solidarity.

An imperative from the roots of the Christian faith

Exodus 20:2-17 and Deuteronomy 5:6-21 enunciate in the same manner the fifth commandment: "You shall not kill!" A simple statement for a complex issue. The study of Jahr on the 5th commandment opens a range of possibilities with respect to the responsibility toward life. In practice, it can be said that Jahr affirmed, with his bioethical imperative, in a positive way, what is secured in the Holy Scriptures of the Judeo-Christian faith. If, on the one hand, the interpretation of the various Christian denominations already contemplated human life in a broader way, that is, included issues of health and healthy life style as an object in this area, not all of them concerned themselves in the same way with the other forms of life on the planet.

Jahr's proposal confirms and broadens the conception of this commandment through the sensitivity towards animals as well, according to his Christian origin, which is Luther's legacy. If we take the example of the Catholic Church, the current Catechism (1983), although it does not contemplate the life of animals in the fifth commandment, it is incorporated, however, in the 7th commandment. (Chapter II, Article 7) – "You shall not steal" (Exod. 20: 15) – "the respect for the integrity of the creation". (cf. 4:624) Although here they are dealing with a different focus, in various aspects, there are similarities with Jahr's thinking. "It is contrary to human dignity to cause animals to suffer or die needlessly." (4:625) Both foci are interesting and can be provoked reciprocally in the sense of their differences and complements.

Jahr takes care always to affirm the unity and the diversity of the various forms of life, all deserving the same level of respect, but in differentiated ways, according to their function. This is why, the application of the fifth commandment to other living beings, besides the human being, is not only opportune, but also recommended, so that, each in its way, has guaranteed full respect for its life. It is not about attributing scales, but about accepting the real differences and understanding that all have the right that their own way of being in the world be respected. Jahr also points out that the struggle for survival is a condition specifically of the living being – *Kampf ums Dasein* – (cf.8:5). This condition today challenges the human being in a special way to rethink his/her relations with the biosphere in a context where consumerist and hedonist tendencies are increasingly being accentuated which appeal to comfort and pleasure without limits. To abdicate, even partially, of the excessive quest for goods and comfort in favor of the other – of the animal and plants as well – seems to be a difficult task, although one is aware that, yes, there are limits dictated by Nature itself. The protection of the animals and plants, that is, of the creation of God, demands of the Christian communities concrete efforts, but also a spirituality of the creation which recognizes the space of the other – of the animal and plant as well – its reason for being, and consequently, knows how to *empty oneself* of the excesses – of self will and personal whims – in solidarity with all living beings.

The human being: looking toward oneself

In an integral view of the relations of the human being with life, it is necessary, even if briefly, to highlight the relation of the human being with his/her own life. Jahr's work has as its hue his beliefs and convictions about human life in the light of his Christian faith. That is why he bases his reflection, frequently, on the Sacred Bible. He accentuates, therefore, the dignity of the human being, affirmed in biblical passages, which distinguishes his work from the current bioethical tendencies, which relate human dignity, especially, with autonomy and liberty, without presenting secure references to practice. In this way, there is no answer to issues such as: Is it possible to be free in a time and space where there are so many external determinations and appellations? Is it possible to be autonomous, that is, to make secure decisions, in this turmoil of possibilities, promises, uncertainties, especially with regard to boundary issues of life?

Jahr's reading of the 5th commandment with respect to human life extends itself to all forms of care. The same precepts which define fraternal relationships are also valid for oneself. The body, in the Christian perspective, is sacred and as such, one cannot attack it. Jahr deduces the respect for one's own life, including through a healthy life style, from Holy Scriptures: "Do you not realize that you are a temple of God with the Spirit of God living in you?" (1 Cor 3:16). This means, the commandment is not reduced to the act of killing, but also has to do with practical issues which interfere in some way with life, exemplified thus by Jahr: "By not taking one's own life, by not shortening it, by not harming or endangering it, by not weakening one's own health by unchastity, excesses in eating and drinking, heavy anger, frivolous foolhardiness and claredevilry, etc." (8:24)

The concepts as well as their practical application dealt with by Jahr in the context of the 5th commandment, still guide the Christian communities today when dealing with the human life. Except for small differences between the various Christian denominations, the essence remains the same and is defended when the Christians participate in public debates. They are provocations for the society as a whole since it demands overcoming a utilitarian view of the person. For the Christians "every human life, (...) is sacred because the human person has been willed for its own sake in the image and likeness of the living and holy God" (4:603)

Respect living beings, that is, all of the divine creation, become indignant faced with all types of destructive power exerted over other living beings, prophetically denounce such acts and take on, with responsibility, the commitment in favor of all life forms on the planet, constitutes the noble task which makes the presence of Christians significant for the whole planet.

2. The challenges for the praxis of the Christian communities

The specific way of being in the world of a Christian community is characterized by the communion of faith and life in Christ and with Christ, therefore the name Christian. A communion which also includes all of creation in a perspective of the edification of the Kingdom of God, here and now, and also in an eschatological view aiming at "a new heaven and a new earth" (Cf. Rev 21:1), characterized by the harmony and solidarity of all creation, in a unison praise of the creator. "There is solidarity among all creatures arising from the fact that all have the same Creator and are all ordered to his glory." (4:100). The universal communion of destinies suggests the integration of human beings with the whole biosphere, where the care or the exploitation of some has an impact on the others. Therefore, it is about respecting the designs of the Creator, which can be read, with the help of the Sciences, in the dynamic itself of life on the planet. "It was the Creator's will that the human being should communicate with nature as an intelligent and noble 'master' and 'guardian', and not as a heedless 'exploiter' and 'destroyer'". (6).

The responsibility of the human being as a thinking being is undeniable, since s/he is capable of *making decisions* based on the knowledge and understanding of the biosphere in all of its complexity, and of *assuming responsibilities* for his/her acts, being this also the most eloquent way of *praising and blessing* the Creator of the marvelous creation. "Bless the Lord, all his works, everywhere in his domain. Bless the Lord, o my soul!" (Ps. 102: 22).

The reading of Jahr's articles relative to the conception of bioethics permits, in the confrontation with the current challenges related to the life on the planet, to deduce important actions to be embraced by the Christian communities, making them agents of social transformation in the range of life in all its forms. It is equally a challenge for the Christian community to integrate itself in education and environmental protection projects promoted by the various sectors of civil society:

> This is about the ecological awakening of the religions: assuming their responsibilities for the human and planetary life, based on their own religious sources (sacred texts, mysticism, theology, morale) and in view of the ecological involvement alongside the sciences, the governments, the civil institutions as well as the other religions. (9:149).

The intention is not to elaborate a complete panorama, but to highlight relevant consequential initiatives for the praxis of the Christian communities with the goal of them becoming significant spaces for all of society:

— To recover the enchantment for creation based on a plausible Creation Theology, in other words, that is able to dialog with the Natural Sciences. The relation with Nature based on a religious profession includes chances and risks. Chances, because in the religious space all things created have a reason for being according to the Creator's designs. The human being, therefore, is called to care for the things of God. Besides this, the Christian communities are places of individual and collective sensitization about respect for life on the planet. The risks come from the fact that the happenings in the world and the manifestations of Nature can be attributed to the *divine will,* stripping the believer of his/her ethical responsibility, care and protagonism in favor of the planet.

— Broaden and deepen the dialog and the joint action in favor of life on the planet with the different religious and philosophical currents and segments of society. "The contemporary religions have a lot to say about the relation between humanity and the earth". (1:223) *Learn to learn* with the different, overcoming in this field an experience of ghetto community concerned only with its dogmas, with little concrete repercussion in society. Be a part of society and not a society apart. Important initiatives are being taken toward this end, but they are still reserved to small groups – religious leaders and representatives – and thus with very little social impact. Finally, break down paradigms and develop a culture of universal care. "We need to more insistently include ecology on the agenda of the religions and religions on the agenda of ecology." (9:150). It is in their midst that the world, the creation, is interpreted aiming toward a greater meaning for existence.

— Become indignant and react confronted with any form of suffering provoked unnecessarily. Jahr's insistence in preserving living beings from unnecessary suffering is one of the greatest challenges of current times. This is about suffering whose causing agent is an individual person or a collective, who voluntarily could choose less aggressive measures, or simply has the condition to avoid suffering. Being that all living beings are created by the divine creator, with God's own purpose, it is up to the human being to respect the purposes of God with regard to God's works. Following are some critical situations which demand of the Christians special involvement to eliminate or minimize suffering:

- Research with animals without following established protocols which grant the procedures that permit greatest possible comfort;
- Butchering of animals with inadequate aggressive methods;
- Abandonment of domestic animals, depriving them of needed food and care;
- Diversion and games where animals are used and these are subjected to innumerable forms of physical suffering, exhaustion, at times, death;

- Taking the animals out of their natural habitat and commercializing them, increasing the suffering caused by inadequate clandestine transportation;
- Containment of animals, thus accelerating their development for obtaining immediate profits;
- Burning of extensive areas of forests, many times caused criminally, and indiscriminate destruction of trees for mere economic ends, both with significant impact on the habitat of the animals;
- Disrespect of urban green areas where many times only special protection can guarantee the survival of the plants.
- ...

— Deepen the relations of life in the light of the Christian faith. In the Christian perspective, as Jahr reminds us, the body is conceived as the *temple of the Holy Spirit* (Cf. I Cor. 3:16-17), therefore, a sacred place of the presence of God. From this belief can be deduced ways of consideration for one's own life, including the limitation of power over our life, that is, that one is not to attempt against one's own life. Understand that not caring for your health, taking unnecessary risks, are also ways of disrespecting life.

— Broaden charity acts, diaconia as service to people, working for policies which guarantee universal accessibility to health care is a way of exercising citizenship as well as participation in the respective political instances. This involvement is especially necessary in regions of poverty where treatments, especially of high complexity, are not accessible for the common folk.

Finally, it is up to the Christian communities to develop a heightened sense of value for life itself, to re-enchant themselves with the creation as a divine gift, to develop an elevated sense of responsibility for the biosphere and also to form protagonists for the protection and respect of all forms of life, who may involve themselves with other sectors of society as co-actors in the changes and defenders of conditions of dignified and perennial life on the planet.

Final considerations

In pointing with such vehemence to the challenges, which, since then, are becoming ever more urgent, Jahr placed himself as a vanguard of a time which can no longer ignore the need to care for the macro and micro life systems on the planet. This marvelous vital structure, in constant interconnection and reciprocity, is, for the Christians, a marvelous work of the Creator and, because of this, deserving respect and care, a place of praise and thanksgiving.

Jahr has as background the network of relations, integrated and integrating, where there is no space for isolated and disconnected affirmations. He overcomes the myth of the freedom and autonomy of post modernity, which sees in the other the limit maker of one's own freedoms. Made possible by modern technologies, global networks of relations are created at the same time as egotism and isolation are increasingly accentuated.

For the Christians, the way of being in the world is in communion. Freedom and autonomy are lived out in the dimension of fraternity and universal solidarity. The other – understood in the conception of Jahr as animals and plants as well – does not limit freedom / autonomy but shares it. It is the task of the Christian communities to contribute to an experience of freedom and autonomy in the way of being in the world in communion with all living beings. This communion in itself implies respect and enchantment, esteem and praise of the Creator.

The important aspects pointed out by Jahr relative to an ecological formation are urgent challenges for the Christian communities to deepen their creational faith in the sense of integrating into the world to be reconciled not only the human beings, "yes, we know that all creation groans and is in agony even until now." (Rom. 8:22).

Of equal importance and need is the dialog with the Natural Sciences to effectively get to know God's creation and deduce adequate actions to treat it with the dignity that is inherent in it according to the design of the Creator.
The coherency between faith and life advocated by the Christian environments should expand itself to all forms of life. For the Christian, the motivation for a responsible relation to the biosphere should be based on his/her faith. This responsibility being that of living in a shared way with all the segments of society concerned with the destiny of all the forms of life on the planet and with the life of the planet.

An ethical relation with the biosphere based on the Christian values is fundamental not only for the intra-ecclesial life of the Christian communities. There are ever more defenders of the faith and spirituality for its contribution to all of society, especially there where the human values of love, solidarity… are in decadence. This is shown when authors such as Edmund O'Sullivan, of broad recognition in the Education sector, dedicate an important space for this dimension of life. "The newly born ecological community needs a mysticism which can offer the great exaltation appropriate to the existence of such a stupendous universe as ours and of such a glorious planet as this one on which we live." (11:376)

Jahr's Bioethic Imperative is not a command to be followed blindly. He questions and demands a decision based on inner values and convictions. This is the true challenge for the Christian communities.

Note
Translation of text and quotes not available in English: Marie Krahn.

References

1. Al Gore AA (2008) A Terra em balanço: ecologia e o espírito humano. São Paulo: Gaia.
2. Bingemer MC (2011) Deus, segredo escondido em sua criação, in: da Cruz ER Teologia e Ciências Naturais. São Paulo: Paulinas.
3. Boff L (2000) Saber cuidar: ética do humano, compaixão pela terra. Petrópolis: Vozes.
4. Catecismo da Igreja Católica (1999) Catecismo da Igreja Católica. São Paulo: Loyola.
5. Catholic Biblical Association of America (1970) The new american Bible. New York: P. J. Kennedy.
6. Catholic Church, John Paul II (1979) Encyclical Letter Redemptor Hominis. Disponível em: http://www.vatican.va/edocs/ENG0218/__PG.HTM. Acessado em: 30/08/2011.
7. Jahr F (1927) Bio-Ethik. Eine Umschau ueber die ethischen Beziehungen des Menschen zu Tier und Pflanze, in: Kosmos. Handweiser fuer Naturfreunde 24(1)2-4.
8. Jahr F (2011) Essays in Bioethics and Ethics 1927-1947. Bochum: Zentrum für Medizinische Ethik (Medizinethische Materialien 188). See also in this volume.
9. Maçaneiro M (2011) Religiões e Ecologia – cosmovisão, valores, tarefas. São Paulo: Paulinas.
10. Moltmann J (1993) Doutrina ecológica da criação. Petrópolis: Vozes.
11. O'Sullivan E (2004) Aprendizagem Transformadora, São Paulo: Cortez.
12. Second Vatican Council, 1962-1965 (1971). Pastoral constitution Gaudium et spes Vatican: Typis Polyglottis Vaticanis.

The Earth and the Multiverse are Living Beings: Let's treat them as such!

Hans-Martin Sass[1]

> *„The guiding rule for our actions may be the Bioethical Demand: Respect every Living Being on principle as an end in itself and treat it, if possible, as such!"*
> Fritz Jahr, 1927

The Integrative Dimension of the Bioethical Imperative

The Bioethical Imperative as a term, an academic discipline and as a virtue and principle was coined by Fritz Jahr, a Protestant pastor and teacher in Halle an der Saale, 1927 in the annual editorial of the leading German language science journal *Kosmos*. Influenced by Buddhist and Hindu thought via Schopenhauer and by Wilhelm Wundt's empirical comparative studies in physiology and psychology of plants, animals and humans, he formulated the Bioethical Imperative "*Respect every living being on principle as an end in itself and treat it, if possible, as such*"[2]. In presenting a new Imperative for moral orientation and action, he deliberately and expressively criticized Immanuel Kant's Categorical Imperative, which had called for the exclusive recognition of fellow humans as ends in themselves: "The moral law is sacred (inviolable). The person is not sacred, but humankind in his person must be recognized as sacred. Everything in the entire creation, if one wants and has power over it, can be used as a means only; only the human person and with it every intelligent being *is an end in himself.* He is the subject of the moral law, which is sacred, based on the autonomy of his will" (Kant: A156). For Jahr the "*sanctity of life*" is the foundation of the 1927 Bioethical Imperative, while for Kant the "sanctity of the moral law" is the foundation of the 1788 Categorical Imperative. Modern quantum mechanics and our new and growing understanding of the multiworlds of living proteins, cells, plant, animals, environments and symbiosis gives us even deeper insights into the multitude of life forms and life interactions.

---

[1] Hans-Martin Sass, hsass@potomacinstitute.org, Academic Fellow at the Potomac Institute of Policy Studies, in Arlington, Virginia, USA, and Founding Member of the Bochum Zentrum fuer Medizinische Ethik, is em. Prof fuer Philosophie at Ruhr University, Bochum, Germany, em. Senior Research Scholar at the Kennedy Institute of Ethics at Georgetown University, Washington DC, and Honorary Professor at Peking Union Medical College, Beijing, China.

[2] Fritz Jahr 'Bio-Ethics. Reviewing the ethical relations of humans towards animals and plants' (1927), also in this volume.

The Bioethical Imperative changes the primary focus of European philosophy and ethics since the Age of Enlightment away from an anthropocentric focus towards a cosmocentric vision and strategy. Such a necessary change was influenced in the midst of the 19th century by translations of classical Asian literature into German and English and by new scientific knowledge of similar neuronal, psychological, and physiological reactions in plants, animals and plants. Jahr also was aware of the interdependence of forms of life, habitats, struggle for life, good life, and survival[3]; he thus called for balancing values, visions, and interests among living beings. Jahr could not have known that modern quantum physics and most recent knowledge in cell, plant and animal biology would have strongly supported his vision of an integrative ethics of life[4]. Most diverse forms of life have to eat and breathe, live together, share the environment, survive and enjoy life and living. Thus the former formal rigorist ethics of Kant becomes situational, content-rich and integrative in vision and action.

The most basic moral intuitions for the Bioethical Imperative are compassion and solidarity, living together in interaction, integration, and harmony. Jahr strongly supports legal protection of animals and rare species of plants and animals; he also voices concern about breaking plants and flowers just for fun and without a civilized and morally justified purpose. In regard to eating animal protein, he observes that in colder climates people eat animal protein, but they should raise and slaughter animals in a respectful way. He explains the interaction of egoism and altruism also for social policy: "For example, what is spend on social welfare and in support and improvement of national competitiveness, comes back with interest income, since the state and the economy have a greatest interest to have trustworthy public servants, good workers, financially well to do consumers, good development of the youth, in general the wellbeing of the entire nation"[5]

The Earth as a Living Being

Cosmos and earth are not just physical preconditions for life and living environments on this globe. Cosmos and earth are living entities themselves. Sakyo Komatsu, influenced by the great earthquake of 1923 on Honshu Island which resulted in over 140.000 death and severe destructions in

[3] Jahr 'Egoism and Altruism. Two moral problems, their contradiction and unification in social life' (1929), also in this volume.
[4] Frank L (2011) My beautiful Genome. Exposing our Genetic Future. One Quirk at a time. London: Oneworld. - Barrow JD (2011) The Book of Universes. - Carr B (2007) Universe or Multiverse? Cambride U Press.
[5] Jahr 'Egoismus and Altruismus', 1928, in this volume

Yokohama and Tokyo, wrote a thrilling novel "Japan sinks"[6]. Onodera, the major figure in this thriller asks his wife while escaping by ship: Can you see Japan? No, she answers. It must have sunk. Can you see smoke? he asks. No, she answers, I cannot see anything. - Since March 11, 2011, we have experienced in Japan that the earth as a living entity, mostly friendly and supportive, sometimes wild, extremely wild, inhuman, cruel killing is real, not just the material for a thriller story.

The earth is dark at night and light during the day, cold in the winter and warm in the summer. Some areas such as deserts or the poles are hostile and not supportive of most forms of life; other zones are full of life and living environments of mutual support, fight for life, cooperation and consumption. Climate changes are occurring over decades and centuries and millennia. We seem to be in a long term warming period; there were warmer periods in Europe in the early 19th century, colder times during the Protestant reformation in the 16th century. We had ice ages and ten thousands of years of hot tropical conditions in Europe. Elephants were roaming where there is permafrost in Siberia now; coal deposits in Europe and North America remind us of millions of years of tropical plant life in these areas during earlier life ages of the earth.

The Bioethics Imperative translates scientific knowledge about life and life cycles into behavioral and attitudinal moral and cultural responses, i.e. into respect and compassion and solidarity with other forms of life. This includes, of course, the Kantian position, respecting fellow humans and sentient and (hopefully) responsible beings. But Jahr goes far beyond those limits of inter-human morality. The recognition of nature and earth as a living being calls for moral protection and cultivation primarily. But recognizing the living nature of plants, animals, environments and the globe itself also calls for accepting naturalness where it cannot be changed. Do we want to ride wild tigers? Do we want to build houses on sand? Do we want to hike lightly closed in icy weather? Do we want to produce enormous and dangerous amounts of alpha, beta and gamma radiation which we might not be able to fully control: radiation of iodine 131 with a half-life period of 8 days, cesium 137 with 30 years and plutonium 239 with 24.390 years? Do we want to venture leisurely into unknown jungle territories or unsafe and unknown social environments? Do we want to grow cultivated crops in unsupportive soil or climate? Do we want to build nuclear reactors on geological fault lines? Do we want to produce pollution, which the globe cannot handle, which will make us sick and the environment suffer? Do we want to build megacities, which during biological or other emergencies we will not be able to be kept alive and livable?

[6] Sakyo Komatsu (1976) Japan sinks, New York: Harper & Row 1976.

It is in recognition and respect to the living powers of the earth, that we will not be able to change the seasons of the year or a global warming or cooling over decades or centuries or millennia, if this is the fate and life cycle of the living earth. However, we can change the pollution levels of our cars, avoid the construction of genetically modified plants and animals which might do harm to our health and the health and harmony of the environment; we are challenged to do it for the protection and cultivation of livable natural, social, and cultivated environments. As bioethics per se is integrative, we might add another field of bioethics not yet seen by Jahr and others: geo-ethics or earth-ethics. A geoethical version of a content-rich Bioethical Imperative, i.e. the Geoethical Imperative in the Kantian tradition would read: "*Respect mother earth with all her forms of life, whether natural or man-made, basically as goals in themselves and treat them, if possible, as such*."

Philosophical and religious traditions of various cultures go even further and interpret the three-plus-one dimensions of our visible universe in multiverse perspective. Stars, fish, insects, babies and the double helix of DNA are visible entities in our three-plus-one world continuum, but there is more. The Bible is filled with invisible worlds, powers and dimensions: 'In the beginning God created the heavens and the earth' (Gen 1); the term 'ha'shamaim' for Heaven is a plural form, ha'arez for Earth is singular, Luther translates both in the singular 'Himmel und Erde', but the King James version uses the plural. The bible continues to only outline the creation of the visible physical world, but recognizes those other worlds as demonstrated in Lucifer talking to Eve in Paradise; also the recognized active role of angels, good and bad spirits, and other invisible powers presuppose multiworlds. Early medieval Islam commentaries ask whether it would better to talk about 'worlds' in plural, as written in the Qur'an 'all praise belongs to God, Lord of the Worlds'. Arab oral and written narratives describe seven heavens, the third full of innumerable angels having seventy heads each with seventy tongues in each head, the fourth with angels knowing and teaching math, biology and other sciences, the fifth full of gold. Habibi in the recent novel by Craig Thompson meanders easily between various worlds seen and unseen bringing Jewish, Christian and Muslim traditions into the 21th century.[7]

Similarly detailed is the concept of multiverse in Hindu tradition: 'Because You are unlimited, neither the lords of heaven nor even You Yourself can ever reach the end of Your glories. The countless universes, each enveloped in its shell, are compelled by the wheel of time to wander within You, particles of dust blowing in the sky' (Bhagavata Purana 10.87.41). Hindu cosmology discusses an infinite cycle of births and deaths and an infinite

[7] Thompson C (2011) 'Habibi', www.pantheonbooks.com

number of universes with each cycle lasting 8.4 billion years. The worlds came out of a dream of Vishnu, who in avatars such as fish, boar, or turtle, involves himself in endless tasks of protecting and saving and in other cosmic actions. The Bhagavata Purana mentions twenty-four different avatars, incarnations or manifestations of Vishnu, but at other places it suggests that the number of his avatars is innumerable. Here is a related Hindu Q&A narrative: 'Where is the world situated? On the back of an elephant! Where is the elephant standing on? On the back of turtle!' If this turtle is just another avatar of Vishnu, than the Vishnu dream story is quite elegant and more in line with quantum physics than other deficient big bang urknall models, which still leave the question unanswered: from what and from where and why at all did the big bang happen?

The Mandeans, followers of John the Baptist then and today, believe that our souls come from a supreme dimension of Light into our bodies and into this world of darkness. Since the times of Jesus and John, and still today they transcend this three-plus-one dimensional world by longing into the dimension of supreme and divine Light and getting recharged for continued battle against the powers of darkness in those moments of frequent and repeated baptism in running water. They believe in a four-piece multiverse, two celestial or spiritual ones and two earthly ones. One of the spiritual worlds sometimes – but not in all books – is understood as non-eternal and an ante-room for post-mortem souls, among them Adam and his descendents and the Mandeans of the past. In the earthly worlds the dimensions of darkness and light are mixed and interacting, struggling to overcome each other; one is our visible one; the other is invisible and filled with good and evil spirits, beasts and other powers. All four worlds interact with each other as interpenetrating dimensions under the most sublime dimension of Light, understood as the omnipotent omniverse[8]. John and Jesus were educated in Ptolemaic cosmology; they would not have imagined the light-centered Copernican model and definitely not post-Einstein quantum mechanics of multiverse. They would found both of these models to be much richer and closer to their vision of the supreme light dimension and to the kerygma of God 'in our midst' and worlds and lives thereafter.

Kant in his 'Critique of Pure Reason' 1781 had demonstrated that our understanding and recognition of the world is based on the categories of 3-dimensional space and time, beyond which we can neither prove, deny or even dispute anything. Thus, he considers the existence of 'God, Liberty, Immortality' as 'regulative ideas' for living a morally and culturally good

[8] Lupieri E (2002) The Mandeans. Cambridge: Eerdmans Publ.- Cf. also Teresi D (2002) Lost Discoveries. The Ancient Roots of Modern Science from the Babylonians to the Maya.

life. Modern cosmologists prove Kant's reasoning true and argue that multiverse theories lack scientific testability but are logically most reasonable, simple and elegant; their foundation is 'hypothetico-deductive logic which permits a theory to propose unobservable entities if these help to explain observable outcomes'. Max Tegmark, a leading multiverse experts, sees the validity of multiverse theory in its simplicity: 'A common feature of all four multiverse levels is that the simplest and arguably most elegant theory involves parallel universes by default. To deny the existence of those universes, one needs to complicate the theory by adding experimentally unsupported processes and ad hoc postulates: finite space, wave function collapse and ontological asymmetry. Our judgment therefore comes down to which we find more wasteful and inelegant: many worlds or many words. Perhaps we will gradually get used to the weird ways of our cosmos and find its strangeness to be part of its charm'[9]. Neutrino particles accelerated by CERN's Hadron supercollider in Geneva and sent to Northern Italy moved faster than light, a phenomenon which cannot be explained by Kant's or Einstein's models but suggests other dimensions known to and used by these particles as some physicists suggest.[10]

Differentiations and Transitions in Living Multiworlds

Visible life already in our own physical three-dimensional world 'sees' this single one geographical universe – our earth – in different world-'views'. We humans have ophthalmological recognition of environment and space and a visual memory for orientation and understanding of life, culture and goal; we have maps, we like oilpaintings, movies and photos. Bats understand space acoustically by communicating in an echo-sound structured world of echolation; what type of echolot music-sound interaction might bats find to be more enjoyable than a Mozart or a Vienna waltz? Dogs orient themselves by sense of smell in olfactory knowledge by olfaction; they 'know' places you or I have walked on and where they have been; they find the whereabouts of people by following these people's odor from yesterday and most likely will enjoy olfactorial ouvres or processes of art as we enjoy movies or paintings. The starfish has a circular brain, different from our two-sided brain and body. Is the starfish superior in reasoning, Ernst Juenger has asked? We humans only differentiate between right and left, good and bad, gauchism and droit, while the starfish would have a more complex (superior?) reasoning in a more complex body and more complex biological interactions with its environment; he might find our distinction between right and wrong, left and

[9] Tegmark M (2003) in: Scientific American 288(5)40-51. – Cf also Randell L (2005) Warped Passages. Unraveling the Mysteries of the Universe's Hidden Dimensions. NY: HarperCollins.
[10] Cookson C (2011) Scientist shed light on findings, 'Financial Times', Sept 24/25.

right quite simplistic. Natural capacities of dogs, bats, starfish and nearly all other creatures are far beyond our human capacities, to recognize, to understand, and to interact with those worlds. They, however - we might assume but don't really know - do not have concepts or models of 'other worlds' or multiworlds, even though they definitely live in, reach out and act in 'universes' unreachable and inaccessible to us. The Book of Revelation goes even further and envisions a 'new world' coming after this one and describes in detail the multiverse interpenetrations and interactions during the last days of birthing the new world. Paul stresses that 'the creation itself will be set free from its bondage to decay and obtain the glorious liberty of the children of God' (Rom 8:21); thus one multiworld will change into another multiworld. Isaiah as well envisioned other dimensions where lions and lambs play together, and small children and snakes in pastoral settings with plants and trees (Is 11:6f)[11]

Quite a number of multiverse modifications occur already in our visible worlds. Many forms of life enjoy different properties, modifications, lifestyles, experiences, environments, some much more diverse than ours. Do they live in their own worlds, even though they share the same space-time dimensions with us? Does the butterfly, in its earlier stages of an egg and a larva, knows about becoming a free-flying butterfly later? Most likely, it does not. Does the egg or the larva aspire to become something else and more free in the future, – in the same life or another life? Cicadas have different forms and shapes of life: the egg hatches into a nymph, which drops to the ground and burrows itself; then it lives underground for a specific time of 17, 13, or 5 years and feeds on roots before in their final stage as an instar it digs a tunnel to the surface and sheds her skin, called molting, and abandons it on the bark of trees. How many forms and stages in the life of one individual cicada would these be altogether? Does the small mustard seed knows, that it will eventually become a huge mustard tree? It has, as Jesus put it, the capacity and potential, and maybe even the ingrained goal and vision to grow into whatever form this vision or potential may be: a huge tree.

And where is 'paradise' in multiworld worlds? Is the paradise another world to come or an already existing dimension in multiverse? Comenius, the great mystic and teacher, wrote a book 'Lux in Tenebris' (Light in the Dark) describing visions of people, who have reportedly seen those other worlds. He also elaborates on three stages in human life: embryonic life, individual physical life, post-mortem life. Jahr was influenced by Comenius[12]. These are the three stages in multiverse life: the first two each are ending in stress,

---

[11] Cf. Jahr's biblical sources in 'Death and the Animals' (1928), also in this volume.

[12] Jahr ‚Das Leben nach dem Tode. Aus J. A. Comenius Didactica Magna zusammengestellt', Ethik. Sexual- und Gesellschaftsethik 1933, 10:50-51

despair, pain and angst and in a natural biological feeling of 'this is the end!'. Do embryos know – in the strict sense of knowledge as empirically proven and provable evidence – that at the end of pain, angst and stress, there will be a new life – 'the real life', – most likely, they do not know. But we know very well, that there is life after birth, indeed a much richer and more complete life. Do we know – in the strict sense of knowledge –, that there will be life after death? No, we don't have that knowledge empirically or scientifically. As already mentioned, Kant has demonstrated that our intelligence and body is bound into a three-dimensional space on a temporal scale. Simply: we don't know for sure, whether it is 'real', whatever modern science explores about a multiverse and multiple worlds, even though the assumption of a multiverse is more simple and easier than that of a universe. Jahr's and Comenius' narrative presents a suggestive probability. They are exercises in plausibility theory and practice similar to contemporary quantum mechanics. As these models and methodologies of quantum mechanics suggest, Jahr's case study could be read as to be more elegant and eventually simpler than models of an abrupt life ending in physical death. Apostle Paul (Hebr 11:1ff) explains that only 'faith is the assurance of things hoped for, the conviction of things not seen … what is seen was made out of things which do not appear'. Pastor Jahr, in reference to Paul (2. Cor.5:1) suggests that dying might be compared to changing clothes: 'Dying always appears like a catastrophic event – at least for him who has not yet experienced it before. And this includes all humankind on this earth. Bu whoever would be able to regard dying in retrospect, might perhaps find that it is not such a catastrophe, not nearly as forceful as it seemed! … In retrospect, we might find that, when our soul departs our body it may be no more uncomfortable than taking off a coat or clothing, which perhaps were rather uncomfortable and too small and now we feel much better.'[13] It is in these arguments, where the theologian and believer Pastor Jahr meets Fritz Jahr, the biophilosopher and bioethicist.

Cultural Geography and Land Ethics: Respecting, Interacting, Cultivating

What we today may call Geoethics, in the 19th century was called Cultural Geography by Carl Ritter and his younger contemporary Ernst Kapp, a Hegelian scholar and liberal democrat, who had to emigrate to Texas after his involvement in the German revolution 1848. Cultural geography was intended to bridge the gap between classical geography and human interaction with land and environment. Philosophical geoscience (Erdkunde)[14] was described as a necessary new philosophy of science and a

---

[13] Jahr ,Three tages in life' 1938, in this volume

[14] Kapp E (1845) Philosophische oder vergleichende allgemeine Erdkunde als wissenschaftliche Darstellung der Erdverhältnisse und des Menschenlebens nach ihrem

useful analytical tool in science, ethics, and politics: "Geoscience alike history can be dealt with in a philosophical manner. Philosophy from Vico to Hegel has produced some remarkable books in philosophy of history. They have earned a special prize for that, as they themselves have been recognized as historical powers, a praise which could have even been higher if they had given more attention to the *geographical existence of the nations*. This deficit is their weakness… *Geography* is ingrained into *every place* of history, into every action of human will in his special spatial limits towards its potential realization... *Therefore philosophical geoscience is the indispensable condition of all true history science.* History in its highest form is philosophy of history or politics in a broader sense. Philosophy of earth science therefore can be defined as a *preparatory school for politics*. All roots of political formation are in man; they are developed by man; the process of this development is history. Geography, however, as with and under history developing is *anthropological*. As such, she is naturally very close to man, because the destination of man is the liberation of his spirit by overcoming nature. We call this process of emancipation history, and via this process the *education* of humankind completes itself as well. The anthropological aspect of philosophical geoscience in this sense, therefore, relates to the task of self-recognition of humankind – via the conscience embodied in history and always renewing itself in it. From a philosophical point of view, geoscience, therefore, is political science as well".

A century after Kapp's vision of cultural geography and 20 years after Jahr's formulation of the bioethics imperative, Aldo Leopold envisioned a new cultural and moral understanding and interacting with nature in its entirety, land, plants, animals, environments, - man not as dominant exploiter and conqueror but as partner, interactive, "an individual as a member of a community of interdependent parts".[15] If Fritz Jahr would have been familiar with Kapp's earlier concept and ethics of cultural geosciences and his humanist vision of transfiguration of nature, and with Leopold's Imperative, that we humans should not consider ourselves to be conquerors of the earth, rather plain members and citizens of the land, just similar to sands, soils, waters, plants and animals in huge land community, he most likely would have made cultural geoscience or land ethics one of the other fields universal and integrative bioethics. Fritz Jahr, of course, also would have strongly supported and signed the *Eubios Declaration* of March 1, 2002, in particular the new and broad description of bioethics, the role and importance of personal commitment, and the protection and interaction of life as a whole[16].

---

inneren Zusammenhange. Brauschweig: Westermann, p. VIff. – Cf. Sass HM (1973) Die philosophische Erdkunde des Hegelianers Ernst Kapp. Hegelstudien 8:163-181

[15] Aldo Leopold (1949) A Sand County Almanac, Oxford: Oxford U Press.

[16] Eubios Declaration for International Bioethics [http://eubios.info/eeidec.html] "(1) Bioethics is an interdisciplinary field that needs to be nourished by debate among all

The Imperative 'Do No Harm' the Lives of the Lands and the Multiworlds

The *primum nil nocere* principle and virtue in Hippocratic medicine of doing no harm in the first place, i.e. balancing minimal or low risk with good success and supporting or healing outcome, can and must be extended to all forms of life in the global dimension of land ethics, cultural geography and integrative bioethics. Forms of life live together and depend on each other; humans are no exemption from this rule. Each of us is home to some 100 trillion microbes and we could not live without them, nor would they live without us. Summarizes Rosamond Rhodes 'we used to think of ourselves as separate from nature … now it's not just us, it is us and them'.[17] We only now become aware of the essential symbiosis between our microbiomes and ourselves. Of course, we consciously or accidentally modify, manipulate, or kill those forms of co-life interdependent with us, but we do so not necessary to our advantage, as we only start to learn today.

The earth in general can and has been hurt by humans; goat herding has changed the Mediterranean vegetation for millennia, shipbuilding by the Romans and particularly the Venetians indiscriminately cutting trees along the Istrian coastline of the Adria not followed by reforestation has washed out most soil over the centuries and resulted in meager vegetation. Earth and land are strong and can take quite some abuse and exploitation. Earth and land can and have recovered from severe natural disasters such as meteoric impacts, severe earthquakes, fires, newly modified microbes and predators changing established and well integrated and interacting environments.

The genetic code of life forms has modified itself accidentally and uncontrolled, subsequently changing the check-and balance of survival and interaction of various forms of life, as Darwin has described. Genetic codes of

---

disciplines and people, not limited to any academic specialty or professionals .. (7) Every person has a lifelong responsibility to develop his or her own bioethical maturity and values. We could define bioethical maturity as the ability to balance the benefits and risks of ethical choices, considering the parties involved and the consequences. At the societal level, public policy and law need to be developed, which requires a social mechanism for balancing conflicting ethical principles .. (8) International cross-cultural bioethics should be developed, including studies and discussions, which respect individual cultures as long as they do not conflict with fundamental human rights .. (13) We recognize the dependence of all life (biota) on intact, functioning ecosystems, and the essential services that ecosystems provide. We urge action to halt environmental damage by humans that reduces biodiversity or degrades ecosystem processes."

[17] quoted by Carl Zimmer 'Our Microbiomes, Ourselves', The New York Times, Dec. 4, 2011

'cultured' plants and trees and of 'cultured' animals, hybrids and crossbreeds produced indirectly by selective breeding, have also changed plant life and animal life together with entire agricultural landscapes. Microbes and retroviruses have been bred strategically in order eradicate deadly diseases or improve therapy. Hybrid plants, some still controversial, will increase food supply and might or might not be more friendly to the environment. The term *culture* originally comes from the Latin word *cultivare*, i.e. working the ground, weeding out unwanted growth, supporting the good and edible and healthy crops and fruits, selectively breeding for even better use. One of the first moral cases related to the culture of deliberate breeding is reported in the Old Testament[18]. Jacob was herding the herds of his father-in-law Laban for no pay, but he requested to be given all crossbreeds after a year or two; Jacob then let the purebred herds be together at the watering places and for longer times, so they could breed across breeding lines. This 'unprofessional' behavior of Jacob was not part of the oral contract between the two partners and was considered immoral by Laban, of course. The bible reports that Jacob 'grew exceedingly rich, and had large flocks, maidenservants and menservants, ands camels and asses', – an indication that genetic manipulation and re-manipulation was well known and practiced and caused ethical issues already thousands of years ago.

Jacob did no harm to the animals and their offsprings; but he was unfair to the father of his wife, who had exploited him unfairly earlier. Modern forms of genetic manipulation are more controversial in regard to life and happiness of those new forms. Purebred pigs having one rib more, do produce more and better meat, but are said to be extreme nervous and scared; thoroughbred milking cows cannot give birth anymore and have severe pain if not milked in short intervals; dachshunds get vertebra pain early in their life from walking staircases in private homes. We don't know if hybrid corn, having different biochemical processes than traditional corn crop or even uncultivated corn, might have a different plant psychology, and would suffer or 'scream' differently. Enormous powers of radiation unintentionally set free, uncontrolled or uncontrollable nuclear warfare, biological and other disasters will cause major changes in genetic change in all forms of life. Thus, the extreme risk of increased radiation is not only related to unfortunate death and cancer of those exposed, but even more so to real changes in the DNA setup of all forms of life. This in turn will change the interaction of individual and species life, of biotopes and balanced or slowly changing environments. Some new forms of life will survive or even dominate, others will do harm, will suffer themselves and will hurt others; long dormant DNA might reactivate. Those uncontrollable events will cause a new high-speed struggle for life, never envisioned by Darwin or anyone else before.

---

[18] Genesis 30: 25-36, 43

To expose all life, the land and the earth in general to an unimaginable increase in genetic modification runs against the visions and virtues of respecting life, making this earth the home and house of human civilization, of cultivation and of stewarding, - against the bioethical imperative. To play with fire and with radiation in an uncontrolled and uncontrollable way harms all forms of life, not only fellow humans. It is the opposite of cultivation; it is uncontrollable destruction and severe irresponsibility. Do recent experiences with nuclear energy disasters meet those standards of irresponsibility and a quest for radical change? *The 'do no harm' imperative, virtue and principle is a central component of the Bioethical Imperative, and as such in instrumental for the protection and respect of all forms of life on the land and on this earth.*

Bioethics of Land and Earth: Respect and Cultivate the Earth as a Living Being

The bioethical imperative calls for respecting all forms of life as ends in themselves, i.e. recognizing and respecting their individuality, including strengths and weaknesses, limits and capabilities. Such a respect does not exclude to use, to manipulate and to cultivate co-life for human and cultural purposes. But there are limits to manipulation and cultivation which are related to the limits of every form of life to change or be changed, to feel pain, to be degraded or to be extinguished for no reasonable and morally defendable purpose. In regard to most animals and plants and environments universal bioethics the Bioethical Imperative calls for respect, for stewardship, and for cultivation[19]. But in regard to mother earth and to wild animals, it also calls for accepting what we cannot change, accepting the uncontrollable capacity and unpredictability of all forms of life.. In regard to deadly microbes in hospitals and houses, the bioethical imperative calls for killing and eradicating.

The future of the cosmos in general and of the earth in particular is unpredictable and far beyond our powers of manipulation and cultivation. Of course, we have the powers to harm and to kill many fields of this earth, even make the earth in its entirety uninhabitable for humans and many species. But the respect for the earth as a living being and for all what is living on and in it calls for good protection from harm. It is definitely a prudent aspect of the bioethical imperative to respect powers, which we cannot change. The moral imperative in those situations requires staying clear form danger and risk in a

[19] John Passmore used the term 'stewardship' in 'Man's Responsibility for Nature' (1974) to describe what Potter, Leopold, Jahr and contemporary positions of deep ecology had in mind.

similar way as we would not ride wild tigers or hug polar bears or infect ourselves with deadly viruses. Rather, the prudent and ethical response is to reduce risk and exposure. Respecting the earth as a strong and powerful living being includes to not build nuclear reactors on geographical fault lines or in other risky places, to discontinue technologies with low probability but highest risk features, to cultivate and not to destroy natural and agricultural environment, to limit genetic and other forms of manipulation to lowest possible risk including the risk of wrong risk prediction under uncertainty.

The *Bioethical Imperative* in its most universal and integrative form is a good instrument to not only respect and cultivate natural and social environments, microbes, plants and animals, but the earth as well in its individuality, its seasons and ages, as a home and support of all forms of live, in its unpredictability and danger. Our interaction with earth and cosmos includes respectful and careful recognition of powers beyond our influence and to adjust and to act prudently and morally for our own protection and cultivation and for the protection and cultivation of our natural and social environments. Riding wild tigers and playing with nuclear radiation contains extreme risks and dangers and runs counter to the Bioethical Imperative to respect and to support all forms of life and to prepare for dangers and disasters caused by living volcanoes and hurricanes, by earthquakes and the shifting of continental plates, by new or old deadly viruses attacking in pandemic proportions, – for the protection of fellow humans, for protection of the land and for protection of the earth as our house and home. Following Jahr's argument that 'if someone does not accept the validity' of the bioethical imperative, 'as far as it is concerned with animals and plants, then, in repeating what already was said, one nevertheless should follow it in recognition of the moral obligation toward human society in general'[20], we may conclude: *This earth is the only one we have, so if we do not respect her for what she is, we should do so for our own sake and for the wellbeing of our fellow humans, communities, and cultures.*

[20] Jahr 'Animal Protection and Ethics' 1928, also in this volume.

BIOETHICS, PHILOSOPHY AND PSYCHOANALYSIS*

Natacha Salomé Lima[1]

The beginning of a conversation: Jahr, Schweitzer, Freud.

In mid 1927, the German magazine *Kosmos* publishes an article written by a then unknown author. His name was Fritz Jahr, who under the title *Bio-Ethik. Eine Umschau über die ethischen Beziehungen des Menschen zu Tier und Pflanze* [Bioethics: a Perspective of the Ethical relationship between the Human Being to animals and plants] for the first time introduces the term and the concept of bioethics. A Lutheran pastor, philosopher and educator who lived in the German city of *Halle an der Saale,* presents, in a recently discovered short text, what is considered to be the inaugural act of a discipline which today is of enormous relevance worldwide. Jahr's perspective stems from a *Bioethical Imperative,* which extends and modifies the Kantian moral imperative, giving way to a more comprehensive and flexible content which includes every form of life: "*respect every living being on principle as an end in itself and treat it, if possible, as such.*"[2]

We will, therefore, hold that bioethics begins to be thought of as a discipline, a principle and a virtue from the very start. The relevance of the context tells of a situational model of equilibrium between moral obligations, rights and cosmovisions. In Jahr's pioneer article this concept has the backing of important philosophers of European Romanticism such as Theodor Fechner[3], Rudolf Eisler and Arthur Schopenhauer. It also recalls the teachings of Saint Francis of Assisi who advocated for care and respect towards animals and plants. This precept is shared with Albert Schweitzer's (1875-1965) ethical thinking which was being developed at the time.

---

* The following paper was written within the framework of the project UBACyT P006: Psychology and Bioethics: Principles of the Universal Declaration (UNESCO 2005) through the use of dilemmatic cases in the practice of Mental Health. Psychology, Ethics and Human Rights, Department, under direction of Juan Jorge Michel Fariña.
The paper was presented at 1st International Conference „Fritz Jahr and the Foundations of European Bioethics", Rijeka/Opatija, March 11-12, 2011, and published in Lima N, Fariña, J J M (2011) *Jahr* 2(4) 565-573.
[1] Natacha Salomé Lima, nlima@psi.uba.ar, University of Buenos Aires, Department of Psychology, Ethics and Human Rights, Buenos Aires, Argentina
[2] Sass, H.-M. (2011): *The bioethics thinking of Fritz Jahr (1927).1934.* A preliminary version was published under the title Selected Essays in Bioethics 1927-1934 Fritz Jahr: Postscipt and References, Medizinhethische Materialien, Heft 186. Zentrum für Medizinische Ethik, Ruhr-Universität, Bochum, Germany. See also in this volume. Translation and technical editing by Natacha Salomé Lima, Susana Gurovich and Juan Jorge Michel Fariña (Science and Technology Program, University of Buenos Aires).
[3] Ellenberger, H. (1956): *Fechner and Freud.* www.elseminario.com.ar.

Fritz Jahr receives artistic stimulus and in this case from Richard Wagner's Parsifal, for his conceptualization of bioethics. The inclusion of art as a resource, opens the possibility of reflection on fiction, where what is important is the setting in motion of a longing or desire in the cosmovision of the time. These are ideas that contain the potential strength, the latent seeds of the bio-ethics concept of 1927.

That same year, Albert Schweitzer was nominated for the Goethe Prize for his literary and philosophical work, award which he received the following year. He had published essays on theology, philosophy, music and humanitarian medicine. In 1908 he finishes his last version of his book *Johann Sebastian Bach* whose worked he had studied in depth. He was also an organist and organ builder and writes on this also. We can, therefore, say that once again the expression of art supplements or completes the conceptual development of ethics.

In 1923 Schweitzer publishes his book *Kulfur und Ethik,* which corresponds to the second volume of *Kulturphilosophie,* where he postulates ethics based on solid ground: the idea of reverence for life[4]. Different paths traverse this idea but according to Schweitzer it is what supports and gives structure to our life in the world. If ethics is sustained on reflection and the application of this universal principle (reverence for life) on situational particularities, the context acquires a distinctive value in Schweitzer's conceptualization of ethics. As each *situation* will intervene its complexity, it does not refer to a universalizable model of approach, but of one thought at a time.

In conversation with Hegel, Schweitzer holds that philosophy has lost its bearings when it begins to act as a classifying instrument at the service of science. Schweitzer prioritizes ethics over epistemology; he considers that (rational) knowledge is not sufficient to reflect on the enigmas of the world. The ultimate knowledge to which we will gain access, will be that all that surrounds us is an enigma.

According to Schweitzer this is why we need to return to the discussion of ethics, from the position of doubt, which excites the action of thought. The 20th century has become the century of specialized knowledge; Schweitzer tries to recover a notion of integral approach to the human dilemma, moving away from dualisms. He believes that specialization has come to narrow the

[4] This principle does not establish a system of values or hierarchies. Life designates all that lives, all the living organisms that make up the cosmos. Schweitzer's ethics is not utilitarian, moral or ethical "progress" cannot be measured either by means or by ends. How then, does one gain access to this principle without becoming an instrumental model?

perspectives, while producing the degradation of manual work. Intellectual aspiration (understanding rationality as an ultimate end) has contributed to support a fictional world. Specialization forces us to be only that, it forces identification *with only one aspect of my knowledge, of my doings, of my profession.*

Albert Schweitzer's book *Kulturphilosophie* [Philosophy of Culture] was translated into English as *Philosophy of Civilization.* However, the terms "culture" and "civilization", for Schweitzer, are not equivalent. Culture is composed of external aspects (material, scientific, institutional, in other words materializations of human production or *civilization)* and internal (which include spiritual aspects, which Schweitzer conceptualizes in terms of *mysticism).* The internal aspects of culture are those which are in decline.

Nature is the bedrock of our culture, but civilization has turned into the process of domination of nature. Schweitzer understands *cosmos* as a live organism which has become mechanical, when what we only pursue is the satisfaction of human needs.

When knowledge becomes an instrument of power[5] and this power is at the service of relations of domination, our civilization suffers: *"the relations of domination are relations of power which, instead of being mobile and allowing the partners a strategy to modify them, are blocked and frozen. The relations of power, contrary to the states of domination, suppose the exercise of liberty.*[6] Hence ethics which results from this type of civilization is an ethics based on domination and violence – vulneration as a way of life. Schweitzer advocates for the retrieval of the Socratic dimension of *knowledge,* where the *power* at play, is the power that circulates –the power that is capable of transforming us.

Also in 1927, Freud publishes *Future of an Illusion.* And, in contrast to Schweitzer, Freud does no establish a distinction between "culture" and "civilization". Although the scientifisism of the time trickles into his pre-texts, Freud again holds that not only reason drives Man. There are drives that tend towards death and towards the destruction of a certain shared or social ideal. Man is not *constitutively* good, he does not live happily in society - first he must relinquish those impulses and instincts. Supposedly he does this for his own benefit, because in the meantime Civilization manages to protect itself against the forces of Nature. This relinquishment is what sustains a society. And it is Jacques Lacan, who from reading Freud, returns to us the

---

[5] Francis Bacon - father of modern philosophy - affirmed that *knowledge is power.*

[6] Castro, E. (2004) *The vocabulary of Michel Foucault. An alphabetical run through of his subjects, concepts and authors*. Page 100

dilemma of *your money or your life. "If I choose money I lose both. If I choose life, I am left with life without the money, in other words, a severed life"*[7] There is no possible way out without loss; something must be sacrificed in order to build a field of impossibility, from something that cannot be done, to enable the possible.

Freud also advances a genuine doubt regarding science and its future. He differentiates it from the illusionary nature of religious belives[8], even though this differentiation does not imply a diminishment of the discursive or coercive power they represent. Freud says: *We now see that the treasure of religious representations does not only enclose realizations of desire, but also important historical reminiscences, thus becoming a joint action of the past and the future, which will give religion an incomparable abundance of power."*[9]

But in *Thus spoke Zarathustra,* Nietzsche already predicted a certain dimension of a body outside the body of rationalism. It refers to a constructed body, a spoken body, of a body-representation, a suffering body. *"You say 'I' and you are proud of this word. But greater than this- although you will not believe in it - is your body and its great intelligence, which does not say 'I' but performs 'I'." What the sense feels, what the spirit knows, never has its end in itself.*[10] And that which does not say 'I', but does 'I', also makes body. *Always* "*The Selbst*[11] *speaks and seeks: it compares, overpowers, conquers and destroys.* The Selbst dominates and is also the dominator of 'I'. *Behind your thoughts and feelings, my brother, there stands a mighty ruler, an unknown sage whose name is Selbst. It lives in your body, it is your body. There is more reason in your body than in your best wisdom. And who knows why precisely your best wisdom*[12]

Thus we have seen different aspects of a cosmovisión, the cosmovisión of 1927 (Jahr, Schweitzer, Freud) where we find similarities and differences. From the incipient birth of bio-ethics as a discipline in charge of coping with

---

7 Lacan, J.: Seminar XI. *The four fundamental concepts of psychoanalysis.* Class XVI The subject and the Other: alienation. Pages 220-221

8 In The Future of an Illusion, Sigmund Freud claim that "we qualify illusion as a belief when it appears as the result of the impulse to the satisfaction of a desire, disregarding its relation with reality, the same way that illusion disregards real guarantee." Page 256

9 Freud, S.: The Future of an Illusion. Page 245

10 Nietzsche, F.: *Tus spoke Zarathustra. On the Despisers of the Body.* Page 29

11 In our translation of the passage we have kept the original term Selbst, avoiding the 'self' as it is sometimes translated into English. In context Nietzsche uses the Selbst to mean what is not-I, and therefore, it can be thought as a Freudian topic like Es (that) or as the unconscious.

12 Ibíd.

the ethical question of *bios,* reflection on the ethical relation of human beings with animals and plants (the living environment), becomes a must. Or, follow Schweitzer, along the path that introduces the principle of *reverence for life,* which recovers that dimension of Life that is not reduced to the merely organic. It is remarkable that both perspectives warn us of such a present-day concern as is the ongoing debate on ecological and environmental issues -the ecological agenda.[13]

The *bios* of bio-ethics is not merely an organic *bios,* it is not only about "body" but about Life (bios). Under the paradigm of a narrative. Life, always situational and contextualized, read from a bioethical narrative, opens up to the dimension of a body that speaks and from a body that often speaks beyond the self. "*Man is a literary genre, a narrative species. Human life consists of history or biography, as the etymological bios of bioethics reminds us that refers to 'life is good' or the good life (the biotos of classical Greek). In the words of Garcia Márques, "life is not the one we have lived, but the one remembered and how it is remembered in order to tell it*[14]

References

1. Bonilla, A.: *Responsabilidad y ética ambiental.* [Responsibility and Environment] in: http://www.ddhhmigraciones.com.ar/publicaciones/bonilla/Bonilla_Responsabilidad ye ticaambiental.pdf
2. Castro, E. (2004): El vocabulario de Michel Foucault. Un recorrido alfabético por sus temas, conceptos y autores. [*The vocabulary of Michel Foucault. An alphabetical run through of his subjects, concepts and authors.*] Prometeo 3010. Universidad Nacional de Quilmes Editorial.
3. Cicovacki, P. (2009): *Albert Schweitzer's Ethical Vision: A Source Book.* Oxford University Press.
4. Ellenberger, H. (1956): *Fechner and Freud.* Published in: www.elseminario.com.ar
5. Freud, S. (1927): *El porvenir de una ilusión.* [The Future of an Illusion] Biblioteca Nueva.
6. Lacan, J.: Seminario XI. *Los cuatro conceptos fundamentales del psicoanálisis.* Clase XVI El sujeto y el Otro: la alienación. [*Seminar XI. The four fundamental concepts of psychoanalysis. Class XVI The subject and the Other: alienation*] Editorial Paidós.
7. Nietzsche, F. (2008): *Así habló Zarathustra.* [*Tus spoke Zarathustra*] Centro Editor de Cultura. Buenos Aires.

---

[13] In order to appreciate the reach that the matter of ethics has in relation to the environment, see Alcira Bonilla's work "Responsibility and Environment", at: http://www.ddhhmigraciones.com.ar/publicaciones/bonilla/Bonilla_Responsabilidadyetica ambiental.pdf

[14] Mainetti, J.A.: Entry "*Bioethical narrative*" from the Latin American Dictionary of bioethics. Juan Carlos Tealdi, Page 167

8. Sass, H.-M. (2010): *Selected Essays in Bioethics 1927-1934. Fritz Jahr.* Medizinethische Materialien. Heft 186. Zentrum für Medizinische Ethik. Bochum. Germany. See also in this volume.
9. Tealdi, J. C.: (2008) Diccionario latinoamericano de bioética. [*Bioethics Latinamerica Diccionary*] UNESCO, Red Latinoamericana y del caribe de bioética y Universidad Nacional de Colombia.

JAHR'S TRANSLATIONAL ETHICS. HOW TO TRANSLATE TRADITIONS INTO THE PRESENT AND THE FUTURE

Hans-Martin Sass[1]

Translational science and translational humanities

There is a new fashionable word in biomedical science 'translational medicine'. It is a compensatory term to mitigate the ever growing specialization in genetics and other biomedical sciences, a specialization which is supported by third-party and public funding and by impact-factor rewards for scientists. Thus highly specialized results of research have to be 'translated' into other fields of biomedical sciences, clinical medicine, diagnosis, therapy, development and administration of drugs, public health issues, lifestyle modification, health education, and health care insurance and payment schemes. Translational medicine is an emerging new field of evidence based biomedical science and interventional epidemiology and clinical medicine and has become an essential, but only compensatory and also high specialized field of modern biomedicine and interventional epidemiology and clinical research.[2]

A brief look into contemporary fields of applied humanities, in particular into medical ethics shows a similar picture, normally typical only for science at the cutting edge of research: quoted material is less than 5 or 10 years old; very rarely are classical studies or century old authorities discussed and actually would disqualify those papers from a favorable comment for publication. An empirical study on citations and references in the dozen leading journals in medical ethics and bioethics would be needed to demonstrate the role and relationship of tradition in these fields. Biomedical ethics, clinical ethics, public health ethics have become a postmodern science without any or only little contact with traditional sources, arguments,

---

[1] Hans-Martin Sass, hsass@potomacinstitute.org, Academic Fellow at the Potomac Institute of Policy Studies, in Arlington, Virginia, USA, and Founding Member of the Bochum Zentrum fuer Medizinische Ethik, is em. Prof. fuer Philosophie at Ruhr University, Bochum, Germany, em. Senior Research Scholar at the Kennedy Institute of Ethics at Georgetown University, Washington DC, and Honorary Professor at Peking Union Medical College, Beijing, China.

[2] The Journal of Translational Medicine [J Translat Med] and the Clinical and Translational Science Network [CTScNetwork] are leading global players in this new field; the AAAS (American Academy for the Advancement of Sciences) already in 2009 has issued a promotional video describing the field and focus of translational medicine. Referenced material in impact-factor relevant publications of medicine rarely is older than 1 to 5 years.

positions in ethics, philosophy, and religion. The so-called Georgetown Mantra – autonomy, non-maleficence, beneficence, justice – just 50 years old, serves as the coat-of- arms and battle-cry, making reference to other sources obsolete and curious. Thus, different fields of health care ethics – in itself a field of many disciplines – are in need of translational services.[3] 'Modern medical care is increasingly dependent on the application of science to clinical practice, which occurs through clinical or translational research. We propose the concept of translational ethics, which incorporates the contributions of research codes of ethics that involve the protection of human subjects into the ethics of clinical practice. The modern research environment, which has contributed the scientific tools of modern medicine, has also framed the ethical environment in which medicine is practiced. The single most important contribution of research codes for protection of human research subjects to clinical practice is the doctrine of informed consent. Translational ethics, based on autonomy and informed consent, progresses beyond the narrow interpretation of those 2 concepts. It requires consensual understanding of a spectrum of clinical interventions that are increasingly complicated. Translational ethics helps navigate the ethical ramifications of technological and scientific advances that will increasingly challenge the corporate-oriented health system in the new millennium'.[4] This quote makes it clear, that translational ethics is a necessary and complementary postmodern form of traditional humanities and ethics regaining lost ground in harmonizing and reviewing the ethics of life and health from a traditional and broad perspective. Translational fields of science compensate for the lost common ground and overall vision of purpose and utility in the growing fields of more and more highly specialized areas of research. As far as conceptual and moral traditions are concerned, more and more classical sources, which may be helpful, even essential and indispensible for understanding challenges to life and health in the 21th century, are missing and compensated for by new highly specialized fields of applied ethics. Here comes the need for translational efforts to regain access to and integration of classical visions in ethics, attitude, and culture come into play. This quest for translational efforts goes far beyond what has developed so far within the extrafugal powers of highly specialized modern sciences and their applications.

---

[3] The Charles Fairbanks Center for Medical Ethics (FCME) in Indianapolis, Indiana offers a program with special services in research, coordination and education in translational ethics; LABTEC, the London-and-Brighton Translational Ethics Centre offers similar highly specialized services to compensate for highly diversified and separated fields of medical ethics.

[4] Kagarese MJ, Sheldon GF (2000) Translational Ethics. A Perspective for the New Millenium. Arch Surg 135:39. – Cf. Cribbs A (2010) Translational Ethics. The Theory-Practice Gap in Medical Ethics. J Med Eth 36:207-210

Candidates for global translational hermeneutics

What are the sources and authorities of tradition, which might give a hint on how to decipher most crucial issues in bioethics and health care ethics today and in the future? I just give a few brief quotes, which you rarely will find prominently in any of our contemporary professional journals in applied ethics: 'Tat tvan asi: this is you as well' (Vedic tradition, 2000 BC). – 'Love your neighbor as you love yourself' (Moses, other sources, 600 BC, St. Paul 50 BC., other sources). – 'The Sages did not wait till they were thirsty to build a well' (Dr. Huang Di, Western Han Dynasty, 200 BC). – 'A superior doctor takes care of the state, an average doctor takes care of the person, an inferior doctor takes care of the disease' (Dr. Sun Simiao, Tang Dynasty, 700 AD). – 'A person of good virtue will have longevity' (Mencius, 600 BC). – 'Trust only those doctors who have a heart of compassion and humanness, who are clever and wise, sincere and honest' (Dr. Yang Chan, 1600 AD). – ' Primum nil nocere: first of all, do not hurt' (Hippocrates, 600 BC). – 'Non homo universalis curator, set unus, quique, nostrum: we do not treat a universal person, but a unique, a special one, our' (Galen, 50 AD). – 'How can you sell land? you cannot sell the air, the moon, the sun?; When animals and plants would not exist, men would die of loneliness' (Indian Chief Seattle, 1700 AD). – 'Respect every living being on principle as an end in itself and treat it, if possible, as such' (Jahr, 1927 AD).

How do we translate these sources into the new worlds of 21th century cutting-edge research and science in the life sciences, in medicine and health care, and into our individual and communal life cultures? How do we invite millennium-old wisdom and classical authorities into our deliberations, visions, into setting goals for research, application and education? Fritz Jahr, the father of bioethics – who himself had been forgotten fort close to a century, another empirical indication for the forgetfulness and loss of tradition and history in the modern humanities – gives a fascinating and challenging example on how to translate tradition into the future.

Jahr's translational case study 'Thou shalt not kill'!

The so-called Ten Commandments originally were written down by the political and religious leader Moses to a small group of nomadic Semitic tribes in the northern Arabian peninsula on their way to find settlements. These commandments were specific and exclusive for these tribes, to make them different and identifiable from others who served other Gods and had other rules. Thus the biblical narrative underlines that the commandments were dictated to Moses by God. A few of these commandments could be generalized such as the command not to kill or not to lie, others with limited authority such as to respect a one-man-one-woman marriage, others not such as the sanctity of the Shabbat and the exclusiveness of the tribal God Yahweh.

Jahr introduces the 2500 year old commandment of not actively killing another person into actual and most challenging tasks of the 20th century: (1) 'Do not kill' is a Golden Rule for everyone on a global scale, in disregard to all differences in religion, race, culture, tradition. – (2) 'Do not kill' positively expressed is an obligation for everyone to take positively and affirmatively care of his or her life and wellbeing, live a preventive lifestyle and avoid risks to health which might kill or harm oneself. – (3) A generalization of the 'do not kill' principle includes the respect and protection of all forms of life, for which humans can and therefore should feel responsible, thus the golden rule for civilized and cultivated anthropological ethics among humans expands into a global life ethics as a new principle to protect life and earth, wherever and whenever possible, as the new Bioethical Imperative, surpassing the traditional Kantian Imperative to respect exclusively humans as ends in themselves, i.e. creating an entirely new field of academic research and teaching and of public discourse and a new global culture for the field of 'bios', i.e. bioethics. Jahr wrote this translation in a crucial moment in German and European history, in 1933, the year the Nazis took over parliament and government and 6 years before they started World War Two. It was the time when in Germany and elsewhere discussions about 'life unworthy of living' [lebensunwertes Leben] of the severely handicapped and demented and even entire racially defined populations were en vogue and later implemented by the Nazis.

Step one: the general golden rule: do not kill

In his hermeneutics, Jahr goes immediately from the discussion of what should be forbidden to a positive goal, asking positively 'What is the Golden Rule?'

'How do we do good? – The so called 'Golden Rule' gives answer to this question is: All, what you want the people do to you, the same do to them (Matthews 7:12; Luke 6:31). Kant's 'Categorical Imperative': Act only according to that maxim whereby you can, at the same time, will that it should become a universal law, - this basically means the same. – But are these and similar formulations not only just a formal criterion for a 'good' action. The motive, disregarding such a criterion, could just be blatant egotism, a so called contract on reciprocity: Do nothing to me, so that, in return, I will do nothing to you (Schopenhauer hints to that in his 'Grundlage

der Moral'). If we are aware that love is the fulfillment of the moral law (Romans 18:10), then we are one step ahead, indeed: We know the motive. But we do not yet know the concrete content of the moral law, we don't know what to do or not to do specifically. Here, Schopenhauer, serves as a helping hand: He calls as the best, the most concrete specification of the moral act the sentence: Neminem laede, imo omnes, quantum potes juva! (Don't hurt anyone, but help everyone, as far as you possibly can!). – More than two millennia before Schopenhauer, the 5th Commandment already has brought such an insight, and in fact in a broader perspective than benefit or harm, namely under the perspective of sanctity of life and life's manifestations. Therefore the calling: 'You shall not kill!'. We know from Jesus, that the 5th Commandment does not only prohibit killing, but all bad deeds against others, even the bad word, even the bad thought. This means: He not only forbids the malicious or careless destruction of life, but everything as well which, in one way or the other, may influence life in a hampering or troubling way. Luther in his Catechism has made it clear, that the 5th commandment has to be understood not only in a negative, but as well in a positive way. - The consequence of all of this, the 5th commandment is a very good expression of what it means to be morally and practically good.'

Thus, an exclusive rule, given to the elected people, became an including rule for 'all good people', integrating 'good' Jews, Christians, Muslims, Buddhists and others, and making them different, not on the basis of skin but on attitude, from 'bad' Jews, Christians, Muslims, Buddhists and other bad people. Of course, the 'do not kill' rule was not only 'given by the Lord', it was a common sense rule of reciprocity and pragmatics 2500 years ago, also could and can be supported strongly by humanist and legal reasoning. A special rule given exclusively to a few can become and has become a universal rule, a common morality. As a rule guiding personal and professional life, it could read today: do not kill other people; do not harm other people; do not exploit other people; do not degrade other people by character assassination, defamation, exploitation. Also in the more distinct health care setting: do not kill your patient; do not harm your patient; do not exploit your patient; do not keep your patient ignorant; do not disrespect your colleagues and co-workers; do not break or harm the rules and values of the care team. In interpreting Moses' relgious rule, Jahr also makes use of the hermeneutic history of the 5th Commandment, quoting Luther and Schopenhauer, an exemplary model to deal with tradition.

Positively expressing and supporting virtues rather than detesting vices, as Jesus and Luther suggested, the rule could read: do support the life of your fellow human, do support her or him in any way you can, offer help, care for your neighbor, encourage and honor her or him in their particular effort and vision. In the healthcare setting: educate you patient to be health conscious and health responsible, treat your fellow human as a person and not just her or his symptoms, respect her or his wishes and values even if you do not share those, be supportive of your colleagues and coworkers and respect their individuality, cultivate the working environment. Thus, a universal rule can be fine-tuned to specific situations and working environments.

Step two: individual health literacy: do not kill yourself

Jahr addresses moral obligations to oneself, translating the classical 'do not' rule into some of the most pressing lifestyle-related issues of caring for health and good health.

'When talking about moral duties, normally we mean duties towards other people in the first place. Routinely we do not consider that each person has moral duties towards oneself as well, and that those duties are of immense importance. Christian religion expressively mentions those moral duties of everyone towards oneself. That, basically, applies to the 5th commandment as well: 'You shall not kill'. In this sense - 'You shall not harm or hurt anyone's body or life, rather help and support him/her in all distresses of body and life, wherever you can'[5] – in the first place means the life of our 'neighbor'. In a later consequence, however, it means: in Christian perspective every human life as such is morally 'sacred' – including one's own life. Preservation of life – and one's own life not excluded – is a duty. And destruction and harm – again, including one's own life – is a moral sin. 'Don't you not know, that you are God's temple and that God's spirit dwells in you? You shall keep God's temple sacred and not destroy it.' (following 1. Corinthian 3:16-17)

How should these moral duties, as expressed in the 5th commandment towards one's own life, be applied in real life's practice? By not taking one's own life, by not shortening it, by not harm or endangering it, by not weakening one's own health by unchastity, excesses in eating and drinking, heavy anger, frivolous foolhardiness and daredevilry, etc... Particularly important is the protection of sexual virtue and the avoidance of abuse of alcoholic drinks. – As far as the first one is concerned, the judgment of the New Testament is particularly clear: 'If you have loose sex, you sinfully harm your own life' (following 1. Corinthian 6:18). But not only is it a duty to oneself to not fornicate, but also the avoidance of everything, which might lead to unchastity: dishonest looks, unclean or double talk, dancing, dresses etc. – As far as alcoholism is concerned, the Christian attitude is based in recognizing

---

[5] *Cf. Luther's explanation of the 5th Commandment, German and Latin. [This and all other footnotes to the quotes from Jahr's essay are his own.]*

that 'wine kills many people' (Sirach 31:30), i.e. alcohol endangers life and brings great dangers to health.

Are the duties towards one's own life not in conflict with duties towards the neighbor? – That is not necessarily the case. On the contrary: He/she who fulfills his/her duties towards himself/herself, avoids many forms of harm towards other people. That can be shown in regard to the already mentioned issues in sexuality and alcohol: He/she, who falls into dependency and unchastity, endangers and weakens himself/herself physically and spiritually. Venereal diseases threaten as well. Weakness and disease cause the victim to be more and more a burden to the community, harming everyone. If he/she has offspring, they also are harmed, as they may inherit a weak or sick nature, causing additional burdens and harm to the community. Who, however, protects one's own life in this respect, fulfills his/her duty also towards the community. Similar with alcohol: Those, who are dependent on consuming alcohol, may eventually expose themselves to severest physical and spiritual dangers. And thus does not only harm himself/herself, but his/her family as well, his/her offspring, his/her country, his/her race[6]. And again: If one protects oneself in this regard against harm, one does, at the same time, good to one's neighbor, actually to one's entire country.'

It is surprising at first glance, that Jahr applies the 'do not kill', 'protect, support and save-guard life' rule in a second step to the personal situation and the 'work environment' of caring for one's own health. He gives the commandment to care for the neighbor the same value as the care for oneself. There is nothing selfish about caring for one self's health and wellbeing, - on the contrary: those who do not care for themselves will become a burden on others. He particularly addresses health risks associated with poor lifestyle choices and that leads directly from the care for individual health to public health care. Drug addictions and, what Jahr considers, immoral lifestyles do not only harm the addict and the immoral person; they make the community unhealthy and sick. The 1920's and 1930's in Europe witnessed a rapid breakdown from old conservative bourgeois traditions to sexual libertinage and free consummation of various drugs and existentialist lifestyle experiments. Jahr, a Protestant pastor, feels obligated to point out the risks to one self and to the community associated with these cultural changes and value modifications; he, thus, emphasizes the relatedness of individual health and public health as a responsibility issue in public health of a civil society.

---

[6] *Alcohol is 'A Mean Enemy of Our Race', cf. the brochure with this title by Wilhelm John, reviewed in no. 2 of 'Ethik'.*

Jahr, however, does not involve himself into the eugenic debates of his times.[7]

Modern health care and health care ethics so far have not put a prime emphasis on health education and health encouragement, on supporting health literacy and health responsibility[8], even though today we know much more about lifestyle risks to health and wellbeing than previous generations. Also the interaction between public health care and a change in individual health care culture has not been made as clear as Jahr does it in this article. Many cultures had a strong tradition in encouraging and supporting lay health culture and responsibility, but the successes of interventional medicines seem to be too powerful today so that we use medicine as a repair facility like we use car shops for the repair of our automobiles.

Would it be advisable to not only develop more and better internet sites in the support of individual health and health care competence and also to use the vast treasures in Asian and European health literacy education?[9] See, for example, the 17th century enlightened rules by Friedrich Hoffmann[10], also

---

[7] It must be mentioned, that Jahr in this article does not use the opportunity to involve himself into the academic and public eugenic debates in Britain, Germany, Japan and elsewhere; nor did he mention Francis Galton, Margaret Saenger, Julian Huxley or J. B. Shaw and others as proponents of 'mercikilling'. The German "Law for the Avoidance of Genetically Sick Offsprings' [Gesetz zur Verhinderung erbkranken Nachwuchses] was passed July 14, 1933 by the Reichstag. Would a German religious or academic journal in 1933 have accepted an article criticizing strategies of eugenic killings or ending lives 'unworthy to live'? Would Jahr have been able to bring his message - support for life as a golden rule, as individual and public health and as respect for life in all forms - across, if he would have gotten involved in the eugenic debate?

[8] Sass HM (2006) Gesundheitskulturen im Internet. E-Health: Möglichkeiten, Leistungen und Risiken, Bochum: Zentrum für medizinische Ethik, Heft 166. – Sass HM (2008) Public-Health-Ethik ist Partnerschaftsethik. Bundesgesundheitsblatt 51:164-174. – Sass HM (2007) Wellness durch E-Health? Die vier K von E-Health: Kultur, Kommunikation, Kooperation, Kompetenz, E-Health und technisierte Medizin, hg D. Gross, EA Jakobs, Münster: Lit, 29-49. – Sass HM (2004) E-Health: Health Promotion and Wellness Communities Eubios Journal of Asian and Intern Bioethics 14:170-174.

[9] See the interactive rules for the lay and the experts by Confucian doctor Gong Tingxian and based on those my own sets of interactive rules: Sass HM (2006) Bioethics and Biopolitics. Xian: 4th Military U Publ, p. 12f, 212f for physician-patient interaction; 146-148, 378-381 for public health; 62f, 274f for e-health. - See also Sass HM (1994) Formulating Global Post-Hippocratic Health Care Virtues. European Philosophy of Medicine and Health Care, 2(1), 1994, 6-10; also in: Eubios Ethics Institute. Newsletter, 4 (1+2), 1994; Japan. transl: Journal of Health Care, Medicine and Community, 5, 1994, 3-6; Chinese Transl: Newsletter. Centre for Applied Ethics, Hongkong Baptist College, 2(1), 1994, 8-15; Italian: Kos, 118, 1995: 46-51; French: Journal International de Bioethique / International Journal of Bioethics, 1995, 6(1), 5-7.

[10] Müller IW (1991) Iatromechanische Theorie und ärztliche Praxis im Vergleich zur galenistischen Medizin (Friedrich Hoffmann - Pieter van Foreest - Jan van Heurne)

from Halle an der Saale, a physician and pharmacist, whose 'Dr Hoffmann Tropfen' are still sold in German pharmacies today to cure headache and stomach pain: *'1) Stay away from everything which is unnatural. - 2) Be careful with changes as routine often becomes our second nature.- 3) Be happy and balanced, that is the best remedy. - 4) Stay in clean air, well-tempered, as long as possible.- 5) Buy the best nutrition which goes easily in out of the body.- 6) Choose foods according to your bodily activity and relaxation.- 7) When you love to be healthy, run away from physicians and from all drugs.'* Could and should we formulate similar guides today for internet-literate people including a 'be careful' advice regarding charlatanry and incompetency, doctors and drugs?

Step three: the bioethical imperative to respect, to protect, to not kill

Finally Jahr expands the 'do not' rule into an even wider context, the respect and protection of the entire world of life, animals, plants, the globe, thus defining the new rule for his 20th century and beyond.

''Thou shalt not kill' admonishes the 5th Commandment. Now, the term killing always means killing something which is alive. Living entities, however, are not only humans, but animals and plants as well. Because the 5th commandment does not expressively prohibits the killings of humans exclusively, should it not be applied towards animals and plants analogously?

But are animals and plants so close to us, that we must recognize and treat them actually as our neighbors? – When we review publications in modern science, we find immediately similar studies of humans and animals as subjects in research, not only in physiology, but also in psychology. Such an equality in treatment today is not reserved, as already mentioned, for humans, as similar methods are applied in the field of animals, and - as there is a comparative anatomical-zootomic research - similarly very interesting comparisons are made between human soul and animal soul[11]. Yes, even the beginnings of plant psychology are recognizable – the most well-known among them are G. Th. Fechner[12] in the past, DE. H. France[13], and Ad.

(=Historische Forschungen im Auftrag der Historischen Kommission der Wissenschaften und Literatur Mainz) Hrsg. von K.E. Born u. H. Zimmermann). Stuttgart 1991, p. 260.

11 *Among recent publication in animal psychology especially recommendable are: Sommer, Tierpsychologie, Leipzig 1925. – Alverdes, Tierpsychologie, Leipzig 1925*

12 *G. Th. Fechner, Nanna oder das Seelenleben der Pflanze [1848; 5th ed. 1921]*

13 *R. H. France, Pflanzenpsychologie als Arbeitshypothese der Pflanzenphysiologie, Stuttgart 1909*

Wagner[14] *at present – thus modern psychology includes all living beings in its research. Given this, it is only consequent, that E. Eisler*[15]*, in summarizing, speaks of a Bio-Psychik.*

From Bio-Psychik, there is only a small step to Bio-Ethik, i.e. to the assumption of moral duties not only towards humans, but towards all living beings. In fact, bioethics is not a discovery of today. Montaigne[16] *already grants - as the only early representative of modern ethics of sentiment - all living being an entitlement of being treated based on moral principles: We owe justice to humans; mildness and mercy towards all living beings, capable of having a benefit from that. Similarly, Herder*[17] *requires that humans - following the model of God in their sentiments - to put themselves into the place of every living being and to feel with it, as much as it requires. Those lines of reasoning are continued by the theologian Schleiermacher*[18]*, who calls it immoral, to destroy life and formation - wherever they are, i.e. including animals and plants - without a reasonable argument for doing so. Therefore philosopher Krause*[19]*, a contemporary of Schleiermacher, requests that every living being has to be valued as such and not be destroyed without reason. Because they all, plants and animals alike humans, have an equal right; but not totally equal, each only as a precondition to reach its destiny. Schopenhauer*[20]*, in particular refers to the Indian realm of reasoning, stresses compassion as the most important motive of his ethics, and requests it also for animals. It was Richard Wagner, strongly influenced by Schopenhauer and a passionate animal friend, who made those thoughts commonly known.*

As far as animals are concerned, the moral request has become self-understanding for a long time[21]*, at least in that form, not to harm animals without purpose. With plants it is different. However, in regard to new biological and biopsychic knowledge (see above) and regarding the circles of thought which I mentioned from Montaigne, Herder, Schleiermacher and Krause, moral duties towards plants become visible. For purely sentimental-poetic argumentation such recognition is nothing new. One only has to think of Goethe, who has Faust call plants his brothers, or of Richard Wagner's Parsival: In pious devotion people, at least on Good Friday, protect weeds and flowers in the meadow by walking carefully, in order not to hurt them.*

---

[14] *Ad. Wagner, Die Vernunft der Pflanze, Dresden 1928*
[15] *E. Eisler, Das Wirken der Seele, Stuttgart 1908*
[16] *Montaigne, Essays*
[17] *Herder, Ideen zur Geschichte der Philosophie der Menschheit*
[18] *Schleiermacher, Philosophische Sittenlehre, Kirchmann 1870*
[19] *K. Chr. Fr. Krause, Das System der Rechtsphilosophie, Roeder, Leipzig 1874*
[20] *Schopenhauer, Über das Fundament der Moral*
[21] *The most comprehensive book in this area still is Bregenzer, Tierethik, Bamberg 1894*

More seriously we have to take plant-ethical reflections of a quite matter-of-fact Ed. Von Hartmann[22]*. In an article on flower luxury he writes about a picked blossom: 'She is a deadly wounded organism, the colors of which are not harmed yet, a still living and smiling head, separated from its stem. – When, however, I put the rose into a glass of water, I cannot help myself but fighting the thought, that man has murdered a flower life, in order to enjoy the dying process by an eye, heartless enough to not sense the unnatural death under the appearance of life*[23]*. The plant-moral requirements leading to such recognition are quite clear.*

As far as the potential realization of such moral duties towards all living beings is concerned, it might seem like utopian. But we may not ignore that moral obligations towards a living being relate to its 'need' (Herder), respectively to its 'destiny' (Krause). So, it seems, that needs of animals are much less in number, and their content less complex than those of people. This applies even more so to plants, so that moral obligations towards them should produce less complications than those towards animals, as they are lower (if not conceptually, so nevertheless practically). Here also comes the principle of struggle for life into play, a principle which also modifies our moral obligations towards fellow humans at no low scale. Within these limits there always will be enough possibilities for bioethical actions. Paragraphs for animal protection in penal codes of various cultivated nations[24] *give guidance in this regard. Confer in particular the new German Reich Animal Law. As far as plant ethics is concerned, we are guided by our sentiment; so it will hinder us to pick flowers and then throw them away carelessly shortly thereafter, or to behead plants with a walking stick, or when we find it disgusting to recognize the blind destructive impulse of rowdy lads in breaking the heads of small trees along the road. Also, excessive flower luxury – in learning from Ed. Von Hartmann – is not morally refined and can be avoided.*

In sum, the universal realm of authority of the 5th *Commandment shows itself and demands to be applied to all forms of life. A transcription of the 5*th *Commandment results in the Bioethical Imperative: 'Respect every living being in principle as an end in itself and treat it, if possible, as such!"*

Given his concept of integrative ethics in a multiworld fashion, it is only consequent for Jahr to extend the protection and care to the entire realm of

---

[22] *Psychological preconditions are discussed in W. von Schnehen, Ed. Von Hartmann und die Pflanzenpsychologie, Stuttgart 1908*

[23] *Ed. Von Hartmann, Der Blumenluxus, 1885*

[24] *For the first time, material has been extensively collected and reviewed in R. von Hippel, Die Tierquälerei in der Strafgesetzgebung des In- und Auslandes, Berlin 1891*

life. He refers to most recent biological and psychological research to argue, that such a scientific biological insight needs to be accompanied by an ethics counterpart. Thus, the wider concept of bioethics as an integrative scientific endeavor and as a personal and communal culture comes from a moral reflection on new responsibilities which are associated with new knowledge. Only in this last section does Jahr quote recent scientific publications of his time, thus providing for a 'translational' service from the sciences to the humanities in general and to bioethics in particular. Potter had argued similarly later in the 1970's. The traditional rule of caring for and protecting life extends to all forms of life, the similarity and equality of which has been strongly confirmed by modern science. This third and last section of Jahr's article is well documented in order to demonstrate that these reflections are based on ethical and cultural reasoning concerning most recent scientific publications and the consequences educated people and communities should draw from there.

Different to the categorical imperative of Kant, Jahr's bioethical imperative is not categorical, but pragmatic, mediated and guided by the 'will to life', the struggle for survival. Humans have to eat in order to survive; humans have to defend themselves against aggressors: animals, microbes, plants, aggressive environments. Situational ethics will have to decide in specific cases where individual or human survival and life will.

It is Jahr's argument, that new fields of scientific knowledge will require new and specialized research in ethics and new attitudes. So, one could extend Jahr's reasoning into at least one new field of research and everyday experience: global warming. Global climate change remind us that the globe itself is a living being with its own seasons, long-term and short-term chances, developments and modifications. Major changes, such as potentially new ice ages or global warming, are beyond human control. But some causes, such as industrial pollution and environmental destruction, can and should be mitigated in order to allow for sustained human development and continued human culture and cultivation. Thus, we could add to the common morality command, to the health responsibility command, and to the bio-ethical command, a new command of 'do not kill' and 'do protect': a geo-ethical command, which would read ' respect mother earth with all her forms of life, whether natural or man-made, basically as goals in themselves and treat them, if possible, as such.'

Conclusion

Fritz Jahr, in his interpretation of a millennium-old commandment, gives a good hermeneutical example of how to introduce and to include traditions into the modern debates and solutions of integrative bioethics in the 21th century, – a truly translational service, far beyond the limited bridge building services in existing fields of translational medicine and ethics into fields of therapeutic or diagnostic application or into ethics review, guidance, and strategy. More than ever before, there is a challenge to European and other traditions to introduce their specific heritage into the modern global and integrative dialogue in the many fields of applied ethics. Jahr, thus, opens a promising field for European and global dialogues in integrative bioethics as an individual and collective attitude and virtue, a new academic discipline, and a new global culture in the 21th century and beyond.

RIJEKA DECLARATION
ON THE FUTURE OF BIOETHICS

Fritz Jahr used the term "bioethics" ("*Bio=Ethik*") as early as 1927. His "bioethical imperative" (*Respect every living being as an end in itself, and treat it, if possible, as such!*) should guide personal, professional, cultural, social, and political life, as well as the development and application of science and technology. In order to promote the Bioethical Imperative and the future of integrative bioethics, the participants of the Rijeka symposium "Fritz Jahr and European roots of bioethics: Establishing an international scholars' network (EuroBio*N*ethics)", wish to highlight the following:

1. Contemporary bioethics quite often has been narrowed down to issues of informed consent and liability in medical ethics, whereas the practical impact of general ethical principles has been minimal.
2. It is necessary that bioethics be substantially broadened and conceptually and methodologically transformed so that it may consider different cultural, scientific, philosophical, and ethical perspectives (*pluriperspective approach*), integrating those perspectives into orientational knowledge and practical action (*integrative approach*).
3. Such *Integrative Bioethics* will have to harmonize, respect, and learn from the rich plurality of individual and communal perspectives and cultures of the global community.
4. Recognising the inexhaustible source of relevant perspectives for *Integrative Bioethics* in the works of authors and teachings using the term and the concept of bioethics, but also of the other „precursors" of integrative bioethical and deontological ideas since antiquity, we strongly call upon the study of classical works and teachings.
5. Respect for life, the considerate treatment of all life forms, need to be supported by all citizens, public discourse and the media, and by educational programs at all levels.
6. If these ideas are successful, bioethics will become a truly open field of meeting and dialogue of various sciences and professions, visions and worldviews, that have been gathered to articulate, to discuss, and to solve ethical issues related to life as a whole and each of its parts, life in all its forms, shapes, stages, and manifestations, as well as to life conditions in general.
7. If these ideas are successful, bioethics will become the basis for the development and implementation of law, nationally and internationally.

8. If these ideas are successful, the recognition and implementation of bioethics will become the „bridge to the future," a „science of survival," and wisdom as „knowledge of how to use knowledge" (as Van Rensselaer Potter defined it in the 1970s) of modern medicine and technology.

The EuroBio*N*ethics International Scholars' Network, promoting the above ideas, will organise further conferences and establish a website to improve global intercultural communication and cooperation. A Fritz-Jahr Award for the Research and Promotion of European Roots of Bioethics will soon be announced. The Network invites scientists and ethicists for communication and cooperation in implementing these ideas of the *Rijeka Declaration*.

Signed in Rijeka/Opatija (Croatia), on March 12, 2011, by:

Christian Byk
Judge of the Court of Appeal, Paris; Secretary General of International Association of Law, Ethics and Science; Editor-in-Chief of *International Journal of Bioethics*

Ante Čović
Full Professor and Head, Department of Ethics, Faculty of Humanities and Social *Sciences*, University of Zagreb; President of Croatian Bioethics Society; Editor-in-Chief of *Synthesis philosophica* and *Filozofska istraživanja*

Eve-Marie Engels
Full Professor and Head of the Chair for Ethics in the Biosciences, Department of Biology, Faculty of Science; associated member of the Institute of Philosophy; spokesperson of the DFG-Research Training Group Bioethics (*Graduiertenkolleg Bioethik*) at the International Centre for Ethics in the Sciences and Humanities (IZEW) of Eberhard Karls University of Tübingen

Igor Eterović
PhD candidate, Department of Social Sciences and Medical Humanities, Faculty of Medicine, University of Rijeka

Márcia Santana Fernandes
Professor, Centro Universitário Ritter dos Reis, Porto Alegre

José Roberto Goldim
Associate Professor, Pontifícia Universidade Católica do Rio Grande do Sul; Professor, University of Extremo Sul Catarinense, Criciúma; Lecturer, Federal University of Rio Grande do Sul; Clinical Hospital of Porto Alegre

Nada Gosić
Associate Professor, Department of Social Sciences and Medical Humanities, Faculty of Medicine, University of Rijeka

Hrvoje Jurić
Assistant Professor, Department of Ethics, Faculty of Humanities and Social Sciences, University of Zagreb

Eleni Kalokairinou
Assistant Professor, Department of Classics and Philosophy, University of Cyprus, Nicosia

Tomislav Krznar
Assistant Lecturer, Department of Gamekeeping and Environmental Protection, University of Applied Sciences, Karlovac

Natacha Salomé Lima
PhD candidate, Department of Psychology, Ethics, and Human Rights, University of Buenos Aires

Amir Muzur
Associate Professor and Head, Department of Social Sciences and Medical Humanities, Faculty of Medicine, University of Rijeka; Editor-in-Chief of *Jahr*

Iva Rinčić
Senior Lecturer and Research Assistant, Department of Social Sciences and Medical Humanities, Faculty of Medicine, University of Rijeka

Ricardo Andrés Roa-Castellanos
Lecturer-Researcher, Institute of Bioethics, Pontifical Xavierian University, Bogotá DC (Colombia)

Hans-Martin Sass
Professor, Ruhr University, Bochum; Senior Research Scholar, Kennedy Institute of Ethics, Georgetown University, Washington DC; Lecturer, People's University of China and Peking Union Medical College, Beijing

Marija Selak
PhD candidate, Department of Ethics, Faculty of Humanities and Social Sciences, University of Zagreb

Ivana Zagorac
PhD candidate, Department of Ethics, Faculty of Humanities and Social Sciences, University of Zagreb